# Technical Topics Scrapbook 2005 - 2008

## Pat Hawker, G3VA

**Published by the Radio Society of Great Britain**
3 Abbey Court, Fraser Road, Priory Business Park, Bedford MK44 3WH
Tel: 01234 832700 Fax: 01234 831496 Website: www.rsgb.org

First published 2008
Reprinted 2008

© Radio Society of Great Britain, 2005-2008. All rights are reserved. No part of this publication may be reproduced, stored in a retrieval system, or transmitted, in any form or by any means, electronic, mechanical, photocopying, recording or otherwise, without the prior written permission of the Radio Society of Great Britain.

ISBN: 9781-9050-8639-9

Editor's note:
When originally published in RadCom some of the images used in this book were produced in colour. Due to the printing process used in this book, the reproduction of these colour images has not been possible. If you find that you require a colour copy of an individual page we suggest that you refer to the attached CD Rom for it. Alternatively you can obtain physical copies from the RSGB Sales Department at a small charge per page.

Cover design: Dorotea Vizer, M3VZR
Desk top publishing: Steve Telenius-Lowe, 9M6DXX
Production: Mark Allgar, M1MPA

Printed in Great Britain by Nuffield Press of Abingdon, Oxon

# Contents

Preface .................................................................................................................... iii

Technical Topics 2005 ............................................................................................. 1

Technical Topics 2006 ........................................................................................... 51

Technical Topics 2007 ......................................................................................... 100

Technical Topics 2008 ......................................................................................... 148

Index .................................................................................................................... 172

# Preface

This fifth of the series of *Technical Topics Scrapbooks* includes the final 40 monthly columns of the regular 'Technical Topics" ('TT'), as published in *RadCom*, the Journal of the Radio Society of Great Britain, between January 2005 and April 2008, thus marking the completion of 50 years of 'TT' that began in a modest way in April 1958 and progressed to an overall total of over 600 articles.

The first collected edition of items culled and collated from 'TT' appeared in 1965 under the title *Technical Topics for the Radio Amateur*. An enlarged second edition followed in 1968 with a new title *Amateur Radio Techniques* ('ART'). Further revised and progressively enlarged collated editions were published in 1970, 1972, 1974 (reprinted 1978) and 1978 with a final (seventh) edition in 1980 (reprinted 1991). A total of some 40,000 books sold, although now all are out of print.

Publication in book form began again in 1993 following a suggestion by Mike Dennison, G3XDV, then the RSGB's Publications Manager, with the first of the quinquennial volumes, each providing the complete pages of 60 monthly 'TT' articles reproduced as originally published in the journal, but each volume with a Preface and Index. The title became *Technical Topics Scrapbook* plus the years covered: 1985-89 appeared in 1993; 1990-94 in 1999; 1995-99 in 2002; and 2000-2004 in 2005.

A collected edition of 'TT' antenna material appearing in the journal between April 1958 and December 1999 was published in 2002 under the title *Antenna Topics*.

These *Scrapbooks* do not aim to compete with or displace the standard amateur radio *Handbooks* but rather extend the reader's awareness of and interest in trends, useful new circuits, new components and devices, and new and updated antennas - what makes and what unmakes a satisfactorily working amateur station. They are thus intended to provide ideas and source books for experimentally-minded radio amateurs and all those who seek to learn more about how the technology has developed.

This volume contains a wide variety of new or updated established topics ranging from crystal receivers to state-of-the-art H-mode mixers, the progress of stable oscillators from Clapp and Vackar to DDS synthesisers, the importance of reciprocal mixers and low-phase-noise oscillators, the birth and development of transistors and other semiconductors, another look at Marconi's 1901 transatlantic experiments, 7360-valve type mixers and product detectors, ergonomics of receiver and transceiver controls, a look at the experimental opportunities offered by the 500kHz assignments, a reprise of parametric up-converters, a novel mixer developed by RZ4HK and many, many other items. I also take further looks at the ZS6BKW (G0GSF) version of the G5RV antenna and the classic but often misunderstood Zepp antenna, and the developing interest in antennas for near vertical incidence sky-wave (NVIS) propagation of particular interest for emergency working on 5MHz. Interest in electrically short antennas continues and although I, as well as the vast majority of professional antenna engineers, have long shown scepticism of some of the claims made for the radiation efficiency of small transmitting loops, there is no doubt that small loops have a role to play in restricted spaces. Experimental opportunities on 500kHz are also explored.

When, 50 years ago in 1958, 'TT' began to appear in the Society's journal (then called the *RSGB Bulletin*) it soon became evident that the regular bringing together of a wide-ranging collection of new and some older but often overlooked circuit ideas, antennas, test instruments, together with general hints and bright ideas for running an effective radio station on a modest budget - having in common only that they were practical, reasonably safe and relevant to day-to-day operation - fulfilled a real need. Until then, technical information in the *RSGB Bulletin* had tended to be presented in a formal, rather forbidding, manner with a heavy lacing of design mathematics. In contrast, a readable KISS ('Keep It Simple, Stupid') approach has always been a prime aim of 'TT', even when dealing with complex new technologies.

Let there be no mistake, amateur radio still has much to offer and many threats to overcome. In the 1950s, the most pressing problem was how to reduce interference from amateur transmitters to the reception of VHF television. Today and in the future we shall have to face up to the problem of high receiver noise levels resulting from the widespread operation of high-speed, broadband transmission of digital data over telephone lines and over external and in-house electricity power lines. The myriad of low and high power switched mode power units found in domestic equipment is adding to the burden of combating locally-generated electrical noise.

Reduction of such interference seems likely to require amateurs with a good grasp of both RF analogue and digital technology. Electrical noise is becoming an increasing problem for many amateurs calling for new trial-and-error experimental approaches. The more relaxed attitude to spectrum utilisation now displayed by Ofcom and other regulatory authorities must be of concern to the DX operator. Even in these days of factory built equipment and the much-maligned 'Consumer Appliance Operators', amateur radio is inherently different from 'fit and forget' domestic electronic products.

50 years ago, in the first 'TT', I vowed that the column would survey ideas appearing in the technical press as well as providing hints and tips coming to my notice, with an occasional comment thrown in for good measure. It is up to you, as a reader, to decide how well, or how poorly, these promises have been met. For my part, writing the 600 columns of 'TT' has been part chore, part pleasure and has resulted in my making friends in many parts of the world by radio or correspondence or, in some cases, personal meetings. Much of the effort has not been producing the words but the long hours spent in seeking out new material; dealing with the correspondence engendered; and checking out references in my chaotic filing arrangements.

I was surprised but pleased to learn in 2006 that I had been recommended (apparently by a group of 'TT' enthusiasts) to receive an MBE in the Queen's 80th Birthday Honours List "for Services to Radio Communications". I duly received my gong from the Prince of Wales at an investiture at Buckingham Palace on 14 December 2006. My grateful thanks to those concerned in bringing this about!

I must also record my thanks to all those who have contributed to 'TT' during the period covered by this fifth volume of *TTS* and to the RSGB editorial staff who handled my monthly contributions, including Steve Telenius-Lowe, G4JVG; Dr George Brown, G1VCY / M5ACN, and Giles Read, G1MFG; to those responsible for the technical illustrations, and also to Mark Allgar, M1MPA, the RSGB's Commercial Manager responsible for book production.

*Pat Hawker, MBE, G3VA*
*London, March 2008*

**Pat Hawker,** G3VA

37 Dovercourt Road, Dulwich,
London SE22 8SS.

# TECHNICAL TOPICS

# TT

G3VA takes a careful look at product and envelope detectors and highlights their performances ♦ Using super-bright blue LEDs in high-level mixers ♦ German and Russian receivers under the spotlight

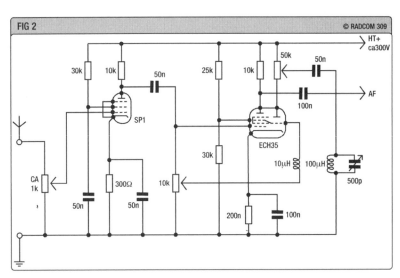

Fig 2: Very simple form of Synchrodyne MW broadcast receiver as described by Prof D G Tucker in 1947. He also described a rather more complex receiver having higher gain.

**DEMODULATION - PRODUCT AND ENVELOPE**
'TT' January, 2004, page 45, included details of a pentode (6SJ7) product detector with in-built carrier oscillator that Peter Chadwick, G3RZP, fitted to his rebuilt HRO to improve reception of SSB: **Fig 1**. Subsequently, Jan-Martin Noeding, LA8AK, sent me information on some of the results of his delving into when product-type demodulation came into use by amateurs for SSB/CW reception.

But first it seems opportune to stress that the technique of using an inserted signal at carrier frequency to provide a 'mixer' (frequency converter) with an IF of 0kHz to form an RF detector was known many years ago, long before the term 'product detector' came into use in the early 1950s.

In the 1920s and 1930s, when the 'straight' (TRF) receivers still reigned supreme, 'leaky-grid' or 'anode-bend' detectors using triodes or pentodes were the norm, giving way to the diode envelope detector for the new superhets. For CW reception, regeneration (reaction) or a local beat frequency oscillator (BFO) provided an audio heterodyne and increased gain. All these were essentially non-linear RF rectifiers and tended to introduce distortion at low signal levels. But one should not knock the diode 'envelope detector' where both sidebands are transmitted. Philip F Panter in his classic book *Modulation, Noise & Spectral Analysis* (McGraw Hill, 1965), concludes that "for high-input signal-to-noise ratio, the envelope AM detector is as efficient as the product AM detector". If it had not been for the expansion of amateur SSB in the 1950s, diode detectors would have remained the norm despite the fact that for weak signal CW reception the more linear product-type detector offers distinct advantages provided that the inserted carrier is much stronger than the signal input at the detector stage.

Even in the early days, there was the alternative, at least for CW reception, to diode rectification of RF signals. of using the more linear frequency-conversion demodulator (oscillating detector), in what was, in the UK, termed a homodyne receiver: **Fig 2**. For conventional double-sideband-plus-carrier AM reception, this form of demodulation ideally requires that the local oscillator is phase-locked (synchronised) to the incoming carrier. Various forms of exalted-carrier and 'Synchrodyne' receivers were developed in the 1940s which largely overcame the severe distortion on AM signals introduced by frequency-selective fading, but did not take off commercially, Widespread 'consumer' use of fully-synchronous demodulation did not arrive until the coming of colour television where it is used to demodulate the colour sub-carrier.

**Fig 3** shows a very simple synchrodyne receiver front end as described in 1947 in *Electronic Engineering* in 'The 'Synchrodyne' – a New Type of Radio Receiver for AM Signals', by Professor D G Tucker. This had an untuned (aperiodic) RF amplifier followed by a triode-hexode product-type detector to which a portion of the incoming RF signal is fed to lock the variable oscillator. A higher-gain receiver used a balanced ring demodulator to remove the carrier of the AM signal while a newly generated carrier is phase-locked to the original AM carrier using a Goyder-lock type approach. This provides synchronous (coherent) reception of AM broadcasts. It would be an interesting project to develop a solid-state synchrodyne-type direct-conversion receiver for broadcast or amateur AM reception.

It is, of course, possible to transmit double-sideband AM with suppressed carrier (DSB). A J Viterbi in his book *Principles of Coherent Communication* comments: "The only advantage of SSB/AM over DSB/AM is that the transmission bandwidth is reduced by a factor of two. It is a common misconception that the output signal-to-noise ratio for SSB/AM is double that for DSB/AM... coherent demodulation yields the same performance for both systems."

Homodyne reception of CW or SSB signals, as in most modern 'direct-conversion' receivers, does not require phase-locking of the local oscillator. In its simple form, it does not provide 'single-signal' reception, and has the audio image signal present. It is, however, usually possible to demodulate an AM transmission temporarily without true phase locking by careful adjustment of a stable insertion oscillator. Homodyne receivers have a long history and were originally postulated for broadcast AM as well as Morse reception.

A detailed paper 'The History of the Homodyne and Synchrodyne', by Professor Tucker appeared in *J Brit IRE*. April 1964, pp143 – 154 (with 62 references). This credits the original homodyne to F M Colebrook, *Wireless World & Radio Review*, Vol 13, 1934, pp645 – 648: Fig 2. Colebrook pointed out that the benefit obtained from the system was in providing linear (ie distortionless) rectification (demodulation). An exalted-carrier receiver

Fig 1: The simple product detector/insertion oscillator using a single pentode used by G3RZP in his rebuilt HRO, suitable for fitting into many older type receivers designed before the introduction of amateur SSB – see 'TT' January 2004, p45.

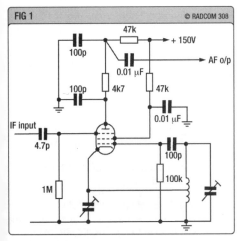

# TECHNICAL TOPICS

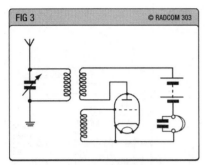

Fig 3: Colebrook's homodyne receiver of 1924.

was patented in 1930 by E Y Robinson using a high-$Q$ crystal filter to remove the sidebands from the carrier which was then amplified and reinserted: **Fig 4**.

However, in a later paper 'The History of Positive Feedback' (up to around 1923), *Radio and Electronic Engineer*, February 1972, pp59 – 79, Prof Tucker notes that the homodyne technique was a logical development of H J Round's autodyne (regenerative) receiver, with Burton W Kendall in 1915 being the first to mention that increased sensitivity could be obtained in a receiver if a local carrier were added to reinforce the incoming wave. Kendall went on to describe a self-oscillating detector.

*QST* in the mid-30s published some simple designs of homodyne-type receivers for HF CW reception (although the term 'homodyne' was not used). An article 'Increased Sensitivity with the Regenerative Detector', by Rinaldo De Cola (*QST*, December, 1934, pp24 – 26), included a suggested circuit for a regenerative but non-oscillating detector with separate heterodyne oscillator: **Fig 5**.

An advantage on CW with any form of linear demodulator is that the resulting audio waveform is ideal for narrow-band audio filtering. With a well-designed but simple direct-conversion receiver, despite the presence of the audio image, the overall selectivity can be excellent; two sharp, clean signals can prove more effective than broader single-signal superhet reception, as well as providing the opportunity to choose the audio signal either side of the carrier frequency, one of which may be subject to less interference.

The use in the 1930s of a triode-hexode mixer valve to form a frequency-conversion demodulator detector has been noted by LA8AK. He draws attention to the 1936 Telefunken Spez 801 design (with only a Spez number it is possible that the receiver never went into full production). The design had two RF (RE|NS 1284 + RES094), ACH1 triode-hexode mixer, and three IF stages (3 x RES094), with an extra RES094 for amplified AGC with four copper-oxide diodes as AGC rectifier in a voltage-doubler circuit. This is followed by another ACH1 triode-hexode in what we would now call a product detector feeding AF to an RE134 output valve via an optional AF filter. IF was 600kHz with five double-tuned IF transformers but no crystal filter. A later (1937) version provided positive and negative feedback in the third IF stage ($Q$-multiplication) to sharpen-up IF selectivity. This technique appeared in the 1939 *ARRL Handbook* as a 'High-Selectivity Regenerative IF-amplifier Circuit'. Regenerative IF or 2nd detector stages were a common feature in the SOE and Anglo-Polish clandestine receivers, and remains an effective approach.

LA8AK points out that the ACH1 'product detector' would not have been intended to demodulate SSB signals as, at that time, the few HF SSB commercial point-to-point links using pilot carriers were received on large 'commercial receivers' rather than on general-purpose communications receivers. The IF output compared with the BFO would have been excessive for SSB although the stage would have functioned well as a product detector for CW reception. As noted above, for optimum SSB demodulation, it is important that the inserted carrier should be very much stronger than the SSB signal presented to the product detector.

It was not until the 1950s after the introduction of single-sideband-suppressed-carrier transmissions (see 'TT' August 2004) that the term 'product detector' began to appear in the amateur journals. LA8AK has traced an article by the late Byron Goodman, W1DZ, 'A Sharp IF Amplifier for Phone and CW', *QST*,

December 1950, using a 6BE6 heptode valve to convert from 455kHz to 50kHz and then a 6BE6 and 6C4 triode in what is termed an 'AF Converter' but is, in essence, a 'product detector'.

I have traced an article by the late O G Villard, W6QYT, 'Selectivity in SSSC Reception' (*QST*, April 1948 pp19 – 22), which discusses 'frequency conversion versus rectification' demodulation in respect of SSB reception. This includes circuit details of a balanced frequency converter for single-sideband reception using two 6L7 valves in a parallel-input, balanced-output, fed with balanced input from a BFO and intended as a substitute for the conventional diode detector.

Michael, VE2BVW, in correspondence with LA8AK, also cites the W6QYT article as a reprint in the ARRL's *Single Sideband for the Radio Amateur* (1954) with the following page including a circuit labelled 'the product detector' with an editorial note claiming "Here is a useful detector for SSB reception, devised by Murray Crosby, W2CSY. It gets its name from its operation – the output signal is proportional to the 'product' of signals in two channels". This is reprinted from an article 'An All-Purpose Super-Selective IF Amplifier' (*QST*, March, 1953). The book adds a note: "Product versus Envelope Detectors – The interesting and significant thing about a product detector is that there is no output with the BFO turned off. Unlike an envelope detector, where two or more signals coming in will give a beat or beats, the product detector requires that the BFO voltage be present. Thus it is very similar to a mixer or converter stage, which also gives no output unless oscillator voltage is applied. The advantage of the product detector is consequently that the output voltage consists solely of beats with the BFO and not cross-modulation beats between signals."

VE2BYW mentions that Don Stoner in his *New Sideband Handbook* (1958) also credits Murray Crosby as inventor of the 'product detector', but does not give a date or reference. There is no use of the term in the classic SSB issue of *Proc IRE* (December, 1956), although it is clear that, by then, the advantages of product-type SSB demodulation were well-known to both amateur and professional operators. SSB adapters using product-type demodulation that could be added to earlier existing communications receivers were noted by George Grammer, W1DF, of ARRL in his paper in that issue on the increasing use of SSB by radio amateurs: **Fig 6**.

## HIGH-LEVEL MIXERS USING LEDs

D G Phillips, G8AAE, while searching the Internet came across an article

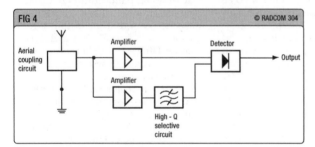

Fig 4: Robinson's carrier-reinforcement (exalted-carrier) system patented (UK) 1930 for which he developed the highly-selective crystal filter, later adopted for a single-signal communications receiver by Lamb of ARRL.

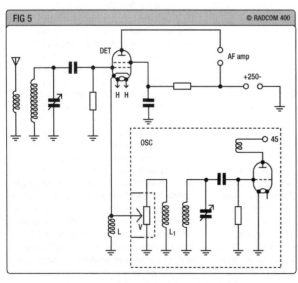

Fig 5: A circuit suggested in QST in 1935 for a regenerative but non-oscillating detector with separate heterodyne oscillator, leading to the description of CW TRF receivers using this form of what was to become known as product-type detection.

# TECHNICAL TOPICS

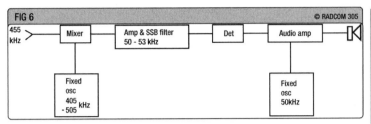

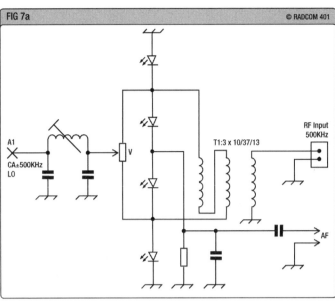

Fig 6: Form of SSB Adapter as used with receivers not intended for SSB reception as noted by George Grammer, W1DF, in the classic 1954 SSB issue of Proc IRE.

Fig 7: Use of GaN LEDs in mixers/product detectors. (a) Singly-balanced detector with potentiometer, V, splitting the BFO signal to the two arms to improve balance of non-matched LEDs and provide better isolation of the BFO. (b) Super-blue LEDs in a doubly-balanced mixer.

'Super-Blue Mixer', by David White, WN5Y, in the March 2004 newsletter of the Flying Pigs QRP Club (new to me). This led him to US Patent 6,111,452 'Wide Dynamic Range RF Mixers Using Wide Bandgap Semiconductors', which names the 'inventors' as Christian Fazi and Philip G Neudeck. The patent, filed February 21, 1997 was issued 29 August, 2000, is assigned to "The USA as represented by the Secretary of the Army".

To quote the patent abstract: "A wide dynamic range RF mixer is shown using wide bandgap semiconductors such as SiC, GaN, AlGaN or diamond instead of conventional narrow bandgap semiconductors. The use of wide bandgap semiconductors will permit RF mixers to operate in higher RF environments, be less susceptible to out-of-band jamming and interference, and be more effective in receiving weak RF signals in the presence of strong unwanted signals. RF receivers can be more closely co-located with transmitters and still receive weak signals without suffering intermodulation-distortion products."

The patent provides an excellent discussion on the operation of solid-state diode switching mixers stressing: "It is important that the mixer should not add an undesirable signal to the incoming wave, since it would be amplified indiscriminately with the desired signal. Low noise, strong nonlinearity, repeatable electrical properties from device to device, and adequate dynamic range are important characteristics in mixer design and selection. The useful dynamic range is bounded by the noise level of the mixer and the level at which the mixer can no longer linearly process the incoming RF waveform... Current methods to increase the saturated output level of mixers and thereby reduce IMD products have focused on increasing the LO power level of the device by increasing the number of diodes used in the mixer circuit... for example by increasing the number of diodes used in the mixer circuit, perhaps via multiple-diode balance circuits. Fabrication and matching of multiple diodes, however, is difficult... increasing the cost of such multiple-diode mixers."

It is noted that while most power semiconductor devices in use today are fabricated in monocrystalline silicon, monocrystalline silicon carbide, SiC, with a wide bandgap (above 3eV) is particularly well suited. In addition to SiC, another wide bandgap material, GaN, is becoming available. Currently GaN (gallium nitride) is being used to make blue-light-emitting diodes and should prove equally effective for the same reasons.

In the *QRP Newsletter*, KN5Y acknowledges that a friend, Pascal Nguyen in Australia, had suggested to him in 2000 that the diodes in a balanced or doubly-balanced (ring) diode mixer should be replaced by super-blue LEDs. KN5Y points out that one of the recognised ways of improving ring mixers using diodes such as 1N4148 or 1N914s (as commonly used in home-brew ring mixers) is to use additional diodes in each leg of the mixer. Since this raises the forward voltage drop in each leg, it shows that a higher voltage drop in a diode string does not stop the mixer from working but does require significantly-increased drive.

Blue LEDs are made of GaN and are now widely available from electronic stores and some surplus outlets: see 'TT' February, 2004. The patent provides a comparison between a conventional silicon diode ring mixer and a mixer using SiC diodes in the same environment. The results are impressive, although it should be noted that the LO power for the SiC mixer is 20dB higher to provide the same 10dB loss because of the higher turn-on voltage needed. KN5Y suggests that results should be similar using super-bright blue LEDs. It is thus not possible simply to replace the diodes in RF mixers or product detectors without providing significantly more LO drive.

KN5Y provides circuit details of his experimental 'Super-Blue Mixer', but I find some aspects rather confusing and feel it better just to show basic single- and doubly-balanced diode mixers fitted with LEDs, **Fig 7**. In forwarding this information on the use of super-blue LEDs in diode switching mixers, G8AAE stresses that the KN5Y mixer is clearly an experimental project with no performance measurements. However, he feels that the idea may be worth exploring by those with the necessary knowledge and test equipment.

While it is always worth pointing out that the issue of a patent is no guarantee that the idea will work as claimed, there seems no reason to doubt the validity of the US patent on wide bandgap semiconductors.

## NO GUARANTEES

'TT' has always aimed at providing a launching pad for controversial experimental ideas and techniques. This inevitably means that some claimed results may not always prove reproducible, although in its 53 years I would claim that there have been very few real booboos in 'TT', although inevitably some minor errors. Reader contributions are not subject to peer review, although I retain the right to comment, sometimes adversely, on any material published in 'TT'.

Antenna topics are particularly difficult to assess, even by professionals. Indeed, of the very few 'TT' items with claims that subsequently proved impossible to reproduce, most have been drawn from professional journals where papers *are* subject to peer review. One such was the paper by American academics that extremely compact meander antennas with a span on 14MHz of only one-metre (August 1999) could have an efficiency and bandwidth virtually equal to that of a conventional half-wave dipole – an item drawn from

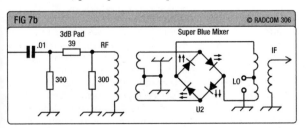

# TECHNICAL TOPICS

*IEEE Trans on Antennas and Propagation.*

I feel relieved that 'TT' has never accepted uncritically the claims made for the CFA, CFL or EH antennas, or the small loops claimed by G3LHZ to have radiation efficiencies >90%. Surely the G3LHZ claims have finally been put to rest by the comments from G3UUR, G8HQP and G0IJZ in the November *RadCom* (p100). If anyone is still in any doubt, there are further critical comments from G0GSF and VE2CV revealing further technical errors in the August and September articles. Personally, I feel happy to stick by my comments in the May 2004 'TT' (p42) in which I concluded that "G3LHZ should be encouraged to continue but to reconsider his >90% claims". Perhaps the most disturbing aspect of this 'controversy' is that his small loop beliefs were accepted without comment for publication in the IEE's *Electronic Letters* and were presented at an IEE Conference (2000) both subject to peer review by professional and academic engineers.

## GERMAN & RUSSIAN HF RECEIVERS

The November notes on the wartime Telefunken E52-series of communication receivers has attracted some valid comments from readers. Richard Walker, G4PRI, points out that the illustration at the top of page 38 was of the E52 owned by Arthur Bauer, PA0AOB, and not, as captioned, an E52a. I had not grasped that the first model in the series was the E52 (U-boat TSK44), with the E52a as well as the E52b, E52c etc, later simplified versions. The lower illustration was of the E52a, although the internal view may well have been that of E52. G4PRI adds that PA0SE contributed a further two-part article on the 'exceptional' E52 receiver in *Radio Bygones*, issues 78 (August/September 2002) and 79 (October/November 2002) using the set owned by PA0CSC for illustration.

G4PRI notes the excellence of the German field radios. He owns (and shows on his QSL card) a Telefunken Tornister (knapsack) portable battery-operated TRF receiver, type Torn E.b. His model is based on the 1836/37 Spez. 976bs design which saw Wehrmacht wartime service as the Torn E.b/24b-305, and was also exported as the AE95. The equipment could be operated from dry batteries or vehicle batteries with vibrator unit EWb carried in the battery compartment.

G4PRI described this elegant receiver in *Radio Bygones*, No 82, April/May 2003, pp4 – 7. It used four similar pentode valves (RV2P800) in a 2-V-1 configuration (2 RF stages, regenerative detector and AF amplifier for use with head-

G4PRI's German Torn E.b TRF mechanically-elegant backpack receiver with large turret coil assembly, as illustrated on his QSL card.

phones including an optional 900Hz filter) and covered 100kHz to 6670kHz in eight bands by means of a large turret coil pack. G4PRI wrote in RB: "It was my humble opinion that the German Torn E.b receiver was probably the most elegant TRF set ever made, if not the most elegant set overall. However, Arthur Bauer, PA0AOB, and Dick Rollema, PA0SE, have been at pains to point out to me that the Lorenz Lo.6K 39 was probably the most elegant TRF set. I also had to agree with them that 'the most elegant overall' was a title to be bestowed on the E52 [1943] series of receivers".

Neil Clyne, G8LIU, noted in the November 'TT' for his *RB* article on the E52 – has written questioning my comment that "the influence of the E52 can be seen in post-war high American and UK professional models". [I had in mind the modular construction and 52in illuminated film scale of the R206 designed by AT&E, although this set was not capable of the performance achieved by the E52 some 20 years earlier].

G8LIU feels it was an opportunity missed: "I would suggest that, for at least several years after WWII, apart from a few scientific establishments such as Farnborough and a handful of fortunate radio amateurs, precious few folks in the UK and USA knew, let alone cared about, the electrical and mechanical qualities of 'enemy' radio equipment. Only one of the former Allies capitalised on the technology of the E52 in a big way after WWII – the USSR".

He notes that several of the desirable features of the E52 "such as the elaborate system of frequency display with coarse and fine tuning scales, the latter being of the illuminated optical projection type enabling clearly-legible scale resolution to 1kHz or better in some cases [calibration reset from internal crystal oscillator]. The bandpass-type single-crystal IF filter was widely adopted; some early receivers (R310, R311) also followed the Wehrmacht technique of using a single valve type for all functions, although later equipment contained several valve types". G8LIU also notes Russian use of the comprehensive metering of the early E52. Again, the modular style of the E52 was widely adopted, with several Russian models using the large coil turrets similar to those found in many German military sets such as the Torn E.b, although not in the E52.

G8LIU believes that much of the Russian equipment may have been made in other Iron Curtain countries, notably Hungary and what was then Czechoslovakia. [High performance communication receivers influenced by the E52 etc also emanated from East Germany – G3VA]. He adds: "Production probably began about 1950 and continued to the end of the valve era in the communist bloc in the late 1970s. At least one E52-inspired Russian design, the massive KROT receiver (allegedly favoured by the KGB) was closely copied by the Chinese for use by their military, as also was the AR88!

"One can only speculate as to the possible results had the technology of the E52 been adopted by the West. A variable-bandpass-type signal-crystal IF filter in the RA17 perhaps, or an optical-projection frequency display with 250Hz resolution in the venerable Collins R390A instead of a Veeder-Root counter? We shall never know."

## HERE & THERE

George Cutsogeorge, W2VJN, recently sent me a copy of his excellent 72-page booklet *Managing Interstation Interference with Coaxial Stubs and Filters*, published in 2003 by International Radio Corporation, 13620 Tyee Road, Umpqua, OR 97486, USA (www.qth.com/inrad). In it, he sets out, in exceptionally clear terms, the basic problem of operating two or more transceivers in close proximity (applicable also when using an amateur transceiver very close to a broadcast or telecommunications transmitter or radar station) where there will usually be some level of interference involved. The level can vary from practically no problem to actually burning up components in the receiving radio (as noted in 'TT' November 2004, p39, in connection with the AR88 and HRO when co-sited with a transmitter). The purpose of W2VJN's book is to identify and quantify the various parameters that create the interference and to show methods involving coaxial stubs and filters that will reduce or eliminate it.

As usual, I am having to hold over a lot of material received from readers, but I must squeeze in a note that the massive (692 pages, A4 format, hard covers) *Wireless for the Warrior – Volume 4 – Clandestine Radio*, by Louis Meulstee, PA0PCR, and Rudolf F Staritz, DL3CS, was finally published in late October, 2004 by *Radio Bygones*. It provides technical information on some 230 sets used by Intelligence services, Special Forces, Partisans, Resistance circuits, Australian & Dutch coast watchers, intercept services etc, in and after WWII. A truly unique collection.

Peter Rovardi, G4HSB, of Rosedale, 8 Cambridge Road, Linthorpe, Middlesbrough, Cleveland TS5 5NQ (e-mail: info@grecobrothers.co.uk), has produced a useful, A4-size, laminated chart providing conversion of RMS voltage to peak-to-peak/dBm; power watts/dBm; and signal (based on 50Ω systems) S-units into dBµV (for under 30MHz and over 30MHz) together with explanatory notes. ♦

Pat Hawker, G3VA

37 Dovercourt Road, Dulwich,
London SE22 8SS.

# TECHNICAL TOPICS

## TT

G3VA looks at how electronic equipment communicates with its operators – ergonomics comes to the fore • pitfalls to be avoided when using vector network analysers • a novel form of micropower transmitter • more information on low dipoles

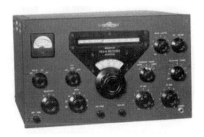

**Collins 75A-4 pre-WARC 'amateur-bands' receiver, often regarded as the best receiver designed purely for amateur radio.**

**The HRO-60 – the final 'valved' version of the late 1950s still bore a family likeness to the original HRO designed in 1934.**

**The classic Racal RA17 professional receiver, featuring the Wadley triple-mix drift-free loop system.**

**Fig 1: Expected relationships of movement between controls and displays etc. Some examples from *Ergonomics for Industry: 2 – Instruments and People*.**

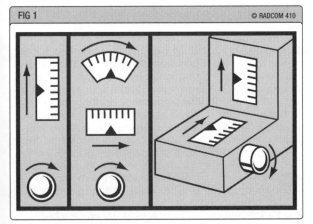

### CONTROLS & ERGOMONOMICS

Correspondence over the past year or so about the merits or otherwise of the now ancient HRO (1934 design) and AR88D (1941 design) as viewed from the 21st century has raised a number of pertinent questions as to what makes a communications receiver memorable – as, in respect of the HRO, Bernard Pettit, G3VD, Peter Watson, G3PEJ and Ian Brown, G3TVU have confirmed.

Clearly, the electrical performance of a receiver is important, but then also is its mechanical construction, including its ergonomics and overall suitability for the required purpose. An amateur whose main operational mode is Morse will not necessarily agree with someone who normally or exclusively uses SSB. Again, data modes put emphasis on stability and a level output that demands a good AGC system – whereas the CW operator can tolerate some drift and can be quite content to have no AGC, manually adjusting the RF/IF gain control as required – a standard practice in older receivers. Modern receivers with good AGC and switchable pre-amplifiers/attenuators often omit any RF/IF gain control.

Both the HRO and the AR88 have excellent, smooth, backlash-free geared drives that have seldom been beaten. But, to my mind, the HRO scores in having a larger, better positioned tuning knob, nicely spaced from the RF/IF gain control. On my AR88, I have replaced the original knob with a larger one, supported by an external flywheel taken from a discarded broadcast receiver. This is fine, but does not make for convenient use of the gain control which is positioned close to the tuning control. I use the knob taken from the AR88 on my old KW2000A transceiver which, again, has its gain control inconveniently placed at a higher level than the tuning knob. The receive/transmit switch is stiff and has been replaced by a larger knob (although for SSB I use the microphone PTT and for CW an external switch). Small matters, perhaps, but they make a noticeable difference.

In 'TT' March 1973, I quoted from an article by Roy Udolf and Irving Gilbert (*Electronics*, 4 December, 1972) and a small booklet *Ergonomics for Industry: 2 – Instruments and People* published by the former Department of Science & Industry Research (DSIR). These sources presented guidelines for the design and layout of front panels, etc, useful for both the home-constructor and when assessing factory-made equipment.

It was stressed that neglect of human factors leads to inefficiency in using instruments and control consoles. Instruments, receivers, etc need to *communicate* with the operator, and the communication is two-way since users generally have to do something, like turning a knob or selecting a switch, to obtain or to respond to the information.

Udolf and Gilbert pointed out that "laying out a display-control panel so that an operator can function efficiently requires more than just making sure that everything fits... not only must the panel layout insure a good man/machine interface, but often the location of certain controls and displays determines the layout of many critical internal components." A system that takes full account of 'human engineering' can be much easier and more pleasant to operate than one that does not.

Such common symbols as 'red for danger/stop' or 'green for OK/go' should be used, but their meanings must never be reversed.

People expect certain relationships between the movement of controls and their associated displays (see **Fig 1** for some examples). Pointers and knobs should be designed to avoid ambiguity (and preferably give some simple '1-9 calibration' that allows an operator to see at a glance whether a gain control is almost fully advanced or nearly minimum. Controls should always operate in the expected manner: fully clockwise for maximum effect, a toggle switch turned down for 'on' etc.

Panels should not be cluttered; unnecessary labelling (legends) should be avoided to eliminate operator confusion due to sensory overload. With panels viewed from the front, labels should preferably be placed *above* the controls to which they refer, and large enough to be read comfortably at the normal operating distance, never less than about 20in (50cm).

For amateur operation of a receiver or transceiver, the most important controls are the tuning knob, the RF/IF gain control (but see above) and the transmit/receive changeover system. Personally, with rather large hands and fingers, I like a smoothly-acting tuning knob to be

# TECHNICAL TOPICS

set centrally at a height of about 4in (9cm) above the desk, fairly substantial (2 – 3in diameter), well clear of all other knobs, providing a tuning rate of preferably not much over 5kHz/revolution. The illustrations show examples of some classic post-war receivers.

In 1972, I wrote: "Operating controls and knobs, far from improving in recent years, seem to have become more fiddling to use; a real problem with miniaturised equipment is that our fingers and thumbs have not been subjected to a similar process. One exception is the modern toggle switch that is slimmer and more elegant and easier to use. Operability is more important than achieving perfect symmetry although panels that are easy to use are usually those that are visually attractive. A lot of this is applied common sense rather than specifically ergonomics."

Thirty-three years later things seem to have got worse rather than better with front panels crowded with umpteen knobs and push-switches. Whereas the AR88 has 11 front-panel controls, my KW2000A 12, PA0SE's transceiver ('TT' October 2004) 24. Modern top-of-the-range transceivers can have over 73 knobs and push-button controls on a crowded front-panel.

The miniature hand-held remote control units for domestic equipment can be even worse with their tiny buttons and illegible legends. My four units (TV; radio/tape-cassette/CD player; Freeview digital adapter with hard disc recording facility; and analogue VCR) together have a total of some 140 push-buttons. One unit is so designed that the curved end is at the *rear*, consistently leading to my picking it up and attempting to use it the wrong way round. Altogether it is not surprising that I often find myself pressing the wrong button on the wrong unit – no wonder that 'technofear' affects all but the young!

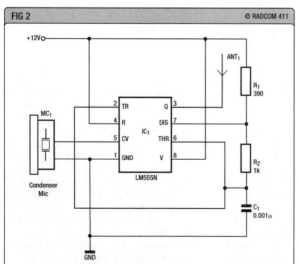

Fig 2: Micropower AM transmitter/signal generator. (Source: Electronics World)

### G3RZP ON VECTOR NETWORK ANALYSERS

The August, 2004 'TT' item on the N2PK vector network analyser (see also 'In Practice' *RadCom*, October, 2004) has prompted Peter Chadwick, G3RZP, to add a warning on some aspects of the use of this type of instrument, whether home-built or a high-cost professional laboratory model. He writes: "One point that seems to get overlooked is the terrible inaccuracy available once the impedance levels being measured move far away from 50Ω. Because the analyser depends, for the measurement of impedance, on measuring the difference between forward and reflected powers, then at impedances around 1000Ω or more, the difference will be very small, and an error of 0.05dB can lead to very large (>30%) errors. A classical example of this can be seen when measuring something like a 1in square patch on a slightly larger ground-plane, and sweeping 400 to 2500MHz.

"Taking all the precautions advised by the manufacturer, you can still end up with the line going outside the boundary of the Smith chart – and this on a pretty expensive professional machine just back from calibration. So, for impedances well away from 50Ω, say SWRs of 10:1 or greater, there are times when the answer from a network analyser needs to be treated with some care, and alternative measurement techniques are desirable.

"Those based on a resonance method have some advantages, since frequency and Q (derived from 3dB bandwidth) can be measured fairly easily, and a change in those parameters allows calculation of the impedance with a reasonable degree of accuracy. Professionally, it's tending to become more important now that a lot of design is done on submicron CMOS and devices having high input impedances. To get round the Miller effect, it is quite usual to use a cascode circuit, just like valves – "*Plus ça change, plus c'est la même chose*" – as Alphonse Karr first put it! Interestingly, just as valves had induced grid noise at UHF, MOSFETs have induced gate noise by pretty much the same mechanism.

"A very good book on measurement techniques is Hartshorn's *Radio Frequency Measurements* – although published in 1940, much of it is basic theory and thus still applicable. Useful too is *Radio Frequency Laboratory Handbook*, by Marcus Scroggie, one-time G(M)6JV. It still has a lot of useful material in it."

### MICRO-POWER AM 'TRANSMITTER'

In the early 1920s, a few amateurs used low-power 'valve-less' transmitters based on relaxation oscillation with large domestic-type neon bulbs used as electronic switches. Apparently, these could be made to oscillate at frequencies up to about 2MHz and were used on the allotted wavelengths of just under 200m, as permitted for some years when post-war amateur licences were issued from 1920 onwards.

This early practice is reflected in a short-range (up to about 25m) simple micropower AM 'transmitter' described by Raj Gorkali of Kathmandu, Nepal in the 'Circuit Design' feature of *Electronics World*, August 2004, p40: **Fig 2**. As described, it was intended for operation on about 600kHz to provide a link with a nearby domestic MW broadcast receiver and, strictly speaking, would contravene UK regulations. However, it might prove interesting to see if component values could be changed to result in a frequency above 1.8MHz.

Alternatively, there is the possibility that the Society may obtain permission for operation in a portion of the old marine band around 600m. Raj Gorkali points out that it could also be used as shown as a form of signal generator in conjunction with AM broadcast receivers.

The only active device is the well-known 555 IC as a free-running multivibrator whose frequency can be set above 540kHz, governed by the values of R1, R2 and C1. Values shown are for about 600kHz (500m) and the frequency can be varied by simply replacing R2 by a variable resistor or C by a ganged variable capacitor (350 or 500pF). A condenser microphone is used and the device operates from a 9V battery. The suggested antenna is a 2-3m wire connected to pin 3.

Might or might-not prove an operational 1.8MHz micropower transmitter even with a good long-wire antenna, but it could be a fun project!

### MORE ON LOW DIPOLES

In 'TT' December 2004, pp33 – 35, Dave Gordon-Smith, G3UUR, contributed some very pertinent remarks on the ground-losses incurred with low antennas, stressing that "most amateurs don't realise that horizontal antennas induce so much loss in the ground when they are relatively low. It's not just antennas fed against ground that suffer from ground loss... Mind you it might take a bit of digging in the literature to find figures for a dipole over real earth... a paper in *Wireless Engineer* by Sommerfeld, about 1942, gave a theoretical treatment of the subject..."

Letters received from John Pegler, G3ENI, and Dr Brian Austin, G0GSF, show that in the years following the appearance of the paper by Sommerfeld, there have been

# TECHNICAL TOPICS

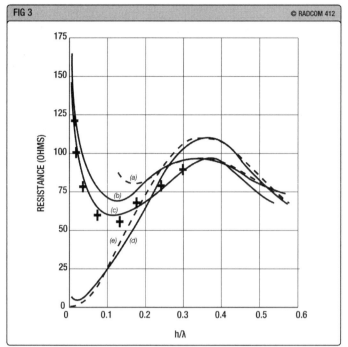

Fig 3: Proctor's 1950-measured input impedance of a resonant half-wave dipole above the ground, with G0GSF's '+' points from EZNEC data ($\sigma = 10^{-3}$S/m; $\varepsilon r = 15$). h = height above ground. (a) Calculated k =5. (b) Dry ground. (c) Wet ground. (d) Conducting mat a/λ + 0.0003. (e) Calculated k = infinity.

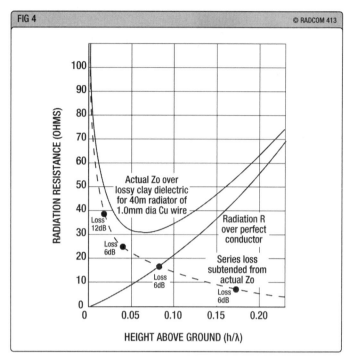

Fig 4: Radiation resistance (driving point impedance) of a half-wave resonant dipole less than a quarter-wave above ground (VK3MI).

several professional and academic research projects aimed at this important topic, including his own research 20 years ago in South Africa.

G3ENI writes: "I also have been frustrated over the years by inaccurate graphs, also several other cases of the 'Father to Son Syndrome'. From my early notes, I see that M J O Strutt wrote papers, published in German, on the subject in 1929 and 1939. K F Niesen followed in 1935 and 1938. In 1942, A S Sommerfeld and F Renner published their 'Radiation Energy and Earth Absorption for Dipole Antennae', with a wartime English translation in *Wireless Engineer*, (Vol XIX, No 227/228/229 August, September, October 1942, pp351 – 59. 409 – 14 and 457 – 62). Sommerfeld had previously published in 1909 the classic treatise on which all our ground-wave propagation theory is based.

In 1949, R F Proctor who was at the Signals Research and Development Establishment (SRDE) published a paper 'Input Impedance of Horizontal Dipole Aerials at Low Heights [above the ground]', (paper No 962, Radio Section) [G0GSF notes that this paper was published (1950) in *Proc IEE* 97, 5, pp188 – 90, see below]. In conjunction with colleagues B J Starkey and E Fitch, he investigated the behaviour of a dipole above dielectric ground at various heights and permittivities. A family of curves in the paper showed the rise of input impedance at different low heights for dry and wet grounds with different conductivity and also over different sized semi-transparent conducting mats. Other interesting findings were the changes in resonant length of the dipole, sometimes up to 10%, with varying height. All the above con- firmed the Sommerfeld and Renner mathematical analysis. All the impedance measurements were carefully taken with a bridge designed by the author with suitable allowance made for the changing reactance components and element lengths."

G3ENI adds: "Finally, please refer to Fig 6.17 (p102) of G6XN's *HF Antennas for all Locations*, (1st edition, 1982) and adjacent text."

G0GSF writes: "My own research of 20 years ago in South Africa touched on this subject of antenna impedance and ground effects, and we made some measurements of the driving point impedance of horizontal wire antennas as they were lowered towards the ground. Its increase at antenna heights of less than λ/10 is quite marked, and this indicates the significant increase in loss resistance caused by the antenna's interaction with the lossy ground. I've added some of the data points we measured in 1984 to **Fig 3** (basically from the 1950 paper by R F Proctor). The effect of this increase in resistance is to cause the 'radiation efficiency' of the antenna *in situ* to decrease, and this can be estimated from these results, as I've indicated in the box alongside. In view of this effect, I question the use of the term 'radiation resistance', as one sometimes sees it, to describe what is essentially the input or driving-point resistance, which consists of radiation resistance *plus* loss resistance. Radiation resistance essentially refers to 'useful' radiation whereas the energy coupled into the ground is usually considered to be lost, except, of course, if the intention is to radiate into the ground (or, indeed, the sea)."

G0GSF points out that we now have the *NEC* code and its many variants that enable us to simulate the situation. He stresses that the agreement between simulation and measurement is very good indeed, a fact shown clearly in some of the 18 professional papers he has listed on the effects of the ground on antenna impedance. He has added some of his own computed data from *EZNEC* to Fig 3. He adds: "Given the good agreement between measurement and simulation I'm a little surprised by G3UUR's assertion that *NEC* (and even *NEC-4D*) show 'next to no induced ground loss' for the case of small loops over average ground whereas measurement (and theory) indicate significant loss."

G0GSF has supplied a detailed list of 19 professional papers on the effects of the ground on antenna impedance from such journals as *IEEE Trans on Antennas & Propagation, Electromagnetics, Proc.IEEE, Radio Science*, etc.

'TT', February, 1995 (see also *Technical Topics Scrapbook*, 1955-1999, pp14 – 15) included a summary of part of a five-page article by William A McLeod, VK3MI, 'Low Radiators and High Ground Planes', reporting his investigation into antenna behaviour "at what seems to be astonishingly small heights above ground" to quote an editorial comment.

To repeat part of the summary given in 'TT' 1995: VK3MI points out that for 7MHz a height of 10m is a bare quarter-wave above ground, on 3.5MHz only an eighth-wave. This raises the question whether, in practice [particularly for NVIS paths], it is worth striving even for this height. What sort of performance can be expected from horizontal antennas only a metre or two above ground? VK3MI summarises

# TECHNICAL TOPICS

the factors involved with low horizontal radiators as follows:

- For low practical heights, the radiation resistance [driving impedance] at the centre of a resonant dipole remains within the 2:1 VSWR range for the usual coaxial cable feeder, so matching procedures are minimal, more so when an electrical half-wave of cable is used to transfer the centre impedance directly to the transmitter: see **Fig 4**.

- Whereas the resonant length of a dipole remote from ground is determined mainly by the length-to-diameter ratio of the conductor, when the ground becomes an increasing part of the dielectric, the length is determined by the height-to-diameter ratio. Due to the wide spread of dielectric constant, no simple formula can determine this ratio.

- The losses increase as height decreases towards ground level, but do not become prohibitive until very low levels are reached; for a 7MHz dipole above common clay, this can be as low as $\lambda/40$ (1m). [In desert conditions, an antenna can be laid directly on the sand, or even buried a few inches below the surface and yet still radiate reasonably well. VK3MI shows that 'saggy' dipoles are better than 'droopy' antennas at low heights – *G3VA*]

- The 'cone' of radiation directed vertically, then reflected back from the ionosphere, can produce non-directional communication with 'no –skip distance' to some 400 – 500km. This is NVIS transmission and is the mode supporting most of those semi-local nets on the 3.5MHz and 7MHz bands. There is usually some fading but, for SSB reception, the long AGC time-constant of the receiver will alleviate this.

- Two- or three-hop transmission can occur where the intermediate reflection points fall at sea, so some long distance working is possible in these favoured directions without low-angle transmission lobes. Land reflection points include greater losses which soon become excessive.

There is a good deal more relevant information in VK3MI's 1994 article but his 'low dipole conclusions' are as follows: "In general, the resonant horizontal dipole is an effective radiator at very low height from ground, particularly for NVIS transmission. Losses increase seriously below $\lambda/30$ (only 1.5m for the 7MHz band) and the high impedance ends of the

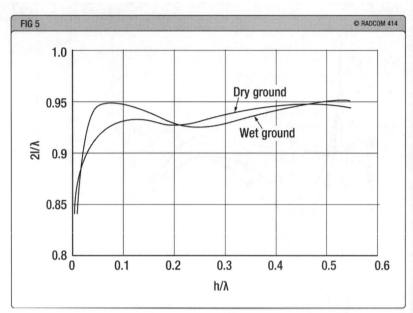

Fig 5: Resonant length of half-wave dipole above ground. h = height above ground. l = length of dipole (Proctor, 1950).

elements should have at least this amount of separation from ground or metallic earthed objects, towers and poles. Kevlar, black Dacron, polypropylene baler twine, and nylon rope are all suitable insulating supports with far less end-effect than the single egg-shaped strain insulator wired back to a steel tower which has been commonly used... With the elements double-insulated inside the popular 132mm polypropylene garden irrigation piping erected at 1.5m on the post side of a suburban wooden fence, a very effective concealed radiator should result. For portable use, a couple of 4m bamboo poles for end supports and a saggy dipole radiator require no apology as to effectiveness for NVIS transmission, but directivity, if any, depends on local obstructions and reflectors."

VK3MI also discusses the use of quarter-wave radials to form an effective ground-plane, not only for the classical 'ground-plane' antenna, but also as an artificial earth for inverted-L antennas and the like. He shows that this can contribute lower losses than an earth rod when elevated say 1m or so above real earth. He notes that the electrical quarter-wave can be significantly less than a physical quarter-wave: see **Fig 5** from the Proctor paper. As in the case of horizontal dipoles, where the radials are at a low height, the ends of the 'radials' should be higher rather than lower than the feed-point, keeping the high RF voltages some 1.5m or so above real earth. This is the opposite to the usual form of ground-plane radials where the base of radiating element is elevated several metres above earth.

VK3MI also stresses that safety is an important consideration for low radiators and for low elevated ground planes. One part is physical... the other aspect is electrical as, even at low power, a nasty sting and RF burn can occur which, for non-technical people or for climbing children, can produce an emotional reaction far in excess of the initial injury. At medium power, around 100W, these effects can be severe and, for greater powers, the effects of corona and irradiation must also be considered.

Readers seeking a copy (two A4 sheets) of G0GSF's list of professional references should send me a stamped addressed envelope plus one stamp to cover photocopying costs. I believe that the IARU journals copyright agreement would also permit photocopying of VK3MI's five-page 1994 *Amateur Radio* article (the SASE plus 2 extra stamps) for those really interested.

### HERE & THERE

The mention of a 24V low-voltage HT in one of the Marconi Marine emergency receivers (August 2004, p47) included an unfortunate error. As G3RZP points out, the pair of PL84s in the output stage provided only 20mW audio output *not* 20W! Hans, PA0TLM adds to the military sets working with low-voltage HT. He writes: "In the 1960s, I served for several years in the Royal Netherlands Air Force as a radio technician. As a navigational aid, the aircraft had on board LW/MW receivers – the American AN-ARN-6. This receiver worked with 28V on the anodes of the valves which were normal high-voltage types 12SK7 (six) 12SX7 (four), 12SY7, 12SW7 and 26A7 (two). Only the two 26A7 were specially intended for 28V. I own one of these receivers and it still works fine, current drain about 3A. This shows once again that normal valves will also work with low anode voltage." True enough, provided you require milliwatts rather than watts of RF or AF output power. ♦

**Pat Hawker,** G3VA

37 Dovercourt Road, Dulwich,
London SE22 8SS.

# TECHNICAL TOPICS

**Pat unfolds more of the small-loops saga •
Wide-band diode and LED mixers • Hybrid
'autodyne' receiver • Premium receivers •**

## BEANO TOPICS

I cannot pretend that the publication of the January 'TT' in mid-December 2004, heralded for me a good start to the festive season.

Due to Christmas postal delays, my corrected proofs did not reach the editor until after the issue had gone to press, only the second time this has occurred in many years of 'TT'. The result was errors in diagrams (mostly minor but in Fig 6 the fixed oscillator block should have been connected to the [product] detector stage rather than to the audio amplifier block); some text errors also went uncorrected.

In the text I had added a note on the proofs of doubts about some of the claims made for the ATUs shown in Figs 3 and 5 of the December 'TT', and corrected the slip of the keyboard that claimed a span of "53 years" for 'TT' rather than the near 47 years since April 1958. Michael O'Beirne, G8MOB, has also pointed out that my reference to the Army 1960s receiver with a 52in film scale and modular construction built by AT & E should have been to the R210, not R206 (see later).

These were minor vexations compared with the receipt of protests from my daring once again to question the obviously sincerely-held belief of Professor Mike Underhill, G3LHZ, that he has shown that the classical antenna theory of small transmitting loops vastly underrates their efficiency (see later also).

But the unkindest cut of all was to see an advance copy of the letter from Andy Talbot, G4JNT ('The Last Word', *RadCom*, February 2005, pp96 – 97), in which he suggests that the contents of 'TT' are reducing *RadCom* to the status of a comic. Andy, it would appear, judges any reference to 'valves' as an unwarranted trip in nostalgia. He is, of course, entitled to his view and to express his views strongly. However, some 95% of the basic circuitry and mechanical facilities used in our equipment were developed in the thermionic era, and any feeling for the development of our technology must reflect this.

## THE G3LHZ LOOP

My continued scepticism (January 'TT') in respect of the radiation efficiency claims of over 90% made by G3LHZ for the small transmitting loops developed in conjunction with his student, Marc Harper, has upset, among others, Tony Wadsworth, MIEE, G3NPF. In a long e-mail, he comments on my belief that a disturbing aspect of the whole affair is that G3LHZ's claims have been published in peer-reviewed professional publications and presented at professional conferences. He considers this can only mean that I believe that "anything that has been professionally peer-reviewed to be technically suspect, but anything that is reviewed only by himself and his attendant sycophants [also condemned by G3NPF as 'so-called experts'] is almost, by definition, guaranteed to be 'the true faith'. Professor Underhill has always maintained that his theories should be proved, or disproved, by practical experiments carried out by others."

Let's get things straight. When G3LHZ submitted his letter to the IEE's *Electronics Letters*, the professional reviewer (a highly respected professional antenna designer) did not accept the efficiency claims as valid. I cannot trace any evidence supporting G3LHZ's claims from any professional antenna engineer. My comments as a technical journalist have been relatively mild, meant to be far from arrogant, and reflecting a high regard for Mike's past contributions to the technology of radio communication.

G3LHZ himself continues the attack on my comments: "Pat Hawker thinks that my 90% small loop efficiency claims should be buried and put to rest. But why? He has given no reasons. Perhaps he takes his lead from the orchestra (or is it choir?) of critics with opinions that do not appear to be supported by any practical efficiency measurements whatsoever. So I repeat my challenge to the critics – please devise a practical experiment that uses traditional physics to demonstrate that the efficiency of a small transmitting tuned loop on its own is less than say 10%. Please ensure that losses from any adjacent environment are not included. Why is there such reluctance to take up this challenge or even refer to it?"

There seem to be several points here. G3UUR in *RadCom* November 2004 showed, to my mind conclusively, that G3LHZ's method of measuring the power losses in his loops is fundamentally flawed, and I have yet to see any attempt by G3LHZ to contradict him. Again, long before G3LHZ launched his University research project, there appeared in *RadCom*, a two-part article 'Loop Antennas – Facts not Fiction', by Tony Henk, FIEE, G4XVF, (September/October 1991). Based on theory and practical measurements, G4XVF concluded: "radiation efficiency as a percentage was often in single figures (significantly below 10 per cent) whereas that of a resonant and well-matched half-wave dipole approaches 100 per cent."

One of the problems in rebutting the G3LHZ hypotheses is that he seems to redefine the rules and IEEE definitions as he goes along. For example, his request above that "losses from any adjacent environment are not included" would suggest that the only way of testing his figures would be to put a small loop in free space – a practical impossibility. I am not going to rehash the many detailed measurements (not only computer simulations) carried out over many years by Dr John Belrose, VE2CV, whom I imagine G3LHZ would define as one of the "orchestra" (though few would think of Jack as an "establishment figure").

Mike believes that I have been persuaded by an orchestrated opposition in a way that could damage my "undoubted reputation". But he does not hesitate to denigrate the classic electromagnetics of Kraus, Wheeler, et al and computer simulations

# TECHNICAL TOPICS

based on *NEC* and the Method of Moments. As a technical journalist and not a professional engineer, I can only play it as I see it, and try not to mislead readers into accepting claims that not only contradict well-established theory but which themselves have shown to be based on error. This, of course, does not mean that small transmitting loops are not of considerable value to amateur radio.

### WIDE-BAND-GAP DIODE & LED MIXERS

If G4JNT had not been put off reading the January 'TT' by the sight of diagrams containing valves, he might have found interesting the item 'High-Level Mixers Using LEDs'. D G Phillips, G8AAE, brought to notice recent developments and ideas relating to the use of wide band-gap semiconductors such as SiC, GaN, AlGaN to form wide dynamic range mixers, including the novel website suggestion stemming from Pascal Nguyen that blue and super-blue GaN LEDs could be used in this way. The item brought to notice the important US Patent 6,111,452 (2000) of Fazi and Neudeck, assigned to the US Military.

Fortunately, Michael O'Beirne, G8MOB, never one to be put off by a triode-mixer symbol, was intrigued and delved further via e-mail and the websites. He was intrigued to find that no member of his elite discussion group 'Premium Radio', comprising mainly American enthusiasts owning 'premium' (high-grade commercial) receivers was aware of this recent development. His posting of the 'TT' item has sparked off considerable interest, with Roger Wehr, W3SZ, digging into the topic and providing a number of relevant references in recent professional journals and the full text of the US Patent by Fazi et al.

In his posting, G8MOB pointed out: "The patent seems well worth reading, though whether one would be able to deliver the required LO power is another matter. We already need about +17dBm to drive a high-level diode mixer, so finding another 20dB will not be easy... I cannot but feel that a better approach to mixer technology is to use the H-Mode mixer developed by Colin Horrabin, G3SBI [first described in 'TT' a decade ago], which uses a high-speed bus switch. The receive section of the CDG 2000 transceiver (*RadCom*, June/August 2002) incorporates such a mixer with an IP3 of +40dBm and a noise figure of 10dB. It also uses a very low-phase-noise LO which achieved −140dBc/Hz at 9kHz offset from the carrier, and −150dBc/Hz at just over 20kHz offset on the 20-metre band. The limitation in dynamic range was, in part, not the mixer but the coils in the bandpass filters, being 13dB better for hand-wound coils than for commercial Toko inductors. This is rather like the old Racal RA1772 'professional' receiver, where the limiting factor in the prototype was the IP3 of the roofing filter and not the then novel switching mixer using a range of four FETs. All the original ferrite material in the filter had to be removed in order to achieve the IP3 spec of +27dBm. In the CDG2000 with its IP3 of +40dBm and a noise floor of −130dB, an IP dynamic range of 113dB was achieved... That's better than many commercial premier receivers."

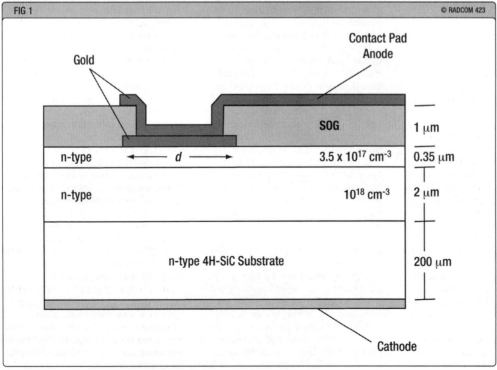

Fig 1: Cross-section of the SiC Schottky-barrier mixer diode as described by Simons and Neudeck of the NASA Glenn Research Center, Cleveland, Ohio.

While I think G3SBI and I7SWX would agree that the H-Mode mixer remains the optimum approach for HF and possibly low VHF receivers, it would appear that wide band-gap semiconductor diode mixers could represent a state-of-the-art approach to UHF mixers.

G8MOB and W3SZ have downloaded *inter alia* two relevant papers: (1) 'Intermodulation-Distortion Performance of Silicon-Carbide Schottky-Barrier RF Mixer Diodes', by Rainee N Simons and Philip G Neudeck (*IEEE Trans on Microwave Theory and Techniques*, February 2003, pp669 – 72), with the work carried out at the NASA Glenn Research Center, Cleveland, Ohio; and (2) 'Resistive SiC-MESFET Mixer', by Kristoffer Andersson et al (Chalmers University of Technology, Göteborg, Sweden), *IEEE Microwave and Wireless Components Letters*, April 2002, pp119 – 21.

While the devices under investigation in these papers are different from the blue LEDs discussed in January, it is clear that the use of wide band-gap semiconductors as high-level mixers represents a new advance in mixer technology. (1) Describes the use of single-balanced mixer circuits with a diode in each arm (two diodes total) at 200MHz and 1.5GHz, achieving conversion loss/input third-order intercept point (IP3) are 8.0dB/+25dBm and 7.5dB/+22dBm at these frequencies, respectively. The measured IP2 over the VHF band is +39dBm. The paper points out that the above conversion-loss values are about the same as that of commercially available single-balanced mixers with silicon Schottky-barrier diodes. However, to achieve a comparable input IP3 performance with silicon Schottky-barrier diodes, a more complex design involving double-balanced mixers with at least eight diodes in a quad configuration is required. **Figs 1 – 3** come from this paper.

The importance of achieving good intermodulation distortion performance at VHF/UHF has been well recognised, not only by amateurs. The paper notes: "Electromagnetic emissions from carry-on electronic devices can potentially degrade the minimum detectable signal level of important RF-based navigational instruments, on board commercial and general aviation aircraft. In some situations, the interfering signal frequencies may not coincide, but may be very close to the navigational frequencies. For example, the lowest frequencies used for VHF omni-directional range |(VOR) and instrument landing systems (ILS) localiser instruments are just above commercial FM broadcast frequencies. In such cases, the IMD products are located adjacent to the desired signal spectrum at the out-

# TECHNICAL TOPICS

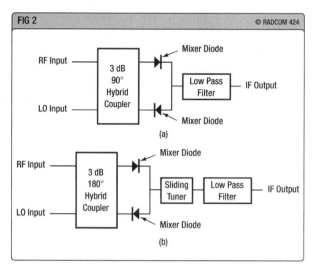

Fig 2: The experimental single-balanced mixer circuits of Simons and Neudeck. (a) For the VHF band with 3dB 90° hybrid coupler. (b) For GPS band with 3dB 180° hybrid coupler.

put... the IM products could be second, third fifth order, etc... In the literature, several schemes have been proposed to suppress diode mixer IMD products, including adding resistance in series with the diodes, or using two or more diodes in series, or using diodes with higher turn-on voltage..."

The short Swedish paper describes the design and characteristics of a single-ended silicon carbide (SiC) resistive MESFET mixer with a minimum conversion loss of 10.2dB and an input third-order intercept point of 35.7dBm at 3.3GHz . "This is, to our knowledge, the best result reported for any resistive FET/HEMT mixer. It is our belief that conversion loss can, by more careful mixer design, be reduced to at least 6dB. The mixer was not optimised for low intermodulation, thus making further improvements possible."

## JA9MAT'S HYBRID 'AUTODYNE' RECEIVER
There is still interest in building simple regenerative HF receivers, with the view quite widely held that for such designs, that the valve can still provide overall better performance that the transistor – though it would be difficult to prove this, and one risks the charge that we should no longer be drilling out the large holes needed for valve sockets.

Be that as it may, several recent 'TT' topics seem, by coincidence, to have come together in an item '12BH7A 12V Separate Autodyne Receiver', by Hidehiko Komachi, JA9MAT (*Sprat* – the Journal of the G QRP Club, Issue 121, Winter 2004/5, p7) with its brief outline of a regenerative 'homodyne-type' receiver using a separate regenerative oscillator, electronic fine tuning and 12V operation of the twin-triode 12BH7A feeding an IC AF amplifier: **Fig 4**. As presented, it covers 600kHz to 12MHz including the 3.5 and 7MHz amateur bands with four plug-in coils (coil data are given for the 7 to 10MHz amateur bands and for 0.5 to 1.6MHz medium-wave broadcast reception).

## PREMIUM RECEIVERS
Although, as emphasised a number of times in 'TT', the future of communications receivers will increasingly depend on software-defined technology and digital signal processing, the performance of such receivers will inevitably be a function of the device designer rather than the circuit designer and the mechanical engineer. There is no doubt that remarkable progress is being made, reflected in lengthy articles in QEX and in the unlimited spaces of the Internet.

But this does not mean that discussion of past analogue designs, valve and solid-state, has become little more than 'nostalgia'. During WWII and the subsequent decades of the Cold War, the importance of signals intelligence (SIGINT) ensured that the leading receiver manufacturers were encouraged to develop designs capable of the highest possible HF performance, flexibility and operability, including remote control over long distances. Even in the 1960s, professional HF receivers could cost around £20,000 equivalent to well over £50,000 in today's 'Monopoly money'.

Top to bottom:
The Russian Volna-K maritime receiver covering 12kHz to 23MHz with gaps in nine bands. Single conversion with IF 85kHz up to 600kHz and double-conversion with IFs of 915kHz and 85kHz above 1.5MHz. Optical projection calibration scale.

The AT & E R210 British Army receiver developed in the 1960s with 52in film scale, modular (hinged) construction, covering 2 to 16MHz.

Internal view showing modular construction of the R210.

The study over the years of British, American, German and Russian 'premium' receivers by Neiol Clyne, G8LIU, Michael O'Beirne, G8MOB, and Dick Rollema, PA0SE, has underlined how little attention the Western powers paid to the excellent German wartime designs, such as the E52 series. G8MOB, for example, has shown how the Russian maritime 'Volna K' receiver (see *Radio Bygones*, Issue 74, Christmas 2001, pp8 – 17 and cover photos) that included an optical projection calibration scale with readout at VLF to 250Hz and better by interpolation: "The scale was also surprisingly linear despite the wide bands. The model purchased by John Midwood, G7PTD, was said to have come from a Russian trawler, but may well have been used for electronic surveillance".

G8MOB has also confirmed that the AT & E R210 (not R206 as stated in the January 'TT') did indeed go into production in the 1960s. He writes: "I used it for many years in Land Rovers, partnered with the powerful C11 AM/CW transmitter. These were usually a Royal Signals responsibility and would form a rear link from battalion to brigade, or other long-range communications. The C11 puts out about 50W of pure AM with high-level anode and screen modulation, and has some voice AGC to keep the level constant. The R210 was extremely reliable and highly sensitive and reasonably selective. Not one ever failed on exercise. The tuning can be a bit stiff, but some are delightful. The film scale takes you to within 5kHz. There is also a normal CW audio filter which is a bit too sharp, and a reasonable noise limiter. The 450kHz IF output can be fed to a box of similar size called the RTA (Radio Telegraph Adapter) to run a teleprinter link with the Siemens T100 teleprinter. These were made by MEL at Crawley and are quite good... One day I will buy a reasonable R210 as a memento of happy days."

To quote further from the letter from Pat McAlister, G3YFK (see 'TT' November 2004, p40), a self-confessed member of the 'high-class HF receiver bunch': "One wonders where all the interesting HF receivers that were easily available (to the cognoscenti) 10 years or so ago have gone? Looking back then I had here, either to repair or purchase, Racal RAs 1171, 1172, 1778, 1779, 1781, 1782 (a rare push-button version of the 1773 specially made for GCHQ), 1784, 1792, 6790, 6793, VHF 1795, etc. Later 3701s, and, in the final two years, DSP 3791s appeared. Eddystone 1650 variants were plentiful, even the rare STR 8212 DSP sets were about! The ultra-rare Drakes like the R4245 and TR4310

# TECHNICAL TOPICS

used as standby by the British Antarctic Survey (with the last production run being bought up by the late JY1 for his army) appeared along with various Collins sets like the 651S1 and the infinitely-better 51S1 – although I have never yet discovered an 851S1A.

"Various Marconi [badged] machines were about, notably the H2540 and H2541 (around in Saudi) and the rebadged Dansk Radio, M5000. Plesseys were plentiful, the PR155, 1553, the later 2280 and 2282 and the better 2250 (mostly ex-Hanslope Park DWS).

"One is always wise after the event. There are few of these that I would bother to keep now! The Collins 51S1 and Racal RA1779 come to mind due to their excellently-clean local oscillators. My later RA3701 (ex Culm Head) sounds awful due to poor audio tailoring, one wonders if these sets were used for data with nobody actually listening on them. I am biased, but none compares with my rebuilt Collins 75A4 amateur bands receiver, a six month task completed in 2002. I incorporated all the Electric Radio modifications and a few of my own resulting in an IP3 of –16dBm as opposed to the normal unmodified-set result of –43dBm.

"German receivers have been notable by their absence. Few R/S, Siemens or Telefunken receivers seem to surface in the UK. Portishead Radio is said to have evaluated R/S receivers but ended up with Racal RA1792s on a cost basis. The main German competitor was the R/S EK070, but I agree with G8MOB that, from the ergonomics view, the best receiver is the RA1792 – totally intuitive! The EK070 does not come close.

"Some time ago, 'TT' mentioned the use of the 4066IC for use as a switching product detector and I note that an FST3125M fast bus version is described by Sergio Cartocet, IK4AUY, in QEX (July/August 2004). R/S seems to have been using 4066 detectors for years in the EK070 and EK085 receivers, and I can vouch for their audio quality. I suspect, but cannot confirm, that the later EK890-series probably also use these or similar devices.

"Finally, I have been thinking of building a home-brew receiver with a 7360 [valve] front end but am now tending towards H-mode technology. I firmly believe that had the American firm Squires Sanders [which introduced the use of the 7360 beam-switching mixer front-end] not been tragically wound up on the death of its founder in an air crash, the firm would have given Collins a run for its money. But I wonder if anyone has any measured IP3 figures for a 7360 switching mixer?"

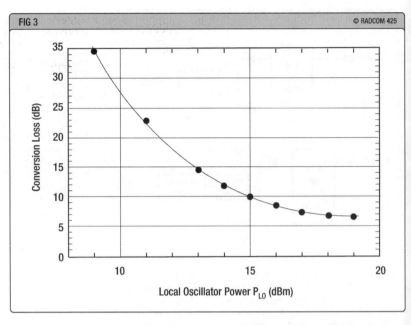

Fig 3: Conversion loss as a function of the local oscillator power. RF input power 11.0dBm at 195MHz, local oscillator frequency 175MHz, d = 50 μm.

Fig 4: The simple autodyne (homodyne) receiver using a 12BH7A twin-triode valve with an HT supply of 12V DC. With four plug-in coils covers 600kHz to 12MHz including 3.5, 7 and 10MHz amateur bands. Coil data for 7 – 10MHz: L1 39 turns, L2 4 turns on T50 – 2 core. For 0.5 to 1.6MHz AM broadcast receiver ferrite slab coil, with IC AF amplifier.

### HERE & THERE

Jim McGowan, M0MAC, believes that some years ago, he read in 'TT' a tip on how to clean surplus variable capacitors possibly using baking soda and hot water although the process used now escapes him. He is anxious to pursue this tip since he has recently acquired some old valved equipment containing various types of variable capacitors.

It was recently suggested to him to use Goddard's Silver Dip and he finds this works really well with the silver-plated types (C804s etc), but doesn't work on other types, and he would like to trace the original tip. Unfortunately, I cannot find it, and wonder if anyone else can either recall the item or provide an effective alternative solution? M0MAC adds: "These old components were made really well and I would like to re-use them in some vintage projects I have planned."

Fuel cell research at St Andrews University is reported to be leading to the development of high-temperature cells with power outputs between 1kW and 5kW with prices of around $150 per kW compared with the current prices of around $2000 per kW. The cells are based on the solid-oxide fuel cell (SOFC) running on natural gas. Markets envisaged include outdoor portable applications at remote sites and for the military. ♦

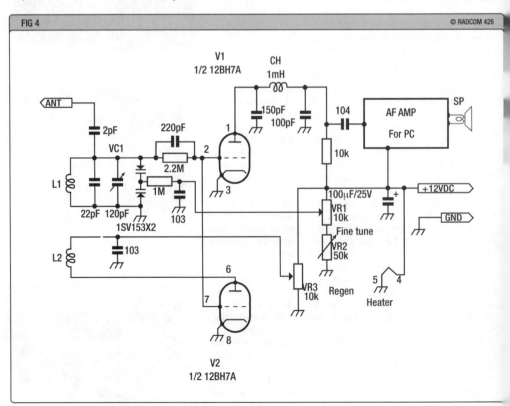

**Pat Hawker,** G3VA

37 Dovercourt Road, Dulwich,
London SE22 8SS.

# TECHNICAL TOPICS

This month, Pat looks at the chequered history of the Wadley Loop, and considers its future ♦ Nylon washers as VHF toroids ♦ Neat earth connections ♦ 2V / 300V DC-DC converter ♦ 3V MOSFET shunt regulator ♦ By-passing: one or multiple capacitors? ♦ Precision time / frequency

### RECEIVERS – A FUTURE FOR THE WADLEY LOOP?

One of the interesting post-war developments in improving frequency stability of receivers and transmitters was the introduction in the 1950s of the Wadley loop (also known as the triple-mix system). The story of how Dr Trevor Wadley developed the system in South Africa and came to the UK in an effort to persuade British communications firms to adopt the system for HF communications receivers is well known – how he failed to convince the major firms that the drift-free triple-mix system provided a feasible and viable system – how he then approached a small and little-known firm, Racal, that had been largely engaged in restoring surplus military receivers for export. The outcome was that Racal took on the task, although their first prototype suffered from self-oscillation until, in desperation, a saw cut was made in the chassis. The requirement for good bandpass filtering and effective screening and bypassing meant that the Wadley loop has never been an easy system to implement. It can be considered almost as a later form (in this case with a crystal-controlled signal stabilising a free-running VFO) of the 'Goyder Lock', pioneered by Cecil Goyder, operator of 2AZ at Mill Hill School, on which he made the first England – New Zealand contact (see the article by John Heys, G3BDQ, RadCom, October 2004, pp 38 – 39).

Dr Brian Austin, G0GSF, believes that the seed-bed of the Wadley Loop can be found in Trevor Wadley's earlier design of 'A Single-Band 0-20MHz Ionosphere Recorder Embodying some New Techniques' (*Journal IEE*, Vol 96, Part III, pp483 – 486). This described a pulsed ionosonde in which the receiver and transmitter swept simultaneously over the spectrum up to 30MHz in a single band. There was only one moving part, namely the rotor of the main sweeping oscillator. Single-span operation was obtained by mixing two high frequencies (one variable, the other fixed) to obtain a beat frequency in a manner very similar to that employed in an ordinary audio BFO, except that the beat is from a radio frequency, variable from zero to 20MHz. The receiver IF is 30MHz. The mechanical simplicity was obtained at the expense of increased electrical complication – a characteristic of the later Wadley loop.

Trevor Wadley added his drift-cancelling technique to the basic single-span up-conversion technique of his ionosonde and developed the triple-mix form of frequency synthesis. In effect, a front-end tuner provides drift-free output over the 1MHz band between 2 and 3MHz, the tuner acting as a crystal-controlled converter with a single 1MHz crystal providing 30 crystal-controlled frequencies by means of a harmonic generator. These signals are locked to the 40 – 70MHz VFO that acts as the band selector for the 30 bands each 1MHz wide. The overall stability of the receiver then depends solely on the stability of the conventional tunable 2 – 3MHz stages of the receiver where it is relatively easy to achieve excellent stability from a free-running oscillator. There will thus be a minimum of three mixers (triple-conversion) in the signal path, plus a further mixer to provide the drift-cancelling facility. The system requires good band-pass filters at 40MHz (±650kHz), 37.5MHz (±150kHz) and 2.5MHz (±150kHz), not easy for a home-constructor to design, build and align.

**Fig 1** shows the basic arrangement of Wadley loop as used in the RA17 front-end, and later in the RA117 and RA171. In 1967, a pre-production Model 919 was also described in which the Wadley principle was used to extend the frequency range into the VHF and UHF bands, although I am not sure whether this design was ever marketed.

The outstanding performance and excellent calibration of the RA17 and particularly the RA17L, with a tuned pre-amplifier, lifted Racal from near obscurity into a major manufacturer of high-grade professional and military communications equipment. This lasted for decades until the firm was sold to a French manufacturer and the subsequent disappearance of the name Racal.

A more complex form of the triple-mix system was also adopted by Marconi to form an analogue frequency synthesiser for transmitters and receivers that avoided the phase-noise problems of PLL synthesisers, but at appreciably higher cost. Marconi frequency-synthesisers were adopted by the Diplomatic Wireless Service for its Piccolo multitone teleprinter links, providing the required frequency stability within just a couple of Hertz. The triple-mix approach was also used in the Wadley-Barlow HF broadcast receiver.

A Wadley-loop design for home constructors appeared in a series of articles by the late Ian Pogson, VK2AZN, in *Radio, Television & Hobbies* (later *Electronics Australia*), including that of a triple-mix 'front-end' valved tuner (converter) in July and August, 1963, later incorporated into a complete 'Deltahet' communications receiver (September and October, 1964). Brief details appeared in 'TT' and *Amateur Radio Techniques* and, over the years, a number of requests for further

Fig 1: The basic Wadley Loop as implemented in the Racal RA17, RA171, RA117 MF/HF communications receivers.

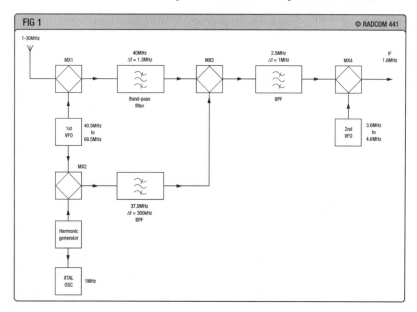

RadCom April 2005

# TECHNICAL TOPICS

| Table 1: Nylon washer dimensions. | | | |
|---|---|---|---|
| (Screw clearance | outer diameter (in) | inner diameter (in) | thickness (in) |
| Number 4 | 0.250 | 0.115 | 0.125 |
| Number 10 | 0.375 | 0.194 | 0.125 |

| Table 2: Inductance windings and measured values. | | | |
|---|---|---|---|
| Toroid diameter (in) | Inductor winding | Max inductance (nH) | Min inductance (nH) |
| 0.250 | 7t / 26AWG | 116 | 86 |
| 0.375 | 7t / 22AWG | 113 | 86 |

details was received and satisfied.

For receivers, the dominance of solid-state technology in the 1960s led to the abandonment of the Wadley loop primarily due to the then poor performance of solid-state mixers. In the late 1960s, Racal produced and marketed a solid-state model, the RA217, using balanced- and single-FET mixers. The model appears to have had limited dynamic range and its performance was much inferior to the original RA17 valved sets. Racal sponsored research at Bradford University on solid-state mixers offering improved dynamic range, but soon abandoned the triple-mix system in favour of PLL frequency synthesis despite the increased phase-noise.

This raises the question: might not an updated triple–mix analogue Wadley loop with the strong-signal performance and high-dynamic range of modern solid-state mixers, combined with careful gain-distribution, outperform PLL and DDS systems with respect to phase-noise and spuriae, at the same time avoiding high-speed digital switching pulses affecting weak signals? For portable operation, such a receiver could probably be less power-hungry than one based on digital technology.

Tony Webb, G4LYF, admits to having long been fascinated by the Wadley-loop as used in the RA17 and Deltahet. He writes: "However, in a home-built version, I would expect a spurious response every 1MHz from the harmonics of the crystal oscillator. It occurs to me that, if a 5MHz crystal were used, the 'birdies' would be at 5, 10, 15, 20, 25 and 30MHz, and at least three of them would coincide with WWV standard frequencies, and so could be used to trim the crystal.

"This system could be used if the 'bandset' oscillator frequency were multiplied by five before mixing with an appropriate harmonic of 5MHz, the filtered VHF mixing product being then divided by five before being used to drive the second mixer in the receiver. It now seems to me that a simpler approach would be to phase-lock the band-set oscillator at 1 MHz intervals. Since its drift would then be very small, the drive for the second mixer in the receiver chain could be taken from a harmonic of the crystal as in **Fig 2**. I wonder if there is any merit in such an idea and whether anyone else has thought of it?"

In Fig 2, M1* acts as a mixer and/or phase detector, so can give a DC or RF output. Appropriate frequencies in the 200 – 350MHz range will mix with the 240 or 320MHz inputs to give outputs in the range 0 – 40MHz, which are mixed with appropriate 5MHz harmonics in the PSD to generate to generate the control signal. The PSD** is arranged so that it can pass a DC input from M1. If this is difficult, any DC from M1* could be added to the control voltage from the PSD using an op-amp, which could also provide an offset voltage.

## NYLON WASHERS AS VHF TOROIDS

Gary Aylward, G0XAN, feels that readers may be interested in an item that appeared a few years ago in the 'Ideas for Design' feature of *Electronic Design* (June 18, 2001, pp104 – 205) by Richard M Kurzrok, 'Low-Cost VHF Inductors use Nylon Toroids'.

This noted that toroidal inductors can be used in passive-filter and equaliser circuits in the VHF range, Although SAW filters have taken over many bandpass filter applications, LC low-pass and high-pass filters are still viable.

Kurzrok added: "When inductance values are smaller than 1µH air cores with unity relative permeability can be used instead of powdered-iron or ferrite cores. For wire size thinner than 20AWG, a coil former is often needed for mechanical support. Nylon 6/6 standard flat washers are usable as low-cost coil forms. The nominal electrical parameters of this material are a dielectric constant of 3.6, a dissipation factor of 0.04 and a dielectric-strength of 385V/mil."

He provides winding data etc based on two standard (American) sizes of nylon washers in **Table 1**.

"Some typical toroidal inductors were wound and tested as shown in **Table 2**. The maximum inductances were obtained with the windings squeezed and minimum with the windings spread. Inductance values were calculated from measured resonant frequencies using a known capacitor. As the winding area is filled, the adjustability of the inductors decreases and there are limitations on the number of turns that can be applied as a single-layer winding.

It is claimed that, despite its dielectric constant and dissipation factor, the quality of nylon toroidal inductors is quite good with typical unloaded Qs in the range 75 to 125, more than adequate for most low-pass and high-pass filters in the range of about 30 to 100MHz. They provide small size, efficiency and some adjustability. With a 0.250in-diameter washer, maximum turns (22AWG) are six and for 0.375in, 15. With 26AWG, minimum turns of 15 and 30 for these diameters.

## NEAT EARTH CONNECTIONS

Bruce Carter, GW8AAG, noted that, in the article on 'The Linear Amp UK Ranger 811 Kit' (*RadCom*. September 2004, p35), Steve Icke,

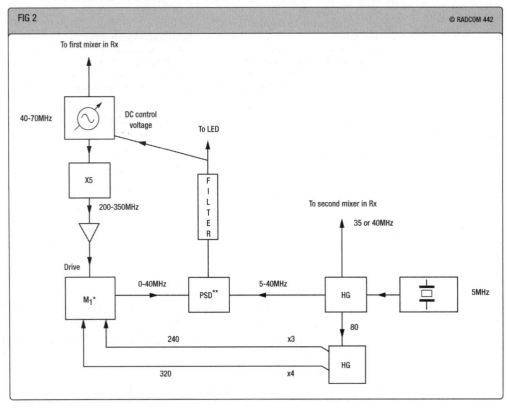

Fig 2: G4LYF's proposed 'bandset' oscillator phase locked at 1MHz intervals replacing the harmonic generator system in a Wadley Loop arrangement.

# TECHNICAL TOPICS

G4ZWY, in the section on 'Valve bases' emphasises the need to "scratch off enough paint to ensure a good earth connection."

GW8AAG comments: "Scratching sounds a bit drastic and probably looks ugly. I had to retro-fit some earth terminals and phono sockets that would remain in full view. To achieve a good earth connection I employed the following method. I measured a 1/4-in wood chisel and found it to be 6.53mm; as a bevelled chisel this was its greatest width. Likewise, a 1/2-in chisel measured 12.67mm. From a sheet of soft grade aluminium, I cut a strip about 18mm wide with a length to suit my particular purpose. At one end I drilled a 7mm-diameter hole and at the other a 13mm hole.

"The 'aluminium hole' was placed centrally over the hole in the unit and held firmly in position. The appropriate chisel was rotated in the aluminium hole until the paint was removed from the unit in a neat circle. The slight discrepancy between the hold diameter and its chisel width made no discernible difference. The chisels only have to remove paint and 'polish' the metal surface slightly so they are easily re-sharpened for their legitimate use."

## 2V / 300V DC-to-DC CONVERTER

A DC/DC converter capable of running a low-energy (9W) light bulb from a single 2V lead-acid cell is described by Paul Bennett of Bristol in the 'Circuit Ideas' feature of *Electronics World*, February 2005, p41: **Fig 3**. It comprises a single-ended forward converter with resonant flyback, with the oscillator frequency selected so that the secondary winding capacitance provides the correct amount of tuning. It is intended to work at a fixed duty cycle, with the output regulation tracking the input, providing an output voltage of about 300VDC determined by the 150:1 turns ratio of the transformer comprising an un-gapped R10 core in 3C85 or similar. Drive for the 5MΩ, 20V IRF3717 FET is provided by the 74HC14 hex Schmitt buffer with a boot-strapped supply. Note that, to guarantee the circuit starts, the 1000μF capacitor is charged from the battery via the switch when the circuit is switched off. Then when first turned on, the capacitor is connected in series with the battery to provide a 4V start-up voltage (logic FETs can have thresholds of 2V). It is also stressed that tight layout and low ESR capacitors are essential for good efficiency.

## 3V MOSFET SHUNT REGULATOR

The increasing use of low-voltage

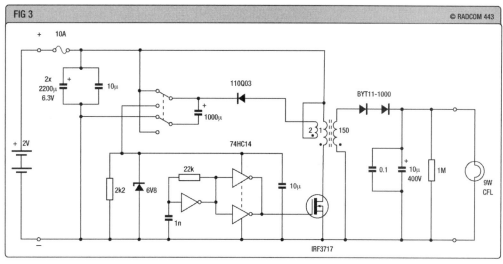

Fig 3: Circuit diagram of the DC-DC 5W converter providing an output voltage of up to 300V from a 2V lead-acid cell. (Source: *EDN*)

Fig 4: A MOSFET used to replace the Zener diode of a shunt regulator provides lower impedance than a diode-based implementation. (Source: *EDN*)

Fig 5: Graph of the key parameters of the MOSFET regulator – gate-to-source voltage and output impedance – versus drain current showing the smoothness of variation over two-and-a-half decades.

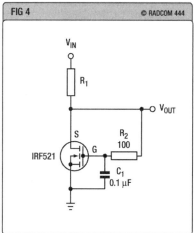

ICs and microprocessors has brought about a demand for 3V DC supplies derived from standard 5.1V supplies. This is usually obtained from a series linear regulator or a DC/DC converter. For prototype circuits etc, it is possible to use a MOSFET shunt regulator as a substitute for a series regulator. Although this has the drawback that, should the MOSFET fail (open circuit), the higher voltage will be applied to the load. With a series regulator a corresponding problem arises if the regulator fails shorted.

A shunt regulator arrangement is described by Stuart Michales in the 'Design Ideas' feature of *EDN* (1 November 2004, pp100 – 101). He points out that the MOSFET in **Fig 4** can replace a Zener diode in a shunt regulator and provides lower output impedance. The MOSFET is self-biased by connecting its drain to its source. The difference between the input voltage and the gate-to-source threshold voltage, $V_{GS}$, sets the current. The IRF521 device has a threshold voltage of 2 to 4V at 250μA, and in this example has a gate-to-source voltage of 3V at 200μA. MOSFETs can vary from device to device but, typically, have thresholds at about the mean between the limits. **Fig 5** shows the output impedance. Although this is near 800Ω at 100μA, it rapidly drops to less than 6Ω at 50mA.

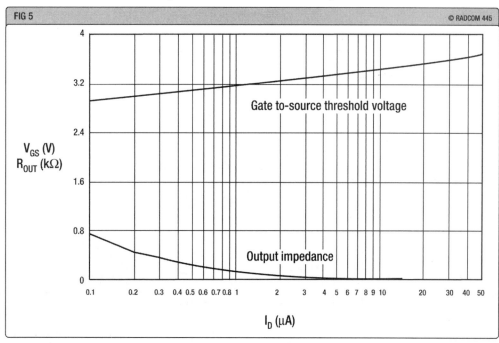

RadCom April 2005

# TECHNICAL TOPICS

## BYPASSING: ONE OR MULTIPLE CAPACITORS?

In 'TT', October 1992 an item 'Decoupling Capacitors – Why Use Two When One Will Do?', based on a paper 'Effectiveness of Multiple Decoupling Capacitors' by Clayton R Paul (University of Kentucky) in *IEEE Trans EMC* (May 1992), called into question the effectiveness of the common practice of using a parallel combination of large value and small value capacitors in order to increase the effectiveness at higher frequencies and to overcome the effect of lead inductance. He concluded that this practice only minimally reduces the high-frequency impedance of either capacitor except over a limited range of frequencies. To quote: "Above the self-resonant frequencies of both capacitors, the impedance of the parallel combination is reduced at most by 6dB. This may not be worth the expense of the additional capacitor or its installation and could be attained by using only the larger value capacitor while simply cutting its lead lengths in half!"

As a follow-up in the December 1992 'TT', Bob Price, GW3ECH, drew attention to circumstances when multiple bypass capacitors may be needed, based on his experience of tuning solid-state VHF power amplifiers. Low-frequency parasitic oscillations may occur. Different small values of impedance in common paths to two stages or in supplies which also feed bias stabilising circuits (to just one stage) can lead to the right conditions for (parasitic) oscillation. One small value but low inductance capacitor is used to decouple the stage at the amplified frequency. A second (and often third) capacitor(s) then decouples the stage at lower frequencies at which parasitic oscillations are likely to occur. He added: "It is not unusual to find a relatively high value polyester capacitor and an electrolytic capacitor used, not to improve the frequency response of the decoupling at signal frequencies but to minimise the risk of oscillation at one or more low frequencies."

In the May 1993 'TT', Brian Bower, G3COJ, noted that the BBC Technical Memorandum R.1027 (90) 'Supply Rail By-Passing in Video Circuitry' included an introductory statement: "In video circuitry (frequencies up to about 6MHz) one sometimes sees supply rails with an electrolytic capacitor used for by-passing together with a smaller capacitor, typically a 22nF ceramic, across it to improve HF performance. The drawback to this has long been known but is not always appreciated."

The Memorandum concluded: "Supply rail by-passing in video circuitry should be by a single capacitor, never by two in parallel. A tantalum capacitor is preferable and arrangements must be made to limit inrush current on switch-on. In more general applications, a parallel combination can provide lower impedance decoupling at higher frequencies than a single electrolytic provided component values are chosen so that the parallel resonance frequency is placed where it will not be a problem – a series element is then optional."

This topic turns up again in the 'Circuit Ideas' feature of *EDN* (April 15, 2004), in an item 'Take Steps to Reduce Anti-Resonance in Decoupling', by Dale Sanders of the firm X2Y Attenuators. This firm manufactures a special type of 'X2Y' capacitor.

The item starts: "To maintain power integrity on PC boards, you need multiple capacitors to decouple the power-distribution system. A typical configuration might comprise five capacitors connected in parallel between the power and the ground traces or planes. To provide broadband decoupling performance, assume the individual values of the parallel capacitors are 470nF, 1nF, 10nF, 100nF, 220nF, as in **Fig 6(a)**, providing a total of 801nF capacitance to the power-distribution system."

With a vector-network analyser,

Fig 6: (a) Typical decoupling arrangement using several multilayer ceramic capacitors connected in parallel and yielding multiple anti-resonances. (b) A 400nF X2Y capacitor connected as shown yields a total decoupling capacitance of 800pF and single anti-resonance. (Source: *EDN*)

each capacitor's self-resonant frequency (SRF) can be identified, each of which can cause anti-resonance. The anti-resonance occurs when one capacitor is still capacitive, while another has become inductive. It is claimed that considerable reduction of the anti-resonance effects is obtained with a single 400nF X2Y capacitor producing, when connected as in **Fig 6(b)**, some 800nF decoupling, with a single anti-resonance frequency of around 10 – 11MHz. It is also claimed that the use of X2Y capacitors saves PCB space and reduces layout complexity. I have no idea whether X2Y capacitors are available in the UK, but is stated that capacitors using X2Y technology are available, for example, from Johanson Dielectrics (www,johansondielectrics.com).

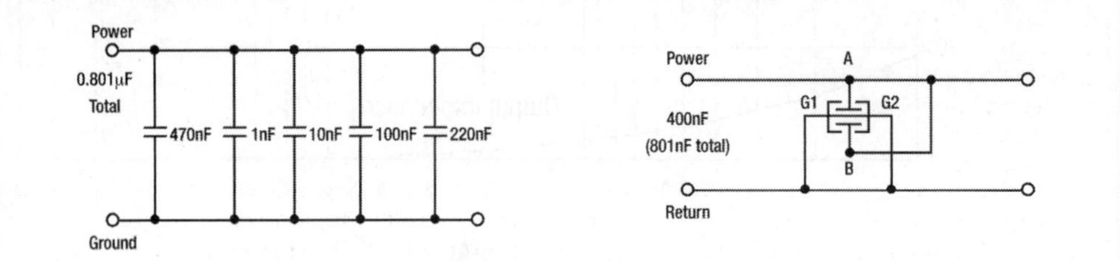

## PRECISION TIME / FREQUENCY

John Osborne, G3HMO, has always been fascinated by clocks, pendulums, hairsprings, quartz crystals, etc, and enjoyed dividing-down frequencies with suitable chips for weather satellite fax and for sync signals for the tape for playback. He draws attention to an article in the October 2004 issue of *Physics World* 'NIST Builds Smallest Atomic Clock'. This reports that John Kitching and colleagues at the National Institute of Standards and Technology (NIST) in Boulder, Colorado have built a 'physics package' with a volume of only 9.5mm$^3$. It consumes only 75mW and its possible applications include frequency control for hand-held radios including mobile phones and GPS receivers. It is claimed to represent a breakthrough in developing a prototype atomic (caesium) clock a hundred times smaller than existing atomic clocks. The clock is stable to one part in 10-billion, equivalent to about 1 second in 300 years The chip is expected to become available in two or three years time.

*Electronics World*, February 2004, p8, reports that UK scientists at the National Physical Laboratory have "found a new way of measuring time that is three times more accurate than any other method in the world." A single ion of strontium is trapped and isolated in a vacuum, and then chilled to near absolute zero (-273°C) with a blue laser beam. It is claimed that it may enable improved levels of accuracy in deep space navigation and exploration. ♦

**Pat Hawker,** G3VA

37 Dovercourt Road, London SE22 8SS

**TECHNICAL TOPICS**

**HF RECEIVER SPECIFICATIONS**

At one time, an amateur preparing to purchase a new HF receiver tended to be influenced by the advertisements, seeing the model at a shop, such as Webb's Radio in London, comparing the price with competitive models, and particularly by recommendations from other amateurs. While all these continue to be used as guides, the role of the detailed technical review, such as that by Peter Hart, G3SJX, in this journal or by the ARRL in *QST,* has assumed especial importance. This, I would suggest, is partly due to the fact that very few amateurs have the complete set of test instruments needed to check for themselves the technical specification or claimed operational features until they have parted with quite substantial amounts of hard-earned cash.

I am not suggesting that current receivers and transceivers do not conform with their advertised performance, although there is a natural inclination for some 'tweaking' in presenting features and figures in the most favourable way. Advertising tends to highlight particular characteristics without making it clear what these mean in operational use. *Caveat emptor* to some degree is often needed.

The usefulness of detailed technical reviews is well illustrated when the model concerned is brought to the market as an improved version of an earlier receiver or transceiver, at a considerable hike in price. It was interesting to read the reviews of the new top-of-the-line Icom IC-756PROIII HF/50MHz transceiver by the ARRL's Rick Lindquist, N1RL, in *QST,* March 2005, pp56 – 59 and earlier by G3SJX in the February *Radcom.*

Both give the new model a thumbs up, but N1RL more specifically questions whether the improvements relative to the PROII are worth the extra $800 cost. The bottom line, he writes, is "If you like the PROII, you'll like the PROIII even more – but you will have a tough choice about whether or not to upgrade"

It is noticeable that Japanese manufacturers including Icom, Yaesu, etc, have in the past few years finally got round to improving

**This month, Pat considers the parameters used for specifying HF receivers ♦ Phase noise and how it may be reduced ♦ More on LED and diode mixers ♦ Cleaning variable capacitors**

the IP3 performance of their receiver front-ends, which for some 30 years have not really been capable of coping with the multitude of strong 7MHz broadcast signals that beset European amateurs during the evenings. This problem goes right back to the early post-war period when so many high-power broadcast stations established themselves in Region 1 between 7100kHz and 7300kHz, although this remained an amateur band in Region 2. With multiple very-strong signals, the effect is largely to produce an artificially-high noise level.

The availability of 7100 to 7200kHz as a shared band in the UK and some other European countries has increased awareness of the shortcomings of most equipment currently in use. A partial solution is either excellent pre-mixer selectivity with *variably tuned* resonant circuits (very expensive with six or more high-Q tuned circuits) or extreme dynamic range (high IP3, IP2 etc). Amateur-bands-only receivers with multi-pole bandpass filters are vulnerable, since these will pass to the mixer any signals within the band 7000 – 7300kHz. It is particularly a problem with relatively low-cost general-coverage receivers with a minimum of pre-mixer selectivity. Remember, the majority of Far-East export sales are to North America where the problems are not so evident.

IP3, IP2 and instantaneous dynamic range figures are thus very important, but they are not the only measures of a receiver's performance. Do not take an advertised IP3 or dynamic range figure as a sole criterion, unless it can be extended by the additional close-in figures provided in the *QST* and *RadCom* reviews. And remember that even then there are other operational features that matter a great deal.

As N1RL puts it: "Icom's advertising for the PROIII trumpets what it calls '+30dBm-class third-order performance on 20 metres'. This would put it on a par with some of the best receivers we've run through the lab. Third-order intercept (TOI) is a number that many like to use as an all-in-one performance benchmark, since its value derives both from the receiver's sensitivity and its front-end selectivity (specifically, two-tone, third-order IMD dynamic range). The more positive the number, the better, and TOI figures can also be negative [ie below 1mV].

"Although the PROIII instruction manual doesn't specify the advertised TOI number, an Icom product guide, originally in Japanese, spells out the measurement conditions: 100kHz spacing (wider than our Lab's widest 20kHz spacing measurement), preamps off and a 2.4kHz filter bandwidth.

"Nonetheless, under the least stringent measurement standard the ARRL lab uses, the PROII came pretty close to meeting the +30dBm mark. At 20kHz spacing, we calculated the TOI at +25dBm on 14MHz with both preamplifiers turned off. That works out [for the PROIII] to a slightly less than 5dBm improvement over the PROII, all other things being equal.

"Under the same conditions at 5 kHz spacing – something much more akin to *real-world* amateur conditions (and this time well within the passband of the receiver's 15kHz roofing filter) we determined the PROIII's TOI to be –17dBm, 1.8 dB better than the –18.8 dBm we calculated for its predecessor. With one pre-amp switched on and 20kHz spacing, the 14MHz figures become +14 and +5dBm, respectively. With 5kHz spacing, the corresponding figures are –16.5 and –29dBm, respectively.

*QST* policy is always to base reviews on models purchased in the market rather than on the less-costly *RadCom* policy of reviewing models on loan from suppliers.

# TECHNICAL TOPICS

Nevertheless, although the measurements by G3SJX on the PROIII cannot always be compared directly with those by N1RL, as they are based on different spacings and omit the case with both pre-amps switched on, they are roughly the same. G3SJX gives for 14MHz, 50kHz spacing +32dBm with pre-amps off, and +21dBm with one pre-amp on. For 5kHz close spacing on 3.5 MHz -16.5dBm and with one pre-amp on, -27.5dB. Peter also gives the 3.5MHz 3kHz spacing as –18dBm and -20dBm.

These are good figures compared with those of a few years ago, even for top-of-the-range models, but it is interesting to note that the IP3 performance of the CDG2000 home-built transceiver prototype described (Part 1) in *RadCom*, June 2002, pp19 – 22 shows that the IP3 figure of approximately +40dBm "is maintained for close-in signals and that this is largely due to the use of the two-tank oscillator circuit as developed by G3SBI and first described in 'TT', January 1995". A two-tank circuit is also used in the AOR7030 receiver which remains an outstanding factory model for its excellent close-in performance.

The CDG2000 offers a spurious-free dynamic range greater than may strictly be needed for general use except on 7MHz, but it shows what can now be achieved. The K2 kit also achieves close-in performance that was shown by *QST* to be better than that of the PROII.

## IMPORTANCE OF PHASE NOISE

The striking difference in most models of the IP3 figures with wide and narrow spacings when tested with two strong input signals underlines the importance of reducing the oscillator phase noise. This problem was discussed at some length in 'TT', December 2002, pp61 – 62, with reference to the excellent close-in performance of the K2 kit transceiver, and the use of a double-tank local oscillator in the AOR7030 receiver and in the CDG2000 home-built transceiver. Incidentally, it is interesting to note that some 250 CDG transceivers have been built by amateurs, with more than half of them built overseas using the information provided on the web.

Peter Chadwick, G3RZP, writes: "As I showed in my *QEX* article 'HF Receiver Dynamic Range: How Much Do We Need?' (*QEX*, May/June 2002. pp36 – 41) IMD isn't potentially as big a problem as phase noise. This is because the IMD from two or three large signals is easier to cope with than the summed phase noise from ten or more signals 10dB lower in level. Thus, phase noise can limit dynamic range to a much greater extent than can IMD. This isn't to say that IMD can be ignored, but that there really isn't the evidence that going from a +20dBm intercept point to +30dBm will make that much difference (if any at all); even going from –10 to +20 probably won't make much difference in practice. With pretty big antennas, my article showed that you don't actually need more than 100dB of instantaneous dynamic range, although you do need a [front end] attenuator and some operating ability. The bottom end is limited by noise, and bigger antennas bring in more signal and noise, so the required dynamic range is determined there. Generally speaking, the presence of the big signals is accompanied by the presence of a higher noise floor, fixing the dynamic range requirement.

"Incidentally, intercept points are notoriously difficult to measure with accuracy. If you take the generator level as ±0.75dB to a 99.5% uncertainty, add the mismatch, combiner and cable loss uncertainties, you can easily get to ±2.5dB overall uncertainty in the absolute level into the receiver, and this reflects a ±7.5dB uncertainty in intercept point. There are a couple of ETSI documents on mobile radio equipment which are well worth reading, although they are fairly heavy going."

## REDUCING PHASE NOISE

Oscillator phase noise affects both transmitted and received signals. Phase noise will continue to represent interference or an artificially-raised noise floor from other people's transmissions even when your own has been reduced to the minimum possible. While close-in sidebands are dominated by oscillator phase noise, far-out sidebands are greatly affected by amplitude noise. Consequently, it is important that all of us, including manufacturers, should try to understand and strive to bring about a general reduction in both phase- and amplitude-noise produced by variable LC oscillators and frequency synthesisers.

It is recognised that quartz-crystal-controlled oscillators, run at reasonable power and using low-noise high-gain semiconductors or valves, produce very little phase-noise. The present target must be to bring LC VFO phase and amplitude noise as close as possible to that of a good crystal oscillator.

As noted in a long paper 'Oscillator Phase Noise: A Tutorial', by Thomas H Lee and Ali Hajimiri (*IEEE Journal of Solid-State Circuits*, March 2000, pp326 – 36), the key parameters include power dissipation, oscillation frequency, resonator Q, and circuit noise power. Another even more relevant source is the paper 'Low-Noise Voltage-Controlled Oscillators Using Enhanced LC-Tanks', by Jan Craninckx and Professor Michiel Steyaert in Belgium (*IEEE Transactions on Circuits and Systems: II Analog and Digital Signal Processing*, December, 1995, pp794 – 804).

One approach to reducing phase noise is to design the LC tank circuit as closely as possible to the equivalent circuit of a quartz crystal, implying that the resonator should be of the highest possible Q. Harold Johnson, W4ZCB, has adopted this. Quite recently, he wrote an (unpublished) article 'Helical Resonator

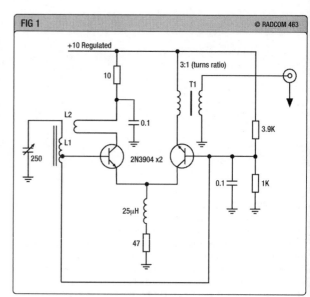

**Fig 1**
K7HFD's low-noise 10MHz oscillator. L1 is 1.2µH (17 turns on a T68-6 toroid core, tap at 1 turn from grounded end while link is 2 turns wound over L1. The link must be properly phased for oscillation. Although not shown, ferrite beads were used on both bases and collectors.
(Source: *Experimental Methods in RF Design*)

**Fig 2**
The low-noise oscillator using a helical resonator (HR-1) as developed by W4ZCB.

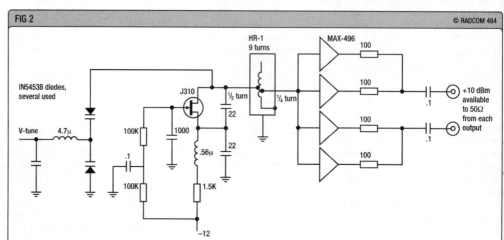

# TECHNICAL TOPICS

Oscillators' from which I will quote briefly.

"After trying a dozen oscillator circuits, from Vackars to Colpitts, Hartleys, Seilers, [Gouriet-] Clapps and even K7HFD's ingenious class-C low-noise 10MHz oscillator (**Fig 1**) described in *Solid State Design for the Radio Amateur*, by Hayward and DeMaw (ARRL 1977, p126) and also in the section on 'Designing Quiet Oscillators' in *Experimental Methods in RF Design*, by Hayward, Campbell and Larkin (ARRL, 2003, pp4.12 – 4.13), it really dawned on me that it's not so much the circuit (though some are better than others) as the components and the way you choose to use them. What was needed was an oscillator with such high tank Q that it would swamp any other characteristics of the circuit.

"Hewlett Packard and its pre-synthesis signal generators pointed the way. Their 8640, one of the quietest oscillators that it ever made, is built around a gorgeous coaxial cavity that tunes 256 to 512MHz and is subsequently divided down to frequencies below that in octave bands."

Coaxial resonators are hard to find. W4ZCB turned to producing helical resonators using copper tubing for an oscillator covering 75 to 105MHz which could be divided down to cover all amateur bands (including WARC) from 1.8 to 30MHz. He designed and built 105MHz helical resonators with Qs in the region of 1000. Various problems were encountered before and after coming to England to check performance with Colin Horrabin, G3SBI, "the one to blame for starting all this".

After describing in detail the constructional details and completing some three dozen helical resonator oscillators (**Fig 2**) for friends, he writes: "What it will give you is an oscillator that is *free running* at 100MHz, and when divided-down to be the local oscillator for a HF sideband receiver, will copy a sideband round-table for an hour without being retuned. It is readily tuned three-quarters of an octave or a shade more. Measured noise sidebands with a divisor of eight (each *synchronous* division by a factor of two provides close to a 6dB improvement in phase noise) are better than -123dBc/Hz at 1kHz and drop to -144dBc/Hz at 5kHz. At 20kHz spacing with divide-by-eight, those numbers are -156dBc/Hz and at 50kHz they exceed my ability to measure them... This is better at *1kHz* spacing from the carrier than a current commercial offering reviewed in one of our periodicals at *any* spacing."

W4ZCB also gives a reminder that one of the things to watch out for in building a low-noise oscillator is the power supply for the oscillator device: "A perfect oscillator will not be perfect with a noisy source of power. A year or so ago, I ran across an article on the Wenzel Associates web page called 'Finessing Power Supply Noise'. Basically, a shunt AC-coupled regulator (**Fig 3**), it functions by shunting the noise to ground through a small series impedance. With no adjustment at all, it is capable of reducing power supply noise by 20dB. With a bit of tweaking, that can be raised to a 40dB improvement, contributing directly to oscillator sideband cleanliness... I don't build oscillators or frequency control circuits without it."

The helical resonator approach is not without its constructional problems, and is not to be tackled lightly. The use of twin- or even multiple-tank coils is still regarded by G3SBI as a better way ahead. His two-coil oscillator was published first in 'TT' February 1996 (see also *TTS 1995-1999*), with a further discussion in 'TT', November 2002, pp77 / 78, noting the use of a similar approach in the still-esteemed AOR7030 receiver (designed in the UK by John Thorpe in the mid-1990s) and is also one of the key features of the CDG2000 transceiver.

The Belgian paper cited above endorses the use of multiple tank coils as an effective means of reducing phase noise in low-voltage oscillators claiming that with $n$ coils, phase noise *decreases* proportionately with $n$, and power and [board] area *increase* proportionately with $n$.

**Fig 3**
W4ZCB recommends the use of this shunt regulator to remove noise on power supplied to oscillators, etc. Design comes from web page of Wenzel Associates.

**Fig 4**
G3UUR's modified diode ring mixer. T1 & T2 – 3 turns primary and 3 turns secondary on Fair-Rite 2843000302 two-hole core. R1, R2, R3, R4 – 100Ω 0.25W carbon film resistor. D1, D2, D3, D4 – matched on forward voltage drop and reverse leakage current.

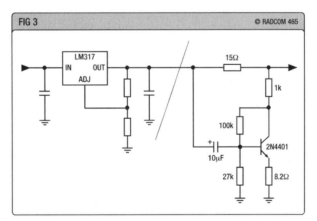

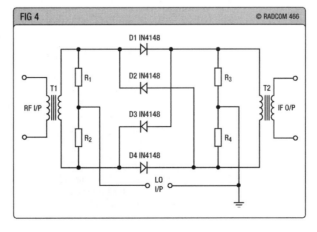

## MORE ON LED & DIODE MIXERS

Dave Gordon-Smith, G3UUR, writes: "I see the blue LED mixer ('TT' January and March, 2005) has created a lot of interest. I wonder whether the fragility of GaInN LEDs will create a reliability problem in mixer applications, though. The original Si microwave diodes used by Fazi and Nuedeck had a reverse breakdown voltage in excess of 50V. The GaInN LEDs have a reverse breakdown voltage of only 5V. In the ring mixer configuration the forward-biased diodes provide some protection for the reverse-biased diodes by clamping the voltage at about 3.2V. However, there is a transformer winding in series with the forward diode drop across each reverse-biased diode that could add a transient voltage from the switching waveform, or RF signals, and exceed the 5V reverse breakdown voltage across the non-conducting part of diodes. Normally, the energy associated with this voltage peak would be tiny for signals received off-air, but locally-generated signals or static could easily provide enough energy to damage the GaInN diodes. It will be interesting to hear how people get on with these LEDs in mixer applications.

On a more general mixer topic, G3UUR writes: "Back in the early 1980s, Wes Hayward, W7ZOI, sent me a copy of a British paper by H P Walker: 'Sources of Intermodulation in Diode-Ring Mixers', *The Radio and Electronic Engineer*, May 1976, pp247 – 55. The author concluded that the time taken to switch stages is a parameter crucial to IMD production in mixers. Thinking about this aspect of IMD performance led me to question the wisdom of applying a fast switching waveform through the inductance of the transformer windings. This arrangement is bound to slow down the switching cross-over and degrade the IMD performance, even with fast-switching LO waveforms, because the current through an inductor can't change quickly. The steady-state resistive nature of loaded transformer windings doesn't apply to fast-switching waveforms. My KISS solution was to feed the switching waveform to the diodes through resistors: **Fig 4**

"The resistors cause some additional insertion loss, but this only amounts to 2dB, or so. The load presented to the LO source is more resistive and, although it absorbs some LO signal, it provides the driver with an almost-constant load – around 100Ω resistive – and this aids faster switching. Another advantage is that the transformers don't have to be centre-tapped. The transformers I used had a 1:1 turns ratio, and presented about 40Ω to the input and output ports. They could easily be changed to provide a

# TECHNICAL TOPICS

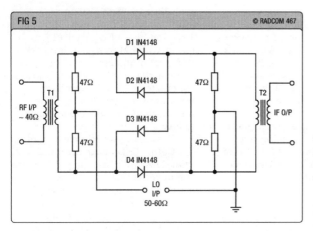

Fig 5
G3UUR's further-improved modified diode ring mixer. T1 – 6t to 4t on Fair-Rite 2865000202 core for 1 – 30MHz. T2 – 4t to 6t on core as for T1.

50Ω input by using an 8:7 turns ratio. To provide a 50Ω input impedance for the LO signal, the input and output transformers could be wound for a 3:2 turns ratio, and the resistors changed to 47Ω, as in **Fig 5**. More LO signal would get to the diodes in this case, so this version ought to be better for strong-signal handling than my first attempt at an improved ring mixer, but with slightly increased insertion loss. The transformer step-down ratio and the LO fixed resistor values can be juggled to give the best compromise between insertion loss and signal handling. I have no figures to back my belief that this mixer circuit is better than the original diode ring mixer, just some subjective tests done on 7MHz years ago with a 3.5Vp-p square wave LO signal. I think it's an improvement, but it requires someone with the right test equipment to determine whether it is, or not."

Incidentally, Colin Horrabin, G3SBI, tells me that he has recently tried the effect of biasing the Fairchild FST3125 fast switching bus in his H-Mode mixer. Preliminary results seem most promising, particularly for up-conversion but they also confirm that present production of these devices do not give quite the performance of the original production devices as switching mixers. It is hoped that the new circuit with biasing will be described before long in 'TT'.

### HERE & THERE

Jan Martin Noeding, LA8AK, writes, re 'TT' January 2005: "I am rather surprised to read – once again – that the German Torn E.b covers 100-6670kHz. I have seen many examples of this receiver, but have never seen a model with this frequency range. The 6670kHz may be right for the 1933 Spez 445 version, but every Torn E.b I've seen covers 98 – 7095kHz. There is another version, with later valves (RV2, 4P700) sold to the Swedish army and covering 195 – 15,410kHz. More details on http://home.online.no/~la8ak/22a.htm

# CLEANING VARIABLE CAPACITORS

As a March 'Here & There' item, M0MAC sought information on a method of cleaning the plates of variable capacitors recovered from old valved equipment that he believed was published some years ago in 'TT' using baking powder and hot water. This has produced a number of varied responses.

John Alford, G3DOE, confirms that some baking powder should be dissolved in hot water. The capacitor is immersed in the solution and left for about an hour. A recent test has confirmed that components come out looking like new. He insists that it must be baking powder and not caustic soda.

George Ashford, G2AOZ, dates relevant 'TT' items to (1) March 1981 p224, with a follow-up note (2) in May 1981. He adds: "I, too, have a recollection that baking powder (sodium bicarbonate) entered the equation".

(1) A tip given by Bill Pickens, WB5NGF (*QST*, December 1980, p54): "For corroded and otherwise dirty variable capacitors, which are almost impossible to clean with a brush, I use a mixture of 4oz (120ml) of concentrated lemon juice in eight to 10oz (240 to 300ml) of water placed in a saucepan. By placing the capacitor in this mixture and boiling for 10 to 15 minutes, the device can be made to look like new. A few drops of liquid detergent might be helpful. A drop or two of oil should be placed on the bearing when dry." The item also contained my comment: "The importance of ensuring that any variable capacitor used in a high-grade VFO is really clean has been stressed on a number of occasions, up to and including the use of an ultrasonic cleaning bath."

(2) WB5NGF's tip stirred R L Halls, G3EIW, to mention that, for the past 30 years, he had used Goddards Silver Dip for cleaning all silver-plated radio components, including variable capacitors. He finds that this transforms blackened silver items into a gleaming 'like new' condition in just a few seconds. Any grease or oil should be removed first with hot detergent. He added a warning: "Goddards Silver Dip can stain 'stainless steel'; any attempt to carry out operations on the stainless kitchen drainer is likely to make you highly unpopular with 'she who must be obeyed'."

Alan Strong, G3WXI, recalls that a colleague used to clean intricate silver components very successfully by standing them on a sheet of aluminium cooking foil placed in the bottom of a polythene washing-up bowl containing a solution of washing soda [sodium carbonate]. He restored a substantial portion of a Chain Home [radar] transmitter that he used in his research work in this way. Unfortunately, I cannot remember whether the component must or must not be in contact with the aluminium foil, but I think he used to cover the foil with a cloth."

I believe that for cleaning fine antique silver objects, cigarette ash is preferable to commercial cleaners such as Silvo that can be rather abrasive. But I don't want to encourage this since the cigarette smoke can be as harmful (long term) to radio equipment as to humans! ◆

Pat Hawker, G3VA

37 Dovercourt Road, London SE22 8SS.

# TECHNICAL TOPICS

## TT

**Pat considers the first software-defined HF transceiver, some handset folded antennas and a 50MHz halo. He assesses the 7360 beam-switching mixer, and ends with the Navy receiver model CJA/CJC**

### DEBUT OF SOFTWARE-DEFINED HF TRANSCEIVERS

The April *QST*'s 'Product Review' (pp73-78) by Steve Ford, WB8IMY, claims to open a new chapter in the history of Amateur Radio. WB8IMY who is the journal's Editor writes: "I'm not indulging in hyperbole, by making such a statement – it is a fact. For the first time in ham history, you can purchase 'off the shelf' an HF and 6m transceiver that uses software to define its functionality – a *software-defined radio.*"

The new product is the FlexRadio SDR-1000, available in various options, including the Model SDR-ASM/TRA (fully assembled transceiver with 100W RF amplifier and RF expansion board) priced at $1325 or as the SDR-ASM/TR with 1W output and RF expansion board priced $875. It is also available in partially-assembled form, with or without enclosure.

The long review makes it clear that the transceiver is still in the course of refinement and may already have additional features, etc. But, before rushing to take advantage of the low cost compared with conventional HF transceivers, remember that it can only be used in association with a PC equipped with a good soundcard, etc. As *QST* puts the 'Bottom Line': "The SDR-1000 may mark the beginning of a new generation of amateur radio equipment, but the pioneers who take it up may need a bit of frontier spirit!" From the review it becomes clear that, at least at this stage, this is equipment that requires a good knowledge of computer and information technology if full advantage is to be taken of the potential advantages and flexibility of this first SDR transceiver to reach the amateur radio marketplace.

As WB8IMY explains, the SRD-1000 is not just a software product: "it most definitely has hardware. If you purchase what I like to call the 'full Monty' version with the 100W PEP RF amplifier and RF expansion board, you are presented with a nondescript 19 x 8½ x 4 inch black box. On the front there is an on/off rocker switch, a four-pin microphone connector and a cooling fan. On the back, you'll find ports for computer connections, DC power unit and, of course, an antenna."

The box contains a few circuit boards. The receiver is an advanced direct-conversion design using direct digital synthesis. It converts RF directly to audio. Separate in-phase and quadrature signals are fed to the computer soundcard for digital signal processing using an innovative in-phase and quadrature (*I and Q*) image-reject approach. For transmission, the hardware is designed to take processed audio from the soundcard and convert it to RF. As WB8IMY puts it: "The true heart is *not* within the black box. *Data* is the lifeblood of this radio; the hardware is just a portal between the analogue and digital worlds… The hardware is like unformed clay on the potter's wheel, waiting for the hands of the artist to shape it into something meaningful. The art – and the artistry – is in the software that runs on your computer. The SDR-1000 software – known as *PowerSDR* – determines how a received signal will be demodulated. It also creates the transmitted signal according to the mode you wish to operate. The SDR-1000 is a software-defined radio in the most literal sense of the term… Don't confuse it with microprocessor-controlled radios that offer firmware updates. The changes implemented by a firmware update are limited in scope because the inflexible hardware defines (and constrains) what can be done… With a software-defined radio, you can make very large changes indeed.

"The SDR-1000 software architecture is completely open. This means that anyone with enough computer savvy can modify the software (and, hence, the radio) to suit his / her individual needs. It also means that hams throughout the world can pool their collective genius and create new software for the SDR-1000. So, rather than a static box full of hardware, the SDR-1000 will evolve through the years as clever hams take up the 'clay' and create new works of engineering art."

The *QST* review is based on the software as it existed in January 2005, and it is pointed out that further facilities were likely even before the review was published. As reviewed, the SDR-1000 receiver had a frequency range of 0.01 to 65MHz with a switchable pre-amp and able to transmit on all amateur bands from 1.8 to 50MHz. Modes with then-existing software included SSB, CW, AM, FM and Digital Radio Mondiale (DRM), although it is pointed out that the CW performance required further refinement. The ARRL Lab measurements indicate that it is capable of good performance although falling some way short of current top-of-the-range conventional transceivers.

It is stressed that the soundcard used in the user's computer is the engine that enables the SDR-1000. "In particular, the dynamic range and distortion performance is *direct-*

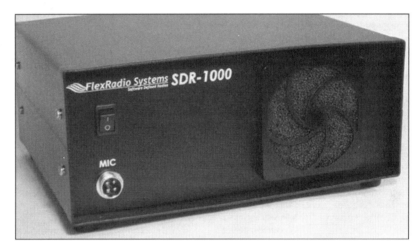

The SDR-1000 Software-Defined 100W HF/VHF transceiver – a truly 'black box' rig with only an on-off switch and microphone socket on the front panel, controlled and defined by software run on a PC. It offers potential for further great flexibility and functionality by updating or developing the software. (Source: www.flex-radio.com)

ly related to the quality of the soundcard. At the time of this writing, FlexRadio Systems officially supported only the SoundBlaster Audigy 2ZS, Audigy 2, Extigy, MP3+ and the Turtle Beach Santa Cruz soundcards. This is not to say that other sound cards cannot be used, but the radio may not perform as specified. And if you run into trouble with a non-supported card, FlexRadio may not be able to help you. In addition, not all soundcards have the separate line input, line output and microphone jacks necessary to work with the SDR-1000. They may also lack mixer controls with independent level adjustments for the line input and microphone input.

"In addition to a quality sound card, you'll need a quality (read 'fast') PC. The minimum requirement is an 800MHz Pentium computer. In our tests, an 800MHz system was *just* adequate. Stepping up to a machine with a clock speed greater than 1GHz makes a substantial difference." At present the SDR-1000 cannot be used with a lap-top computer.

It is clear from the detailed review that the true challenge for a user begins when you install the software: "My first step was to download the latest version of *PowerSDR* from the FlexRadio website (**www.flex-radio.com**). The file is less than 1MB in size, so that step went quickly. When I ran the setup program to install the software, however, it came to an immediate halt and informed that I didn't have the Microsoft 'Net Framework' installed on my PC. Oops!" This was only the first of several problems encountered before the rig came into operation.

To quote briefly from some of WB8IMY's final impressions: "The receive performance of the SDR-1000 was at least comparable to a traditional transceiver in its $1300 price class. Of course, the crucial difference is that it will undergo continuous updating and improvement for years after the initial purchase. A hardware rig remains essentially the same for ever... Is the SDR-1000 a radio for all hams? At this point, probably not. The current incarnation is best suited to the amateur who knows his or her way around a computer. It takes a ham with intermediate or advanced computer skills to get the most out of an SDR-1000 with the least amount of frustration.

"But this is just the first step into a new era. As the SDR-1000 evolves, new versions are likely to emerge that will be 'friendlier' and well within the understanding of any amateur. The manufacturer reports that it plans to offer a turnkey soundcard and radio solution, including a CD with driver and calibration routines by the time you read this. If so, that will be a real plus. With the

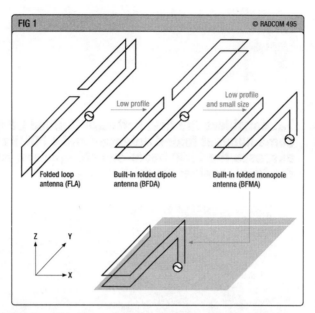

**Fig 1**
Development of the antenna structure leading to the built-in folded monopole antenna for handsets developed by Japanese engineers (Source: Electronics Letters)

collective intelligence of the global amateur radio community at work, the potential of the SDR-1000 is almost limitless."

SDR-1000 manufacturer is FlexRadio Systems, 8900 Marybank Drive, Austin, Texas, TX 78750, USA.

I must confess that as a computer illiterate and someone who still uses a PC primarily as a word-processor, I do not see myself as a potential SDR customer. But, for the computer-cum-radio enthusiast, it opens up interesting new possibilities, although stand-alone HF transceivers with front-panel controls are likely to be still with us for the foreseeable future.

### HANDSET FOLDED ANTENNAS & A 50MHz HALO

Much of the current professional research into antennas is keyed to improving the performance or appearance or convenience of UHF cellular handsets operating at 900 or 2300MHz. But the basic structures can be used at much lower frequencies by scaling up the dimensions. Some of the structures can then be used for amateur mobile operation, etc.

**Fig 1** comes from 'Built-in Folded Monopole Antenna for Handsets', in *Electronics Letters*, 25 November 2004, pp1514 – 5, by S Hayashida and colleagues at the Japanese National Defense Academy.

The synopsis reads: A built-in folded monopole antenna (BFMA) for handsets is introduced and investigated. The characteristics of the BFMA are compared with those of a planar inverted-F antenna (PIFA) which is one of the conventional handset antennas. As a result, it has been confirmed that the BFMA has smaller size and wider bandwidth compared with the PIFA."

In the text, it is noted that a folded loop antenna for handsets (FLA) has already been introduced and shown as one of the balance-fed antennas for handsets, and is very effective in mitigating antenna performance degradation due to body effect. To meet the requirements for the latest handsets, the antenna is modified to have small size and low profile and the performance is analysed. Low profile is achieved by folding a loop element sideways so that the antenna can be placed closely on the ground plane. Small size is achieved by it consisting of half of the built-in folded dipole antenna (BFDA) which has a structure of a folded loop with elements sideways. The antenna characteristics such as VSWR, the current distribution and radiation patterns are compared with those of the PIFA. As a result it has been confirmed that the BFMA has smaller size and wider bandwidth. The Japanese authors describe and analyse a BFMA designed for use at 2250MHz.

Paul Danzer, N1II, in *QST*, September 2004, draws on a 30-year-old *ARRL Handbook* (1975) design (**Fig 2**) to present constructional details of a practical 'Six-metre Halo' antenna for fixed or mobile operation. The 'halo' antenna comprises a half-wave resonant dipole bent into a circle and in this case gamma-matched. N1II lists the following reasons for making this an obvious choice: "Inexpensive (okay... cheap); omni-directional; horizontal polarisation to be able to work locals; only one trip to the nearby home supply store; no exotic components to be ordered; easy to tweak with 6m test equipment."

Wider bandwidth should be obtainable by using the folded dipole configuration shown for the FLA in Fig 1.

### THE 7360 BEAM-SWITCHING MIXER

The February 'TT' mentioned briefly the RCA beam-switching valve type 7360 that was for many years recognised as capable of coping with extremely strong signals, of the order of volts rather than millivolts. But how does its overall performance compare with state-of-the-art solid-state mixers such as an H-mode mixer using the FST3125?

First, a little history. The 7360 was introduced by RCA in 1960 for use as a high-level balanced modulator for SSB transmitters, a receiver product detector or phase splitter. It subsequently attracted considerable amateur interest for use in receivers as a switching mixer (balanced or unbalanced) and for product detectors, soon gaining the reputation of being able to handle extremely strong signals and providing high gains when used as a product detector. I recall the late Reg Cole, G6RC, showing me how well the 7360 per-

# TECHNICAL TOPICS

formed when he used one to modify his Hammarlund Super-Pro receiver.

Its use as a front-end mixer in a factory-made receiver occurred in 1963 with the Squires-Sanders SSR-1 communications receiver. Unfortunately, William Squires, the designer, died in an air crash after just a few models had been produced and I am not aware of any ever reaching the UK. Several designs using the 7360 in home-built receivers were published, for example by W2PUL and by W1DX in his novel 'Miser's Dream' design (*QST*, May 1965, noted with front-end circuit-diagram in 'TT' and subsequently in later editions of *Amateur Radio Techniques*).

The low-cost 'Miser's Dream' featured an unbalanced 7360 mixer with no RF stage, but with an RF Q-multiplier (6C4 triode) to sharpen up the characteristics of the signal input circuit to reduce image response, a 2.8kHz crystal lattice filter immediately following the mixer, plus a 250Hz filter for CW in the subsequent IF stage. This remains a valid approach, even today, particularly as a hybrid design with solid-state devices used for all stages other than the mixer. Costs could be reduced by the use of a home-built crystal ladder filter, a little-known filter technique at that time, and would compensate for the current high cost of a 7360 - if one can be located.

The 7360 was the best, but not the only, RCA beam-switching valve. The earlier lower-cost 6AR8 and the later 6JH7 were developed primarily for use as synchronous demodulators in colour TV receivers but also formed excellent mixers and product detectors. Brian Mitchell, G3HJK (QTHR), used two 6AR8 valves in a unique version of the G2DAF receiver that was built in two separate units. He later sold this receiver to J A Cox, G4AQD who subsequently became a silent key. As a result, G3HJK lost track of the receiver but believes it still exists and is probably somewhere in the Cheshire area. He would welcome any information as to its present whereabouts and ownership.

Beam-switching valves were not the only valve-mixers capable of handling inputs above 1V. Peter Chadwick, G3RZP, writes: "My RCA tube manual lists the 7360 as producing -40dB IMD for 2.8V RMS input in balanced mixer service. One half of a 12AU7 double-triode will produce this level of IMD for 2.1V input, according to Pappenfus et al of Collins Radio. There are no figures given for the 7360 noise [see below- *G3VA*]: it's probably fair to assume a noise figure equivalent to a tetrode, since there will be partition noise. This gives an equivalent noise resistance (ENR) of 2500ohm, or about 0.34µV in a 3kHz bandwidth, if you assume the transconductance from deflector plate to anode is the correct term to use. If you use the main transconductance, it becomes around 17kohm, or 1µV in 3kHz. The 12AU7 double-triode, again according to Pappenfus, is about 20dB worse than the first figure and 10dB worse than the second, so overall is between 12.5 and 22.5dB worse in dynamic range than the 7360. On the other hand, 12AU7s can [still] be purchased without paying an arm and a leg! You will remember the naval receiver that McCormack of GEC described at the 1963 HF Conference at the IEE, which used a 12AU7 mixer and another one as a cascode

### Table 1: Intermodulation measurement 7360 mixer

| Measurement | Rk | Vg1 (V) | Vg2 (V) | Ig2 (mA) | Iat (mA) | Gain (dB) | IP3 (dBm) | CP (dBm) | Noisefloor BW=2.5kHz (dBm) | 3rd order Dynamic (dB) |
|---|---|---|---|---|---|---|---|---|---|---|
| 1 | 47 | −0.6 | 105 | 2.8 | 10 | 18 | −11 | −25 | | |
| 2 | 470 | −2.9 | 178 | 1.2 | 4.7 | 19 | −2 | −17 | | |
| 3 | 170 | −1.6 | 140 | 2.0 | 7.4 | 20 | 0 | −15 | −135 | 90 |
| 4* | 170 | −1.6 | 140 | 2.0 | 7.4 | 14 | 6 | −15 | −131 | 91.3 |

* In measurement 4 the grid connection is tapped halfway at the input parallel circuit

### Table 2: Intermodulation measurements ECF 80 mixer

| Measurement | Rk | Iatr (mA) | Vg2 (V) | Ig2 (mA) | Iap (mA) | Gain (dB) | IP3 (dBm) | Noisefloor BW=2.5kHz (dBm) | 3rd order Dynamic (dB) |
|---|---|---|---|---|---|---|---|---|---|
| 1 | 300 | 10 | 190 | 0.8 | 4.2 | 4 | 6 | −113 | 79.3 |
| 2* | 300 + 100uH | 9.2 | 170 | 1.3 | 5.8 | 2 | 21 | −113 | 89.3 |

* In measurement 2 the result is strongly dependent on oscillator level

Table 1 PA0KDF's intermodulation measurements on the 7360

Table 2 PA0KDF's intermodulation measurements on the ECF80

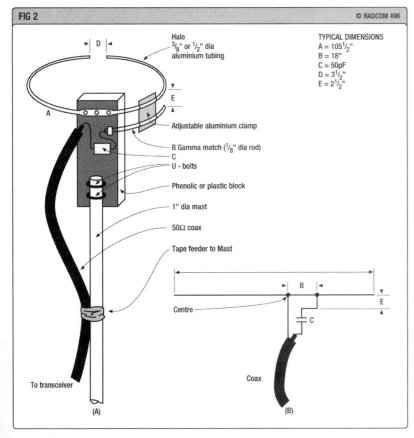

Fig 2 50MHz halo antenna as described in the 1975 ARRL Handbook, but recently revisited by N1II in QST, September 2004.

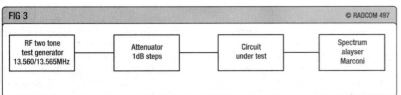

Fig 3 Test set-up used by PA0KDF in 1984 to measure the intermodulation performance of the 7360 beam-switching mixer (see Table 1) and an ECF80 triode-pentode mixer (see Table 2).

# TECHNICAL TOPICS

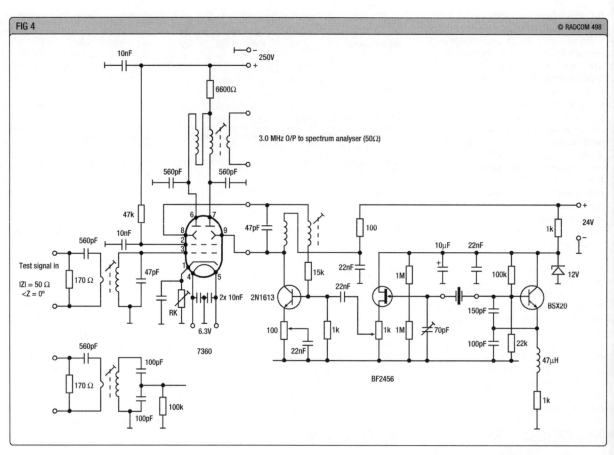

**Fig 4**
Circuit diagram of the tested 7360 mixer with alternative signal input circuits as published in Electron, April 1984.

RF amplifier." See opposite page.

Koos Fockens, PA0KDF, has also responded to the request for information on the performance of the 7360, comparing it with that of an ECF80 triode-pentode mixer. He writes: "In 1984, I carried out a series of measurements on the IP3 and other relevant characteristics of the 7360 and the ECF80; the results were published in Dick Rollema's column 'Reflecties Door PA0SE' (*Electron*, April 1984). I used a test generator specially developed for intermodulation measurements in the arrangement shown in **Fig 3**. This was in connection with an antishoplifting system that I developed professionally in the late 1970s that depended on the detection of intermodulation produced by a security tag. The relevant test circuits are shown in **Figs 4** and **5**.

"The test generator provided two signals: 13,560 and 13,565kHz, each at 0dBm with its own intermodulation products lower than –90dBm, so its IP3 was >45dBm. Measurements for the 7360 are shown in **Table 1**, and for the ECF80 in **Table 2**

"My conclusion is that the 7360 beam-switching mixer was indeed a very good mixer compared with other valve mixers, but lags far behind the best current semiconductor mixers [for example the H-mode mixer – G3VA].

"Incidentally, I used a 7360 as a second mixer in a double-conversion 144MHz converter I built in 1968; this converter was used at PI4THT for some years.

"Finally, I would draw attention to the following: Very often a third-order dynamic range is specified without specifying the receiver bandwidth. Such a specification is useless since because the noise-floor of the receiver is dependent on the bandwidth, the dynamic range will be too. Therefore a third-order dynamic range must specify the receiver bandwidth."

## HERE & THERE

With reference to the 'TT' April item '2V/300V DC-to-DC Converter', shown (Fig 3, p81) powering a 5W CFL (compact fluorescent lamp), Alan Floyd, G3PNQ, points out that fluorescent lamps do not like working on DC as one end of the lamp will blacken and burn out, considerably reducing the service life of the lamp.

Andrew Holme of Brentwood read with interest my comments about the Wadley loop in the April 'TT' but adds: "However, I am not convinced that it offers a phase-noise advantage over PLL designs. Surely the first VFO in the RA17 has phase noise? According to John Wilson's review in *SWM*, December 2000, it may have quite bad phase noise. The question is does the loop cancel the phase noise? The loop cancels slow drift on the first VFO, why not fast 'drift' (phase noise) also?"

Andrew shows diagrammatically that with a Wadley loop, you increase rather than decrease the phase noise, noting that "You cannot remove phase noise by mixing it with itself. It just spreads twice as wide. I am inclined to think that phase noise *is* a problem with a Wadley loop."

Godfrey Manning, G4GLM, remains doubtful about the short-range 'Micro-Power AM 'Transmitter" ('TT', February 2005, p74). He writes: "I'm not sure about [its] function. The microphone, applied to control pin 5 of the 555 timer IC varies the internal reference for the comparators that determine the threshold for switching the output. Normally, the timing capacitor has to charge and discharge by fixed proportions of the supply voltage, triggering a change in the output state as it does so. Altering the internal reference voltages means that the thresholds are reached at a different proportion and so the output state flips over either sooner or later than normal, depending on which way the control pin is pulled. So this would appear to modulate the frequency and not the amplitude, but slope-detection means that a domestic MW broadcast receiver could still resolve the signal if slightly off-tune. Also, as a transmitter, the square-wave output would be the most efficient harmonic generator possible! Very stringent low-pass filtering is needed and the

# TECHNICAL TOPICS

insertion loss would reduce the radiated [micropower] signal." G4GLM would be interested in any thoughts or experience that proves or disproves his views.

According to a report in *New Scientist* (12 March 2005, p17), gigantic solar storms destroyed nearly 60% of the ozone above the Arctic during the spring of 2004 with ozone levels remaining low into July 2004. The ozone which shields us from harmful ultraviolet radiation lies mostly in the lower and mid-stratosphere. Man-made chemicals such as the CFCs have been mainly responsible for the depletion of the ozone layer. Now Cora Randall at the University of Colorado at Boulder and colleagues have shown that a record barrage of charged particles from the sun in October and November 2003 also destroyed large amounts of ozone to a level never before seen in the northern hemisphere. My Kiel beacon records show that in November 2003 the KA ratings exceeded 150, with further strong magnetic storms in July (>120) and November 2004 (>140). An article in the following issue of *NS* (19 March, 2005, p10) 'Superflares Could Kill Unprotected Astronauts', noted that the most powerful solar flare ever recorded was observed by British astronomer Richard Carrington in September 1859. It easily surpassed the monster eruption of March 1989 which knocked out the power grid in Quebec, Canada. The powerful flares of November 2003, July 2004 and November 2004 all had major effects on HF propagation.

## THE GEC NAVY RECEIVER MODEL CJA/CJC

The GEC 1960s naval receiver was briefly noted in 'TT' and in many editions of *Amateur Radio Techniques*. It has been identified by G8MOB as the massive Royal Navy type CJA with its separate GEC frequency synthesiser, also massive, as type CJC. In a letter to *Radio Bygones*, (No 94, April/May 2005, pp34 – 35), Ted Minchin, ZL1MT, reports that he is the fortunate (?) owner of a pair of these receivers and their accompanying synthesisers. Mounted in their rack cabinet, the whole makes an assembly in excess of 1000lb: "Just as well my workshop floor is reinforced concrete!" He adds "the performance of this pair is impressive in SSB. I use a reference signal derived from a rubidium source for the synthesiser and it is possible, using this to tune these receivers to a SW broadcast station and receive it in ISB mode (gives an interesting 'stereo' effect to the recovered audio if you listen to both channels simultaneously, and the receivers stay locked and in perfect sync for days... But it is easy to see why the Navy needed squadrons of technicians to maintain these receivers, it can be a time-consuming job keeping them in tip-top order... I have enjoyed owning them and don't hesitate to say to someone with the room and expertise 'go for it'. They are a lot of fun!"

The 1963 Conference paper stressed three main design features for a high quality receiver: (1) there should be maximum selectivity before the first non-linear stage; (2) The valve [or semiconductor] stages must be designed for maximum linearity; (3) The signal level must be maintained at as low a level as is practicable until the maximum selectivity has been achieved.

The CJA was a single-conversion superhet with an IF of 1.6MHz designed to cope with extremely strong local signals. To achieve satisfactory image response, six tuned circuits are employed before the mixer in the form of three coupled pairs, separated by two RF stages. The gain of these stages is only just enough to maintain the signal above noise level. The design of the RF stages was the result of a study of many different valves and circuits. A cascode amplifier employing a B329 double-triode (equivalents 12AU7, ECC82) was chosen as this gave the best overall compromise between linearity and noise factor. The mixer circuit also uses the B329 with the two halves connected in push-pull forming a single-balanced mixer. The result was such that, when the selectivity required to give the necessary image rejection was achieved, no other spurious response was significant."

The CJA covered 2 to 30MHz with an overall noise factor of 10dB ±2dB. Response to a signal on the image frequency is better than −130dB. Response to a signal at 800kHz (half IF) above wanted frequency: an antenna EMF of 3V will give an output equivalent to an antenna EMF at wanted frequency of less than 0.2µV. Third-order intermodulation: signals to give 0.2µV equivalent antenna EMF – (a) signals near on-tune: 12mV; signals 10% and 20% off-tune 1.7V. IF breakthrough: with receiver tuned to 2MHz, the response to a signal at 1.6MHz will be better than −130dB relative to the wanted response.

It was designed for shipborne use, where several transmitters are likely to be operated simultaneously, and where very small separation between receiving and transmitting antennas is possible, resulting in strong unwanted signals at the input to the receiver. A major requirement is thus extremely good spurious response performance and circuits that give a very wide dynamic range. ◆

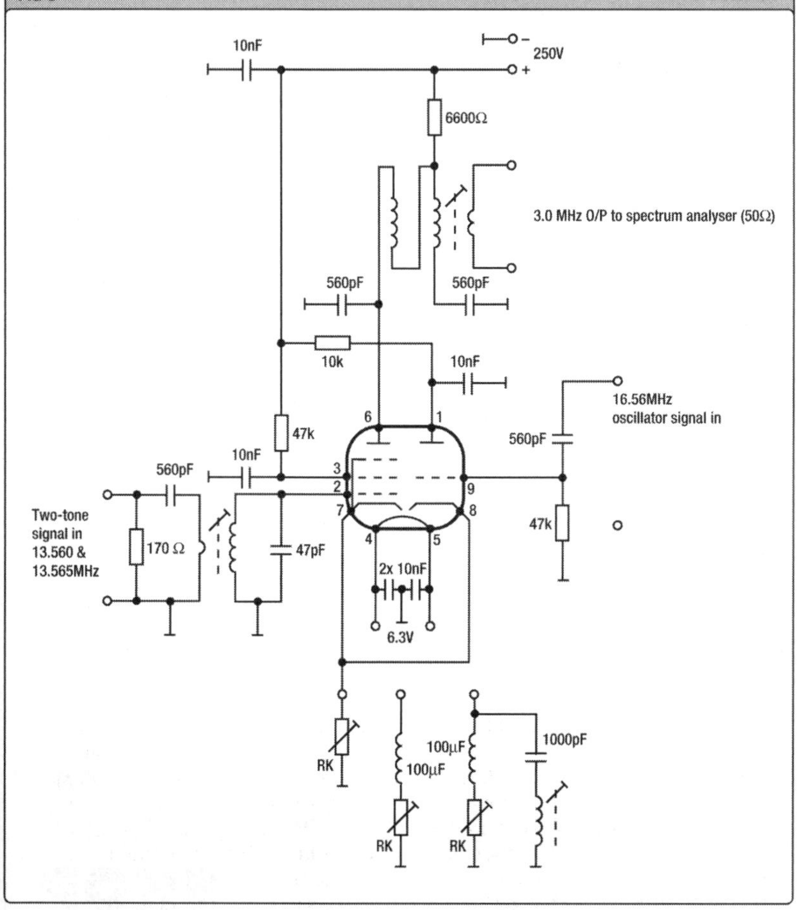

**Fig 5**
Circuit diagram of the tested ECF80 mixer with alternative cathode-bias arrangements as published in Electron, April 1984.

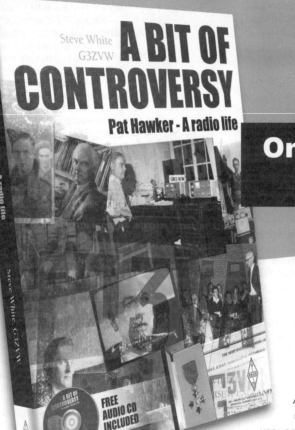

# For the best selection of Amateur Radio books

**Only £13.99 plus p&p**

## A Bit of Controversy
### Pat Hawker - A Radio Life
### By Steve White, G3ZVW

*A Bit of Controversy* details the extraordinary radio life of Pat Hawker MBE, G3VA. From a very early age though to his eighties, Pat has been involved in many aspects of radio, from WW2 with the Britain's military intelligence organisations through to time working for the RSGB, the IBA and as the editor of *Electronics Weekly* and the *Royal Television Society Journal*.

Starting with his early upbringing and his interest in radio from a tender age, *A Bit of Controversy* brings this remarkable man to life. Pat was first licensed with an 'artificial aerial' licence (2BUH) in 1936 and a transmitting licence (G3VA) in 1938. There are details of his WW2 experience when Pat was a member of the Radio Security Service (RSS) with its connections to MI5 and MI6. Pat followed the allied invasion through Europe with the intelligence services, including a spell with the Dutch Bureau of Intelligence. After ending his military service in 1946 the book brings out details of his published work for the RSGB and George Newnes. Never far from a "radio controversy", this book covers the 50 years that Pat wrote the hugely popular column Technical Topics which appeared in the RSGB's journal *RadCom*.

Pat Hawker has without doubt led an extraordinary life and this book details it. If you were a fan of Pat's long running column or just interested in the activity of the security services in WW2 this book is a fascinating insight.

**Free Audio CD**
A special bonus the book includes on a free audio CD a specially commissioned interview with Pat Hawker, MBE. Talking about his wartime experiences, Pat adds an extra dimension to his fascinating story.

Size 174x240mm, 144 pages plus CD   ISBN 9781-9050-8640-5

**Radio Society of Great Britain**
Lambda House, Cranborne Road, Potters Bar, Herts, EN6 3JE  Tel: 0870 904 7373  Fax: 0870 904 737
**www.rsgbshop.org**

Pat Hawker, G3VA

37 Dovercourt Road, London SE22 8SS

# TECHNICAL TOPICS

LA8AK, silent key ♦ HF transmitter limitations ♦ Measurements professional and amateur ♦ CDG2000 H-mode mixer modifications ♦ Why clean variable capacitors? ♦ Compact DC-DC converter ♦ Here & there

**JAN-MARTIN NØDING, LA8AK, SK**
Experimental amateur, VHF DXer and QRP enthusiast, professional engineer, prolific 'TT' contributor for some 30 years, Jan-Martin Nøding, LA8AK (and sometime holder of G5BFV) has fallen silent at the early age of 59 years. In April, he went for a walk in the Norwegian hills close to his home between Søgne and Kristiansand. but did not return. After being missing for just over a week, he was found dead on 27 April, the victim of a massive heart attack.

LA8AK was a notable pioneer on VHF, UHF and microwaves in Norway, but as his many contributions to 'TT' and to the Norwegian journal, *Amatør Radio*, show, his interests were catholic. I can trace his 'TT' contributions back to at least October 1976 (linear CMOS gates). There quickly followed such items as Wien bridge oscillators (August 1977), stable BFO using CMOS (August 1977), IC voltage regulators (July 1978) etc. Over the following years, contributions continued right through to the short Here & There item in this year's May issue commenting on the frequency coverage of the German military receiver type Torn E.b.

LA8AK was always interested in oscillators using crystals or ceramic resonators and in showing how far they could be 'pulled' in VXO designs. I believe that professionally he was concerned with Norwegian television transmitters. In the 1970s he undertook building the amateur beacons LA3UHF and LA3VHF. In this project he found that the noise sidebands of most oscillators – with the exception of one originally developed by DJ2LR ('TT' February 1976) as a 144MHz variable-frequency signal source – resulted in poor CW signals. LA8AK adopted elements of DJ2LR's design to form a VHF source incorporating an IC regulator. He appreciated, even then, that it is important for a low-noise oscillator to have a very 'clean' supply voltage. A reminder of this basic principle turned up again in the May, 2005 'TT' p77, Fig 3, with W4ZCB recommending the use of a Wenzel Associates design of an IC regulator with an additional shunt AC-coupled regulator section. W4ZCB echoed the 1979 comment from LA8AK by stating: "A perfect oscillator will not be perfect with a noisy source of power."

As a reminder of the practical work of LA8AK, who will be so sadly missed, **Fig 1** reproduces his 1979 extremely low-noise crystal oscillator likely to be useful when attempting to modify a transistorised FM or AM transmitter for CW operation.

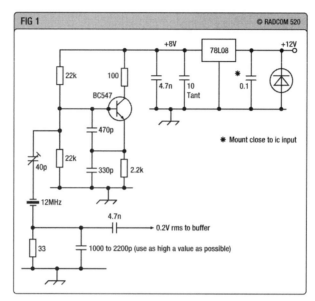

**Fig 1**
LA8AK's version of the DJ2LR extremely low-noise crystal oscillator that he developed in the 1970s for use in the Norwegian VHF and UHF beacon transmitters in order to provide a good CW note.

### HF TRANSMITTER LIMITATIONS
The May 'TT' discussed some important aspects of HF receiver performance and specification. In particular, the question of specifying third order intermodulation and the related instantaneous dynamic range was discussed including the effect of close-in oscillator phase noise as affecting reciprocal mixing. It was shown that there are marked differences in the instantaneous dynamic range of a receiver when two-frequency noise measurements are made with wide or narrow frequency separation of the input signals. Published measurements made on current top-of-the-range transceivers show that the third-order intercept (TOI) can change from better than +30dBm with 20kHz spacing to about -18dBm with 5kHz spacing. But it was also pointed out that oscillator phase noise affects both transmitted and received signals. IP3 from multiple highpower transmitters is received as interference or usually an artificially-raised noise floor.

Receiver close-in TOI characteristics is significantly more important to an operator using CW and narrow-band modes than to the SSB operator. A letter from Karl-Arne Markström, SM0AOM, puts the case clearly for reducing transmitted wideband noise from its effect on adjacent channel interference – although, apparently, he does not consider the case of the cumulative effect of the whole batch of high-power 7MHz broadcast transmissions that can produce an artificially high noise level, even on receivers that have reasonably good IP3 performance. Peter Chadwick, G3RZP, in his *QEX* (May/June 2002) article 'HF Receiver Dynamic Range – How Much do we Need?', emphasised that when considering reciprocal mixing, the assumption that only the strongest incoming signal needs to be considered is invalid: "The phase noise of all [incoming strong] signals will add directly. Ten incoming signals at a level 10dB below the strongest signals will add as much phase noise as one signal in the top signal level." While G3RZP concluded: "Do our receivers have adequate intercept points and ILDR (intermodulation-limited dynamic range)? The answer is apparently 'Yes', but only if you can move the dynamic range up and down to suit conditions. A not-too-distant future job at the G3RZP station is to build a finely-variable step attenuator to go in the antenna line to the receiver." It is his view that once an IP3 of some +20dBm is reached, the main receiver limitation becomes the phase-noise generated in the receiver synthesiser or VFO.

SM0AOM writes: "I have read with interest the May items about HF receiver specifications. I just want to add that, in my opinion, the receiver is no longer the limiting factor when assessing overall HF system performance.

"Current high-end receivers appear to be able to deliver a 90 – 100dB SSB adjacent channel rejection or dynamic range, limited by either IMD or by synthesiser phase noise. However, the chances of finding a current transmitter having a corresponding adjacent channel rejection in order to obtain full advantage from improved receiver performance seems remote. For this reason, I question the emphasis put on receiver IP3 specifications by many manufacturers of amateur radio equipment.

"To obtain full advantage from the now attainable receiver performance, the transmitter in the adjacent channel would be required to have an IMD or noise floor suppression at, say, 5kHz spacing of more than 100dB relative to the on-channel power level, which corresponds to -135dBc/Hz. Even if we could assume noise and IMD-free amplification in the transmitter signals paths, the requirement would be to generate an RF signal with noise and distortion products in the closest adjacent channels at least 100dB down from full power. The effort

# TECHNICAL TOPICS

needed to design and produce such signal quality would be difficult even to imagine.

"Measurements on high-priced commercial HF MOSFET power amplifiers [(a) Nygren, 'A New Generation of HF Power Amplifiers', *Proc HF89 Nordic Shortwave Conference;* (b) Sabin et al, *Single Sideband Systems and Circuits* Chapter 11] point to a *best-case* amplifier noise figure of around 20dB that would correspond to a wideband noise level of -180dBc/Hz for 1kW output power in SSB bandwidths. To this, the noise power contributions of all the translation oscillators, baseband signal processing and power control loops will have to be added.

"The best tunable oscillators currently available appear to have phase noise levels around 150dBc/Hz at 5kHz spacing, which would be further degraded by the noise contributions of the baseband and frequency translation circuits. A recent paper on HF naval system component performance [Hubbard 'A High Dynamic Range VLF to HF Active Receive Whip Antenna', *ProcHF04 Nordic Shortwave Conference,* Faro, Sweden] quotes a wideband noise plateau from a 1kW transmitter of -165dBc/Hz at ±5% frequency offset. It is conceivable that one could design a very low-distortion low-power SSB exciter with a noise-related 5kHz adjacent channel suppression of -140dBc/Hz, or 105dB relative to the output signal in an SSB bandwidth.

"This leaves us with the adjacent channel suppression attainable in the transmitter signal-path due to amplifier non-linearities. A two-tone IM3 ratio of -50dB relative to one-tone can perhaps be obtained in production equipment by using RF negative- or envelope-feedback techniques, and maybe another 10dB by the use of cartesian-feedback linearisation. This would still be 40dB less than the suppression required to co-exist on equal terms with the best HF receivers.

"Another aspect is the dynamic influence on the adjacent channel spectrum that comes from gain control and ALC loops. Current research by Dr Leif Asbrink, SM5BSZ, [www.sm5bsz.com/dynrange/alc.htm] using the Linrad measuring system, points in the direction that ALC actually *worsens* the adjacent channel spectrum due to uncontrolled amplitude and phase modulation of the output signal by the power control loops. The ALC loops also seem to be used in improper ways to set the output power level and transmitter gain across the operating frequency range.

"From my point of view, it appears that the amateur radio community at large is quite uninterested in the actual output spectrum of their transmitters, happily trading increased adjacent channel interference for a S-unit more in the signal reports, or a 'lively' ALC meter indication. I will end this letter with a 'plea' on two counts. First, to the amateur equipment manufacturers – that they invest more engineering man-hours in optimising the transmitter spectrum. Second, to radio amateurs – to take the adjacent channel suppression specifications into serious consideration when making system design and purchase decisions."

It seems to me that SM0AOM makes a number of valid points, and it is much to be hoped that both amateurs and manufacturers will take note of his 'pleas'. Most of us are well aware of the problems caused by 'splatter' resulting from over-driving a linear amplifier, etc but are less aware of the problem of the various forms of wideband noise that emanate from our transmitters. But I would stress that the aim of reducing receiver phase noise and improving TOI is not primarily to cope with the inevitable noise and splatter from a transmitter using the adjacent SSB channel. Rather, it is to cope with the cumulative effects of all the very strong signals reaching the first mixer stage, of especial importance in coping with HF broadcast signals. There is at present, as SM0AOM implies, little expectation that we can, in the foreseeable future, expect interference-free reception of a weak signal when there is a strong local or broadcast transmission in the adjacent channel. SM0AOM is right to emphasise that this represents a limiting factor to overall HF system design.

## MEASUREMENTS PROFESSIONAL & AMATEUR

There is an old but still valid adage that a measuring instrument should have a performance that is an order of magnitude better than that of the device it is measuring. This is increasingly difficult to achieve when making critical measurements on receiver intercept points, oscillator phase noise etc, of high-performance receivers. In the May 'TT', G3RZP suggested that "Intercept points are notoriously difficult to measure with accuracy... you can easily get to ±2.5dB overall uncertainty in the absolute level into the receiver, and this reflects a ±7.5dB uncertainty in intercept point." G3SBI and others feel that this is an unduly pessimistic appraisal. With professional laboratory instruments, they believe, accuracy should be within better than ±1dB.

There have been several mentions in 'TT' of the American firm of Wenzel Associates Inc of Austin, Texas. For example, in April 2001, in connection with a 'Wide-Span Tuned-Toroid VCO' and, more recently, in May 2005 (Fig 3, p77) with regard to the removal of power supply 'noise' from oscillators, as used and recommended by W4ZCB.

Wenzel Associates has become one of the leading American firms in the supply of high-cost, state-of-the-art components for professional and laboratory use. Colin Horrabin, G3SBI, has sent along some details of an 81.25MHz SC Ultra Low Noise Crystal Oscillator unit as purchased recently by the CLRC research laboratory at Daresbury. Such a 'clean' signal source has great potential for signal generation in measuring instruments etc. The plug-in unit illustrates what can now be achieved in reducing the noise output of a professional VHF crystal oscillator having an integral temperature-stabilised 'oven' in a unit measuring 2.94 x 1.75 x 1in. The phase noise (measured by Wenzel Associates) of this oscillator is a remarkable -130dBc/Hz at 100Hz, -158dBc/Hz at 1kHz and -176dBc/Hz at 10 and 20kHz. These figures clearly represent current professional state-of-the-art.

Most of us, of course, have to be satisfied with less-costly devices and test instruments. Dr G L Manning, G4GLM, draws attention to an article 'Passive Component Testing', by Mike Tooley in *Everyday Practical Electronics* (May 2005). G4GLM writes: "Most of the article describes how a palm-sized device (Peak Atlas LCR40 Analyser) rapidly provides L – C – R measurements of components out of circuit, replacing bridge techniques at the touch of a button. As I was lucky to win one in a *Practical Electronics* competition, I can vouch for the device's practical usefulness. The article mentions that a simple calculation will estimate the characteristic impedance of coaxial cable. Two quick measurements are all it takes, provided the cable is reasonably long and there is access to both ends. Furthermore, best-case Q may be estimated for inductors, although the author rightly cautions that RF losses are not taken into account. I suppose it's back to the microwave oven test if in doubt!"

The Atlas LCR40 is available through the usual distributors (Maplin, Farnell, etc) or direct from Peak Electronic Design, Atlas House, Kiln Lane, Harpur Industrial Estate, Buxton, Derbyshire SK17 9JL. Tel: 01298 70012.

## CDG2000 H-MODE MIXER MODIFICATION

Mention was made in the March 'TT' that Colin Horrabin, G3SBI, has found that a simple modification can improve the FST3125M H-mode mixer as used in the home-constructable, world-class CDG2000 transceiver as described in a series of *RadCom* articles during the year 2002. G3SBI writes: "This is a simple modification that gives IP3 performances of 42dBm on 30MHz, 45dBm on 21MHz, 48dBm on 14MHz and 50dBm on 7MHz for an input noise floor of -130dBm, achieving on this band an IP3 dynamic range in the region of 120dB. On 7MHz, the IP3 performance of the Loadstone Pacific coils in the bandpass filter and that of the crystal roofing filters 'bottom out' at the same point.

"The modification involves the use of AC-coupled drive from the AC74 or AC109 flip-flop that provides the local oscillator drive to the Fairchild FST3125 fast bus switch used as the mixer and the use of a balance potentiometer to adjust the switching thresholds between the two pairs of switches in the mixer: **Fig 2**. Full details of the original FST3125 H-Mode Mixer and its fundamental frequency squarer (Fig 6) are given in 'TT', September 1998, pp58 ff] and in *Technical Topics Scrapbook 1995 – 1999*, pp234 – 235.

"One of the nice features about the use of the FST3125M as a mixer is that, with absolutely no circuit adjustments, the mixer delivers an IP3 in the region of 40dBm. I clocked the AC109 at 78MHz so that the receiver was effectively tuned to 30MHz, put in a 9MHz signal and tuned the balance pot for minimum feed-through at 9MHz. This gave a null of 70dB and coincided with maximum IP3. A similar test on 14MHz gave the null as only 50dB but, again, coincided with maximum IP3.

"In the CDG2000 receiver one could link-out one of the bandpass filters to do this and use a 9MHz signal generator as the input and, turning the multiturn pot with a trimming tool set the trimpot for minimum S-meter deflection. It is important to use a trimming tool, since the capacitance of a small screwdriver makes it difficult to set the trimpot accurately. Another method may be to tune the receiver to the 10MHz band where the 9MHz trap is only 50dB, put in a 10dBm signal at 9MHz and tune the trimpot for minimum S-meter deflection. In an ideal situation, this balance adjustment of the mixer should coincide with IP3 maximum and, in practice, this does seem to be the case. Very few radio amateurs are able to measure IP3 so the fact that the feed-through null ties up with maximum IP3 makes it an easy adjustment to make.

"The modification involves the use of two 0.01μF monolithic ceramic capacitors, two 10kΩ 1% fixed resistors, a 20kΩ multiturn trimpot and a 390Ω resistor as a safety precaution in case of failure to set the trimpot to mid-scale before fitting the modification. The 390Ω resistor connected between the trimpot wiper and the 7V supply is to prevent the bistable used as the divide-by-two squarer seeing a short circuit should the trimpot be adjusted to one limit or the other of its track.

"Before fitting the modification, the best approach is to set the trimpot to mid-scale using a digital multimeter. If this is done, the balance point for minimum feed-through will probably be within two turns if the AC109 or AC74 bistable is presenting a 50:50

**Fig 2**
The simple AC-coupled modification from the divide-by-two squarer to the FST3125 H-mode mixer in the CDG2000 or other applications of this state-of-the-art G3SBI mixer originally published in the September 1998 'TT' (see text). The text shows how the balance can be obtained simply by means of presetting the multiturn trimpot.

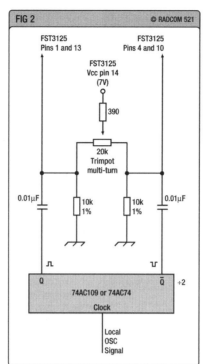

waveform. Apparently some manufacturers of these devices are, in this respect, better than others. I know Harris parts are OK."

## WHY CLEAN VARIABLE CAPACITORS?

Recent 'TT' notes and queries on cleaning variable capacitors in 'TT' March and May 2005 continue to attract comments from readers well-conversant with the chemical effects of the various cleansing agents that have been suggested, as well as adding further suggestions.

But, first, what are the reasons for this concern with cleaning variable capacitors – a class of component that is fast disappearing from modern equipment, other than for ATUs and some high-power linears? For virtually all small-signal applications they have been replaced by electronic tuning diodes (varactors) or, for RF filters, by fixed capacitors. Varactors are low-cost, take up very little board area and are convenient to place, since the variable control takes the form of a potentiometer that can be sited remote from the diode. Progress indeed.

Yet Patrick Hutber once laid down his law: "Progress means deterioration". Certainly the change to electronic tuning diodes rather than using old-fashioned air-dielectric variable capacitors has its disadvantages as well as its advantages, Havelock Ellis once claimed that "What we call progress is the exchange of one nuisance for another nuisance."

Many of the early limitations of electronic tuning diodes have been overcome or reduced by improved devices or by careful design. The use of two diodes instead of one is now widely used. There can still be problems: limited capacitance variation: limited reverse breakdown voltage;

limited $Q$ values; possibility of introducing non-linearity into signal-frequency tuned circuits leading to intermodulation on strong signals; etc. There remain worries about using tuning diodes in oscillatory circuits where a major objective is to achieve the lowest possible phase and amplitude noise.

In such circumstances, there is much to be said for mechanical tuning systems. But variable capacitors are subject to deterioration – poor and variable contact to the rotor vanes leading to ohmic resistance, jerky movements, dirty vanes etc. I recall that, at one time, at least one manufacturer of military radio equipment routinely cleaned tuning capacitors in an ultrasonic bath.

There is also one specific application where variable capacitors are subject to very high voltages and required to pass large RF currents. That is when used as the tuning resonator in small transmitting loops. For this application the requirements are extremely demanding. Professional loops use high-cost vacuum capacitors generally considered too expensive to use in amateur loops.

Looks rather than performance is often the requirement for those restoring old equipment, particularly where these are required only for display.

While the "baking powder in hot water" procedure as outlined in the May 'TT' seems to provide a useful way of cleaning capacitor vanes without causing adverse chemical reactions, etc, Dick Biddulph, M0CGN, writes:

- "'TT' mixes up 'baking powder' with 'baking soda'. It is baking soda, not baking powder, which is sodium bicarbonate whereas the former is a mixture of this with sodium acid tartaric.
- "If you heat baking soda in water, it liberates carbon dioxide and becomes washing soda which will attack aluminium.
- "If you heat soda solution with the component in contact with aluminium foil, the tarnish, which is silver sulphide, will slowly disappear being replaced by silver. 'Silver Dip' dissolves the tarnish (it makes a short circuited cell liberating hydrogen at the silver electrode).
- "I've never heard of the use of lemon juice (citric acid) for this sort of cleaning.
- "In *all* cases the component should be rinsed in distilled or de-ionised water and dried in a low oven.
- "Finally, when dry and warm, a little Vaseline should be applied to bearings and to any rubbing surface."

M0CGN continues: "I have used washing soda solution plus a little washing up liquid or even washing detergent, both hot, followed by rinsing in distilled water and treating as in the last item above. I have also

# TECHNICAL TOPICS

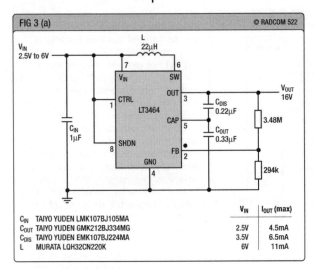

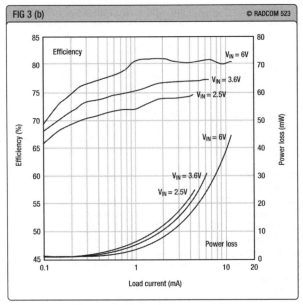

**Fig 3**
(a) Use of the Linear Technology LT3464 to obtain an efficient 16V bias supply using surface-mount components.

(b) Efficiency and power loss of the 16V output for input voltages of 2.5, 2.6 and 6V – maximum outputs 4.5, 6.5 and 11mA respectively. (Source: Linear Technology Design Notes advertisement)

**Fig 4**
Potential damage or weaknesses (a) of conventional lead component after reflow at higher lead-free soldering temperatures; (b) Lead-free component after reflow at lower conventional lead solder temperatures. (Source: Electronics World)

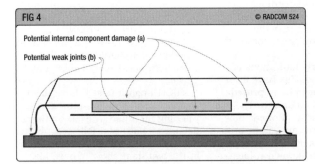

used the water recovered when defrosting the fridge, which is more or less pure, if a little smelly."

Tony Webb, G4LYF, confirms some of the information provided by M0CGN, but also adds additional points. He writes: "Can I warn readers not to use baking powder to clean variable capacitors? Baking soda (sodium bicarbonate) is fine, when used correctly, but baking powder contains other ingredients such as rice flour, which are unlikely to enhance appearance or performance. To be on the safe side, use only a product bought under the names sodium bicarbonate or bicarbonate of soda.

"Tarnished silver or silver-plated articles can be cleaned very effectively by contact with aluminium (foil, or old aluminium pan) in a hot solution of sodium bicarbonate in water. Washing soda may also be used (bicarbonate turns into this in hot water) but definitely *not* caustic soda! In fact, you shouldn't use caustic soda for any purpose unless you are qualified to handle it. It will enthusiastically dissolve aluminium, etch ceramics and destroy many plastics.

"In the cleaning process with bicarbonate, electrical contact between the metal to be cleaned and the aluminium is necessary. This sets up a (short-circuited) cell (the open circuit voltage is about 1.5V) and the tarnish (silver sulphide) acts briefly as a depolariser, being converted to hydrogen sulphide and silver. There is sometimes a delay in the action, while the oxide surface is etched from the aluminium. Any oil or grease on the silver surface (including fingerprints will hinder the action and it is best to clean this off first. A trace of detergent (but *not* soap) in the bicarbonate is also helpful. Boiling water should not be used.

"Any component thus cleaned must be very thoroughly rinsed with distilled or demineralised water. If you can get hold of methanol or isopropanol, these are usually good for a final rinse, as they are non-aggressive solvents, although they are very flammable, and methanol is poisonous, so should be handled appropriately. Isopropanol is known as 'rubbing alcohol' in the USA. I'm doubtful about using methylated spirit, as this may leave a residue of 'pyridine bases' which could encourage copper and some other metals to corrode in air. Don't forget that re-lubrication of bearings may be needed.

"I have tried the same cleaning technique on copper, which usually has a mixture of sulphide and oxide tarnish, but without success."

### COMPACT DC-DC CONVERTERS
There is often a requirement for a DC voltage in the region of 16V or 20V, when the equipment is powered from 2.5 to 6V supplies, eg for electronic tuning diodes, bias supplies etc. Design Note 358 'Compact Step-Up Converter Conserves Battery Power', by Mike Shriver appears as a Linear Technology advertisement in *EDN*, 17 March 2005. This provides an introduction to the surface-mounted LT3468 device, with circuits showing how it can be used as a 16V bias supply, ±20V bias supply, or 34V bias supply.

It is claimed that the LT3464 is an ideal choice for portable devices which require a tiny, efficient and rugged step-up converter. "The device, housed in a low profile (1mm) 8-lead thin SOT package, integrates a Schottky diode, npn main switch and pnp output disconnect switch. For light lead efficiency Burst Mode™ operation is used to deliver power to the load. This results in high efficiency and minimal battery current drawn over a broad range of load current. Quiescent current is only 25µA. While in shut down, the output disconnect switch separates the load from the input, further increasing battery run..."

**Fig 3** shows a 16V bias supply that can provide 6.5mA at an efficiency of 77% from a 3.6V lithium-ion battery. The circuit uses a 22µH surface-mount chip inductor with a 1210 footprint and a 0.33µF output capacitor with 0805 footprint. Data sheet download at **www.linear.com**

### HERE & THERE
In connection with recent 'TT' items on high IP3 valve mixers, André Jamet, F9HX, comments: "One suggestion: why not use vacuum diodes such as the 6AL5 or the old 6H6? Well, we need a 6.3V supply for the heater! But we could expect a very, very high IP3 [if used in switching-type ring mixers]. Not only inductors but also capacitors have to be *linear* – ie to have a constant value even when the applied voltage varies. Mica, polystyrene and NPO ceramic types are useful."

An article "Lead-free – or Not Lead-free?' Asks the US Military', *Electronics World,* May 2005, pp18 – 19, notes that the US military is looking very cautiously at lead-free components: "While the world of electronics manufacturing is turning its attention – if not its enthusiasm – to lead-free processing, the US military is taking a somewhat different view. The US military establishment may, at some point, begin accepting lead-free as a fact of life, but that time is not likely to arrive soon.

The fear is that mixtures of leaded and lead-free components etc may end up on the same board. While leaded components etc are required to pass a heat test of 260°C for 4s during lead-free reflow, the same component might reach 260°C for 30s due to the higher working temperatures of lead-free solder: **Fig 4** shows the potential damage of a lead-free component after reflow at higher and lower temperatures. ♦

Pat Hawker, G3VA

37 Dovercourt Road, London SE22 8SS.

# TECHNICAL TOPICS

### Digits, digits everywhere ♦ Buying overseas ♦ Short-span multiwire folded dipole ♦ The coming of coax

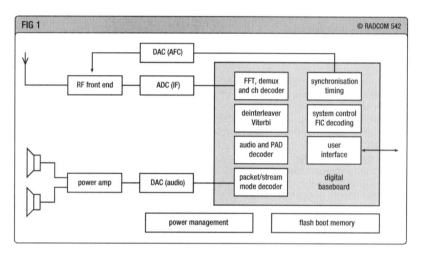

Fig 1
Block diagram of a typical DAB domestic receiver. (Source: Texas Instruments sponsored feature in *New Electronics*)

**WILL DIGITS DOMINATE?**
Like it or lump it, multi-mode high-speed data (including audio) seem set increasingly to take over our bands from HF to EHF, following in the footsteps of other telecommunications services. For broadcasting, there are already DAB (Digital Audio Broadcasting) multiplexes and DRM (Digital Radio Mondiale) presently used only on HF, although theoretically capable of being used on all AM broadcast frequencies up to 30MHz. Both DAB and DRM can also transmit other forms of data. **Fig 1** shows a block outline of a typical DAB receiver

DRM, it is claimed, overcomes the problems on HF of multi-path, Doppler and most forms of electrical interference, although severe fading may still cause more problems than the less dramatic increase in noise on analogue transmissions. On DRM, appreciable amounts of redundancy provide error correction to increase robustness, but there remains a critical signal level below which the audio drops out. However, transmitter power needs be only about a quarter of that required for equivalent AM coverage. Several coding schemes can be used to offer optimum quality at different rates and with different programme content. These include subsets of the MPEG-4 AAC, CELP and HVCX codecs, with optional spectral band replication.

A DRM transmission may contain up to four audio or data components, enabling transmissions in two languages simultaneously, for example. A few DRM receivers are on the market, although there are also software-defined receivers that run in conjunction with PCs. Power consumption is a problem for battery-operated portable receivers. A useful short tutorial article appears as an advertisement feature in *New Electronics,* 26 April 2005: 'Digital Radio Developments', by Les Mable (sponsored by Texas Instruments, with more information on **www.ti.com/digitalradio**).

Digital audio (voice) is creeping into the HF amateur bands,

although it is mutually incompatible in regard to frequency-sharing with both SSB and narrow-band transmissions, including CW and low bit-rate digital modes. ARRL is proposing radical changes to HF band-planning and has submitted a draft petition to FCC seeking to govern the usage of amateur spectrum by emission bandwidth rather than by mode, with segments limited to maximum bandwidths of 200Hz, 500Hz and 3kHz and, on part of the 28MHz band, 16kHz. This proposal would seemingly restrict American NBFM, DSB and AM below 30MHz to only 29.0 – 29.7MHz.

The June *QST* includes an overview article on D-Star digital voice and data systems suitable for use above 30MHz, plus an article on the installation and testing of the new Icom UT-118 digital voice modules as fitted in their IC-V82 and IC-2200H models. The AOR ARD9800 digital HF voice modem using the G4GUO protocol was introduced in 2003, permitting digital voice on at least some HF transceivers..

To quote *QST:* "Interest in digital technology runs deep in amateur radio, back to the early days of radioteletype, or RTTY. Packet radio was created from the Bell 202 and X.25 standards, and more recently there has been an explosion of new modes such as PSK, PACTOR, MFSK and others. Voice transmissions have been slow to be adapted to digital transmission standards, but the G4GUO protocol ['Practical HF Digital Voice', by Charles Brain, G4GUO, *QEX,* May 2000] has been implemented on HF by AOR, and Icom has now released equipment that supports the D-Star standard for transmitting both voice and data.

"D-STAR was the result of three years of research funded by the Japanese Ministry of Posts and Telecommunications to investigate digital technologies for amateur radio. The research was conducted by a committee administered by JARL. Included in the group were representatives of the Japanese amateur radio manufacturers, including Icom, which provided the equipment used for development and testing. The committee produced a standard called D-Star in 2001. Although D-Star is a standard published by JARL, it is available to be implemented by anyone. It is an *open protocol,* meaning that any equipment complying with the published standard can inter-operate with D-Star-equipped radio equipment. Icom is the only manufacturer to date that makes equipment that supports the D-Star standard."

Voice signals are converted to and from a digital stream of data by a codec implementing the AMBE (Advanced Multi-Band Excitation) and connected to a computer with

# TECHNICAL TOPICS

either an RS-232 or USB serial connection for low-speed data, or a high-speed connection via Ethernet. Data speed for the D-Star air link is 4.8Kb/s for voice and 128Kb/s for digital data, sufficient to support communications-quality voice and about twice as fast as good-quality dial-up connections. The data is transmitted over the air in the 0.5GMSK format – Gaussian Minimum Shift Keying.

**BUYING OVERSEAS – CAVEAT EMPTOR**
The May 'TT' noted the appearance on the market of the first software-defined radio amateur transceiver (Flex-Radio SDR-1000) based on the review in *QST*, April 2005. The SDR-1000 comes in a black box containing a few printed circuit boards using miniature surface-mounted devices. The hardware is defined and controlled from a PC equipped with a suitable sound card. At present, the unit offers both analogue and digital (DRM) modes of reception, and the conventional forms of analogue voice and digital transmission with the possibility of additional facilities with further software developments. It is clear that this is a unit that is virtually still under development, and at present is not available through UK agents or retailers.

It thus seems opportune to suggest that any major purchase overseas, except perhaps from a firm with an established reputation in the amateur radio market-place, should be approached with some prudence and a little caution. What is the after-sales service offered? Can the surface-mounted components be serviced, or individual boards be replaced? Can we be certain there are no inherent or hidden glitches in the hardware or software? Does the relatively small amount of RF hardware conform to good analogue design practice? It may seem a simple matter to order equipment over the web from a source 3000 miles away, but the distance seems much greater if you run into the difficulties so often associated with new systems based on a massive amount of software and with novel circuitry.

I must stress that I have had no personal contact with FlexRadio or its equipment or its after-sales service, but simply feel that this is the type of situation that can give rise to the types of problem discussed in the item 'Buying and Selling of Equipment', 'TT' August 2004 (see also *Technical Topics Scrapbook 2000-2004*, pp235 – 236). Among other comments, I wrote of purchases made in the UK or Europe: "A point of considerable importance to amateurs is the supply of spare parts. There is a legal obligation on suppliers, retailers and manufacturers to make spare parts available for at least seven years from the time of selling a product without which a product is unusable... The Sale of Goods and Services Act, 1982, requires a service to be carried out with reasonable care and skill within a reasonable time and, where no price is agreed, the charge should be reasonable. Remember that the buyer's contract is with the retailer or supplier, not [usually] with the manufacturer."

With equipment bought from a UK-accredited agent or firm, this is all fairly straightforward, but it could prove difficult and expensive to attempt to argue a legal case in the USA. For example, to prove an inherent fault, to determine what is a 'spare part' for a faulty board based on tiny surface-mounted components that amounts to a 'throw-away' part rather than repairable. Then again, a glitch in the software may or may not be regarded as an inherent fault, but could cause problems to an operator dependent on his equipment functioning without requiring computer expertise. It would be interesting to hear from anyone who has been an early purchaser of the SDR-1000 of their experience in regard to the hardware, software and after-sales service. The web has opened the way to a global marketplace and competitive pricing and has encouraged the setting up of many enterprising new businesses. But it has also brought in train some problems.

There is much to be said for purchasing complete equipment, wherever it is manufactured, from an accredited UK or even a European firm, as it then comes with the protection of the various Sale of Goods Acts etc discussed at some length in the August 2004 'TT' item. *Caveat Emptor!*

**SHORT-SPAN MULTIWIRE FOLDED DIPOLE**
While 300Ω or 450Ω balanced twin-wire feeder has much to recommend it – it can be used as both matched or resonant feeder – many amateurs remain wedded to the use of coaxial cable as a most convenient low-impedance feeder.

In 1939 – 40, the late, great John Kraus, W8JK, investigated and published details of various forms of folded dipole and monopole (folded unipoles), details of which have been published at various times in 'TT', most recently in April 2004, in an item marking his many contributions to antenna design. The folded antennas shown then had feed impedances of about 350Ω, 900Ω, 450Ω and 230Ω. The item also noted that Walter Roberts, K4EA, and Ham Clark, G6OT, had described how to calculate the feed impedance where wires of different diameter were used.

However, Kraus also published a further, half-sized, four-wire version in the form of a shortened, double-folded dipole in an article entitled 'The Three-band Rotary Antenna' (*Radio*, February 1940). This showed how a driven four-wire folded element with a span of some 34ft could be used on 7, 14 and 28MHz (no pre-war 21MHz band) if reconfigured by means of five switches (remote or manual switches in the element) to reconfigure it as outlined **Fig 2**. Dave Gordon-Smith, G3UUR, has developed the quarter-wave version

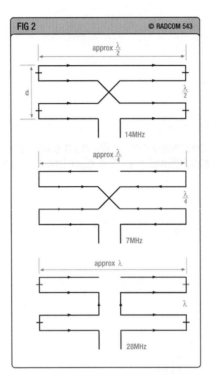

Fig 2 Arrangement of the driven element of the four-wire, three-band doublet described in 1940 in *Radio* by Dr Kraus, W8JK. With a span of approximately 34ft, five (remote-controlled or manual) switches could be used to configure the folded element as shown in (a) as a half-wave element for 14MHz, (b) as a quarter-wave element for 7MHz, and (c) as a full-wave element for 28MHz. It is possible that (b) could also have been used on 21MHz as a 0.75λ element had the 21MHz band been available before WWII.

Fig 3 Short, four-wire, double-folded dipoles. (a) The original W8JK quarter-wave version as shown in Fig 2(b). (b) The low-impedance-feed version as adopted by G3UUR.

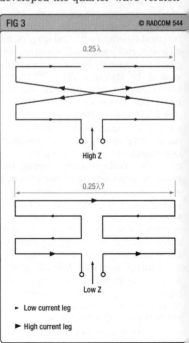

32  RadCom August 2005

to provide a central low-impedance feed point that can be fed directly by coaxial cable without the requirement for an ATU. He has used this arrangement intermittently over several decades. He writes: "The Kraus half-sized, four-wire, folded dipole was fed with open-wire feeder, as shown in **Fig 3(a).** I saw that this could be modified to work with low-impedance coaxial cable quite easily: see **Fig 3(b).** The arrangement is equivalent merely to folding a folded dipole back on itself so that the ends almost touch, and the feed impedance should be close to four times 12Ω. This is a very obvious modification to make, and I'm sure Kraus would have done it if low-impedance cable had been in common use by amateurs in 1940. It is exactly the same as the structure of the FLA and BFDA on which the Japanese engineers based their Built-in Folded Monopole Antenna (BFMA), see 'TT' June 2005, p70.

"Kraus did not mention the reduction in self-inductance caused by folding the element legs back on themselves, and I didn't appreciate what the effect would be at that time. In fact, the resonant frequency increases by about 17% with an average separation of about 1in between the conductors in each folded-back element. My first attempts to make a version that could be fed with low-impedance coaxial cable were very disappointing. W8JK didn't have to worry about the increase in the resonant frequency caused by folding the elements because he used open-wire feeder and a matching unit. I didn't have the knowledge, or equipment to find out where my version was resonant. Later, in the 1970s, when I was better equipped, I was able to make a rough prediction of where the design might be resonate, and got it to work very successfully. Since then, I've used it, on and off, on various bands, as a short rotary dipole and as a sloper. It could also be used as an inverted-V, of course, if the matching and length were tweaked slightly. The change in resonance, compared with the same conductors in a straight line, is very much dependent on the spacing between the two parts of each folded element, and the proximity of the ends of the two legs of the dipole. The reduction in self-inductance is the dominant effect, though, and this varies as the logarithm of the separation between the conductor in the folded-back part of the element and those in the original line of the antennas.

"There are three variations of this short, double-folded dipole, which allow 50Ω (T-match version) or 75Ω coaxial cable (simple version) to be

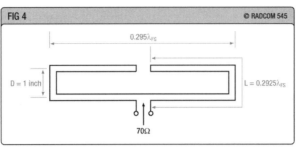

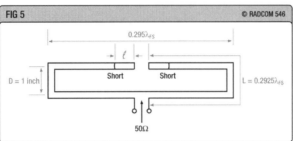

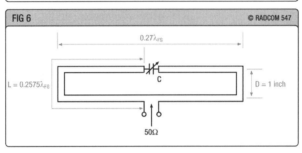

**Fig 4**
Practical dimensions of a simple, short, double-folded dipole intended for 70Ω or 75Ω feed as used by G3UUR. $\lambda_{FS}$ = free-space wavelength.

**Fig 5**
T-matched short, double-folded dipole for 150Ω feed. l = 8% to 9% of L.

**Fig 6**
Tuned, short, double-folded dipole with 50Ω feed. C = 3 – 5pF for 14MHz, 5 – 9pF for 7MHz.

used with it, or 50Ω cable with a tuned version, which may be remotely tuned if desired. The tuned version can be used with a fixed capacitor if the bandwidth is adequate without remote tuning. The bandwidth of the shortened, folded arrangement is about one-third to one-quarter of the full-sized version, which is still considerably wider than an inductively-loaded dipole of the same span. Ground loss will increase the apparent bandwidth of this design if it is used at low heights on the lower-frequency bands, although having the high-voltage ends in the middle of the antenna does reduce the dielectric loss slightly in the inverted-V form. However, it is still preferable to have the tips of the dipole span as high as possible to minimise ground-induced loss, because the RF voltages at these points are still more than 0.7 of the voltages at the conductor ends, and these voltages are much increased compared with a full-sized dipole due to the folding/shortening of the antenna.

"The arrangement and dimensions of these three versions are shown in **Figs 4, 5** and **6.** Generally, I have used 300Ω ribbon feeder supported by bamboo poles to make short, double-folded rotary dipoles, and have used small pieces of the same ribbon feeder to make fixed tuning capacitors for the tuned version. In this case, the exposed ends of the feeder should be sealed with blobs of wax to keep out moisture, and the wax should not be allowed to spread across the polythene dielectric between the conductor ends. PVC tape should not be used as it is very lossy at RF. Small beads of almost any adhesive will do instead of wax, as long as it is restricted to just sealing the conductor ends where they emerge from the polythene, and doesn't start to bridge the gap between them. There is very high voltage across the 300Ω ribbon when used as a fixed tuning capacitor, even at moderate power levels, and also between the two ends of the antenna. This ribbon cable is not the best material to use for either application, but is readily available and convenient. Heavy-gauge, enamelled wire would be much better if a convenient means of separation can be fabricated from readily-available material. Nowadays, fibreglass fishing-poles would be much better than bamboo poles for supporting these antennas.

"Sloping versions can be matched to 50Ω coaxial cable by shortening the tuned version down to something around 0.25 of a free-space wavelength to get a lower feed impedance, and re-resonating the antenna with more capacitance across the ends. Alternatively, the shorting connections on the T-match version can be moved further down the antenna, away from the ends, to achieve a better match. The sloping T-match version can take a bit of trial-and-error to set up properly, because of variations in the effect of the ground below the antenna, the angle of its slope, and its proximity to any conducting support poles. But it is well worth the effort, as these antennas are far better than loaded dipoles."

### THE COMING OF COAX

Amateur radio without the use of flexible coaxial cables would seem almost unthinkable. Yet, in practice, RF coaxial cables were seldom used by amateurs in the pre-WWII era and only just beginning to appear as footnotes in the handbooks... The first time I ever saw such cables being used was in 1942 for the Hanslope Park SCU3 intercept station in conjunction with the wide-band distribution amplifiers of 'Dud' Charman, G6CJ, using the trusty 807 RF power valves to achieve excellent linearity over bandwidths of more than an octave (4 – 8MHz or 8 – 16MHz, if memory serves me) when fed by the multiple strong signals collected by the large rhombics or V-antennas.

I was reminded of this by receiving an enquiry from Stewart Revell, G3PMJ, querying when coaxial (concentric) cable was first used to deliver

## TECHNICAL TOPICS

RF power from a transmitter to an antenna. He had traced early patents, etc, on the use of such cables for submarine cable systems, and US Patent 1,835,031 of December 1931 (filed May 1931) by Lloyd Espenschied *et al*, *Concentric Conducting System* for cable television.

The answer was an easy one: C S Franklin of the Marconi Company, the eminent engineer who was responsible for the building of the Marconi Short Wave Beam System in 1926. To quote from a 1934 Science Museum publication by W T O'Dea on the history and development of radio communications: "In 1926, the first portion of the beam system between Bodmin and Canada was opened for regular service on 16.6 and 32.4 metres... The beam array at Bodmin, designed by C S Franklin, marked a great advance in directional aerial construction. It consisted of a vertical sheet of separate aerials, each with its own feed. The parallel wire feeders formerly employed were replaced by concentric copper air-spaced tubes fitted with expansion joints. The outer tube was earthed and the inner one connected to a suitable tapping on a screened transformer..." A Patent was issued to the Marconi company in 1929.

References to rigid and flexible concentric-line cables began to appear in American amateur radio publications about 1936. By 1938 the first edition of the RSGB's *Amateur Radio Handbook* included a photograph showing a collection of concentric and twin flexible cables, probably manufactured by The Telephone Construction & Maintenance Co Ltd (Telcon). The second edition, published July 1940, included a note: "Concentric lines are difficult to construct, but there are several varieties on the market, chiefly flexible. In particular the 'Teleconax' cables specially manufactured by the Telephone Construction & Maintenance Co Ltd may be used at all frequencies up to 56Mc [*sic*]. They are moulded, some solid, some partially hollow, and various surge impedances are obtainable from 60 to 100 Ωs. Another variety is marketed under the name 'Pyrotenax'. This consists of a copper wire with a ceramic covering fused on, and an outer copper tube drawn overall. It is not strictly flexible, though it may be bent as required without damage. The surge impedance is 50Ω or less, and the loss very low."

Home-constructed twin-wire balanced lines (usually about 600Ω impedance) were widely used by amateurs either as matched or resonant lines. Where low-impedance feeders were required, the usual solution was to use good-quality twisted electric flex (rubber insulated) which was reasonably satisfactory up to about 15MHz, but increasingly lossy at higher frequencies. With the coming of 405-line VHF television, Belling-Lee introduced a low-loss twin line appreciable better than conventional electric flex. I suspect that only a handful of amateurs used the early, expensive coaxial cables pre-war.

The post-war period saw coaxial cable feeders dominate the amateur radio scene, particularly in the UK and Europe, stimulated by the growth of VHF and then UHF television as well as for military communications. A couple of 'TT' items on coax usage appeared in May and July 1984. The May item included advice and suggestions on 'the coaxial jungle' from the late William Orr, W6SAI, that still bear repetition.

*Inter alia*, he noted that the highest quality cables are those manufactured to meet tight military specifications but, contrary to popular belief, not all RG-8/U cable comes into this category. "At one time, this cable (52Ω) was widely used by the US military, but subsequently they standardised on 50Ω cable (eg RG-213/U) manufactured to MIL-C-17D or E specification. New mil-spec cable is not cheap. RG-8/U and RG-8A/U cable is still manufactured in volume, but no longer to American mil-spec standards. The high cost of copper and other materials has led to cost-cutting. Some firms produce very-high-grade cable by normal standards, others have reduced the percentage coverage of the outer copper braiding from 97% to about 60% and use lower-grade materials. The quality of the outer jacket, the dielectric material and the inner conductors can all vary. Lower-grade cables may serve reasonably effectively at least for a time, but their attenuation is likely to increase more quickly than a high-quality cable and is more noticeable at VHF and UHF."

W6SAI listed six factors that can help prolong the useful life of coax:

- Keep the cable off the ground and make sure it can dry off after rain. Because modern outer jackets are slightly hygroscopic, moisture can penetrate the jacket material, reach the outer braid and cause corrosion.
- Try to keep the cable out of direct sunlight; ultraviolet rays (UV) are damaging over time. For prolonged exposure to strong sunlight, the cable outer jacket should be a high-molecular weight polyethylene with imbedded carbon black (expensive).
- Support the cable every 10ft or less. Do not let it sag on a long run.
- Do not let the cable whip around in the wind. Repeated flexing is not conducive to long cable life.
- Seal the ends of the cable. Use type-N (waterproof) fittings rather than the cheap and plentiful PL-259 plugs. Coat the termination with non-acid type silicone rubber sealant. ("If it smells vinegary this indicates acetic acid in the sealant. Don't use it.")
- Do not step on the cable or otherwise flatten it (eg with fixing staples). Do not bend it around a sharp radius. The minimum recommended bending radius is roughly equal to 10 times the outer-diameter of the cable (about a 5in radius for RG-8A/U or RG-213/U).

In the July 1984 'TT', I commented on a number of points after visiting the stands of Raydex International Ltd and Delta Enfield Cables Ltd at a CAI exhibition. These firms were then supplying much of the cable used in TV downleads (mostly 75Ω) and wired distribution systems. I noted the very wide variation of cable attenuation, even within the standard ranges. For example, attenuation in decibels/100m at say, 600MHz, can vary from over 20dB to under 8dB for distribution (relay or cable TV) cables and from about 17.5dB to around 20dB for 'low-loss' downlead cables, and up to 30dB for lower-cost cables (remember these figures are for *new* cables). The percentage coverage of the outer conductor of standard downlead cables was then between 40 and 60% with recently-introduced cables having finer braid wires in the outer conductor to cut costs. Delta cable intended for buried use had the addition of a moisture barrier under the outer polythene sheath; Raydex semi-air-spaced distribution cables include PVC-sheathed cables, polyethylene-sheathed cables, and various forms of bonded shield cables, including some having an aluminium barrier as part of the outer sheath. The outer sheath of standard feeder cables must be considered as providing only limited protection against moisture ingress with Raydex suggesting that "ingress of water in all types of trunk and distributive cables is the primary problem, resulting from sheath damage during installation or as the result of cables being inadequately sealed. Raydex was also then producing 93Ω 'data transmission cable' for video display units, etc.

It would seem that, more recently, some manufacturers of coaxial cable for CATV have substituted aluminium for copper for the outer braid conductor. ♦

**Pat Hawker,** G3VA

37 Dovercourt Road, London SE22 8SS.

# TECHNICAL TOPICS

## TT

**Binaural cocktail parties ♦ Low-cost huff & puff stabilised VFO ♦ Valve diode mixers & product detectors ♦ G3DXZ's ceramic resonator VXO**

### BINAURAL COCKTAIL PARTIES

A *New Scientist* feature by Robert Adler 'Are We On Our Way Back to the Dark Ages?' (2 July 2005, pp26 – 27) discusses the controversial views of Jonathan Huebner of the US Naval Air Warfare Center that, far from enjoying a golden age of human inventiveness, we are fast approaching a new dark age. It is his belief that the rate of technological innovation reached a peak a century ago. Significant breakthroughs flourished between about 1873 and 1915 but have been declining ever since.

This is clearly an unfashionable view. Many futurologists say technology is developing at exponential rates. Nano-technology offers, they say, a fantastic future. Far be it from me to judge between these diametrically-opposite views, but I cannot help feeling that many of what are claimed as new developments in radio communication often represent primarily the utilisation of ever smaller devices in new implementations of ideas, concepts and circuits originally developed many years ago; some stretching back to the thermionic era and even (for antennas) to the Age of Spark.

The *RadCom*, July 2005, pp78 – 83, article 'The Buccaneer', by Phil Harman, VK6APH, and Steve Ireland, VK6VZ, presented a convincing implementation and endorsement of the value of binaural or pseudo-stereo reception of CW signals. It drew attention to the section dedicated to this technique in the excellent ARRL book *Experimental Methods in RF Design*, by Wes Hayward, W7ZOI, Rick Campbell, KK7B, and Bob Larkin, W7PUA. In turn, the book draws on KK7B's article 'A Binaural I-Q Receiver' (*QST*, March 1999, pp44 – 48), with full constructional details of the 'Binaural Weekender'. Of this receiver, an editorial endorsement by Ed Hare, W1RFI, ARRL Lab Supervisor, runs "Once my ears got used to the effect, they had to drag me away from this radio. This is one I gotta have!" **Fig 1** shows the basis of the KK7B binaural I-Q receiver as used also for the Buccaneer.

I cannot resist a wry smile. Some 30-odd years ago, in several items headed 'The cocktail party effect', I drew attention to the potential value of pseudo-stereo reception of CW signals and, in October 1973, revealed for the first time the pioneering work on this technique by F J H ('Dud') Charman, G6CJ. Earlier (August 1973) I had drawn attention to a CCIR investigation into a novel, switchable synchronous (exalted-carrier) binaural detector for AM broadcast receivers that offered greatly improved performance over conventional detectors. Also noted then was a performance improvement and a 'frequency scissors' detector that had been developed in 1958 by Hans Ever, PA0CX, that split the upper and lower sidebands of AM or CW signals and fed them to separate earpieces.

To quote from introductory remarks in the August 1973 'TT': "The human ear can provide a 'filter' bandwidth of around 50Hz with a remarkably large dynamic range (well over 100dB) and the ability to tune from about 200 to over 1000Hz and all without introducing 'ringing'. With such good 'no-cost' filters, the degree of improvement provided by additional electronic filters is limited.

"This naturally raises a further question: are there not other ways in which we could use our ears to better advantage in eliminating signals we do not want to hear? Which bring us, as you may have guessed, to the so-called 'cocktail party' effect. This has been described (for example, in the book *Correlation Techniques* by F H Lange) as follows: 'If a large number of people are in conversation with one another in one room, it should in general be impossible to carry on a conversation with someone even in the immediate vicinity. Nevertheless, experience teaches us that this raises no great difficulty – in fact quite the reverse provided that the listener is 'tuned in' to the partner and the subject of conversations. This implies, therefore, the existence of a tuning (modulation) mechanism in the human ear, certainly of another kind to that used in radio receivers, since all conversations are using the same frequency range... the human ear achieves more than all the methods of analysis hitherto known... Classical filter theory with its band-pass and rejection bands breaks down here.' Undoubtedly, one of the ways in which we can separate wanted and unwanted signals is by means of apparent differences of direction, making use of the fact that we have two ears rather than one. Stereo and more recently quadraphony are examples of how we can take advan-

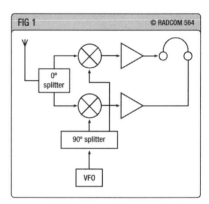

**Fig1:** Block diagram of an I-Q binaural receiver that allows the ear-brain combination to process the detector output resulting in stereo-like reception. As used by KK7B for 'The Mountaineer' (*QST*, March 1999)

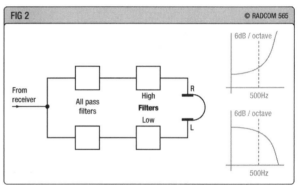

**Fig 2** G3OTK's suggested method of providing a 'stereo' effect for CW reception using AF processing ('TT' August 1973).

tage of the spatial characteristics of our ears. But, in communications, we are normally interested in what can be done with only one channel."

The item also included a suggestion by Richard Harrison, G3OTK, for providing a 'stereo' effect to provide subjective selectivity on CW: **Fig 2**.

This was followed up in the October 1973, 'TT'. To quote: "What we did not know at the time was the pioneering and extensive work in the field of CW subjective selectivity by F J H 'Dud' Charman, G6CJ, and the clear priority established by him in British Patent No 916,843 'Improvements Relating to Radio Telegraph Receivers'. This patent, now expired, was taken out by EMI (formal application date 24 January 1958) naming G6CJ as the inventor... It remains essential reading, describing the arrangements necessary to assist a radio operator to separate wanted CW signals from unwanted transmissions around the same frequency by means of 'psychological or subjective' effects. In practice, these arrangements –which have been used by G6CJ over many years – provide exactly the facilities sought by G3OTK, although they differ in the sense that G6CJ is

# TECHNICAL TOPICS

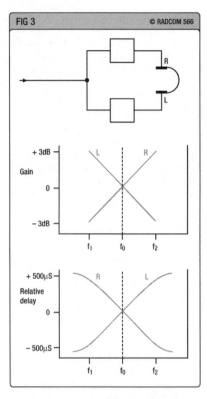

**Fig 3**
Gain and delay dispersion as used in the G6CJ / G3OTK 'Stereocode Processor' (*Radio Communication*, September 1975)

**Fig 4**
G0UPL's two-chip VFO plus Huff & Puff stabiliser (*Sprat*, Spring 2005).

convinced that the stereo effect is achieved by minute time delays rather than phase differences: **Fig 3**.

I continued: "How has it come about that this most interesting and valuable work has remained little known over the years? Well, Dud hinted at the possibilities in his lecture 'The Human Machine as a Radio Operator' in January 1958 – though he frankly admits that quite deliberately he has never described his work in any detail except in the patent application. He says: Well, it's nice to have a private secret, especially in competitive amateur radio operating – but really I was aware of the placebo pitfall of psychology and wanted first to get a proper statistical test done, using operators who did not know what was going on, and once it was widely known this would be impossible. The patent was taken out not for financial gain, but rather to establish priority of the work..."

G6CJ suggested that it was by now evident that we are extremely good at detecting incredibly small differences of *time of arrival* of sounds at our two ears: times so short that it must be done virtually at molecular level. The cocktail party effect thus depends partly on time delays, but also makes use of all other features of the signals, including amplitude, waveforms, room echo and the like.

The 1973 item continued: "Recent work on stereo has brought to light a good deal of fresh information. It is already possible to take a pair of loudspeakers, provide them with identical inputs and then insert effects that make the sound 'walk around' in imitation stereo, successfully fooling the hearing mechanism (if this experiment is repeated with continuous tones, varying only the phase, nothing unusual will be observed). It seems that essentially the ears work on starts and stops (the musician's *attack*) so that Morse code is an almost ideal medium for achieving pseudo stereo selectivity.

"G6CJ recalls how, as long ago as 1925, he was working with phones on a bridge system with two nearly identical 1000Hz inputs, when all of a sudden the beat note between them began wobbling about inside his head. The effect was so startling that the memory persisted although it was many years before a reference in the classic book by Fletcher began to set him wondering whether this effect could possibly be used to turn the output of a receiver into a stereo presentation.

"This was in the mid-30s when, at EMI, G6CJ was working with such people as Blumlein, the father of modern stereo recording. Unfortunately, he never put his ideas to Blumlein, who was later tragically killed during the war while working on airborne radar. Fletcher had shown that the relative time delay for a 'side' signal was about 0.6ms and that there would be an amplitude difference of about 1dB when a signal goes round the outside of the head. It seemed feasible to establish such conditions artificially, but this needed networks of a type then little understood even by the experts in this field.

"By 1945, 'Dud' had made a model using filters with m > 1.0 to get delay differentials, but it was not very satisfactory in spite of being frightfully difficult to realise. However, he then discovered the 'all-pass' network and by 1951 had a two-channel amplifier with a delay cross-over at 700Hz. 'As soon as I put it on the air I knew I was home – though aware that a psychologist would jump in and say – Ah, but you wanted it to work that way, so you haven't proved anything. But I discovered that not only could I concentrate on the signal tuned to the centre (ie the crossover frequency) but also if, for example, a rare callsign turned up on the 'edge' so to speak, I could immediately switch my attention to it."

G6CJ provided an outline of his experimental box of tricks as provided in his Patent 916,843 and this was reproduced in the 1973 'TT' item. He used some 10 valves (six in cascade) giving severe stability problems, an expensive collection of LA3 cores, and problems with decoupling which produced its own brand of delay distortion. However, in the end it did what was intended, providing about a millisecond per octave delay. The original valved unit was used operationally and successfully at G6CJ over a number of years, but a later solid-state design was never completed as it was abandoned when G6CJ began to suffer a hearing loss in one ear.

However the publication of the two 1973 items in 'TT' brought G6CJ in touch with G3OTK and, between them, a solid-state design emerged. The principles and practice of 'stereo' reception of CW signals was extensive described in 'Subjective Selectivity and Stereocode', by F Charman, G6CJ, and R Harris, G3OTK (*Radio Communication*, September 1975, pp674 – 681, with additional notes in January 1976, p23). This included full constructional information on a solid-state stereocode processor using eight integrated circuits, including all-pass filters based on operational amplifiers. For a time, a complete kit of parts was marketed.

Since then, a number of articles on pseudo-stereo has appeared, although the pioneering work of G6CJ seems generally to have passed unnoticed. However, as a result of my referring briefly to the 1975 article in 'TT' in February 2001, I received some comment from Peter Montnéymery, SM7CMY, who had been encouraged to complete the stereocode processor

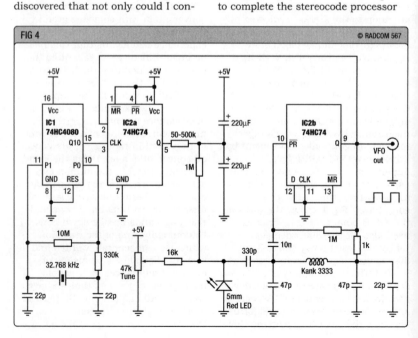

that he had begun in the past. He found that while use of the processor did not *per se* improve recognition, it was more pleasant to listen to CW using the stereo adapter and it results in less fatigue.

He also enclosed his 120-page book *Signal Detection in Noise with Special Reference to Telegraphy*, published in 1994, based on his medical thesis in support of a doctorate at the Department of Otorhinolaryngology, Head and Neck Surgery, University Hospital, Lund, Sweden. His main conclusions, not unsurprisingly, was as he summarised in his letter, that: "(1) the ability to recognise telegraphy masked by noise decreases with increasing telegraphy speed; (2) a low frequency 500 – 700Hz improves recognition; (3) best recognition is obtained with the telegraph signals 180° out-of-phase between the ears, but with the noise 'in-phase' (unfortunately impossible in practical receiver design, as far as I know); (4) a human telegraph operator outperforms the electronic decoders. (This was all before PSK31. On the other hand, I understand PSK31 is more a visual proceeding than an aural proceeding)."

In 'TT' July 1989, I drew attention to some limitations of the human ear in rejecting closely adjacent signals due to the phenomenon 'Zwicker masking'. In his 1967 book on the ear as a receiver of information he showed that there exist, between 30Hz and 20kHz, 24 sub-bands within which the most powerful component conceals (masks) adjacent, less powerful components, including noise, making them imperceptible to the ear. The width of these sub-bands varies from 100Hz in the low-frequency domain up to 2kHz in the high-frequency domain. While this effect became important in the field of digital sound recording and bit-rate reduction, It explains why our ears alone cannot filter out or even detect a weak wanted CW signal if there is at the same instant a more powerful interfering signal *within the same sub-band*, even though, in other circumstances, our ears may have a nominal selectivity of the order of 50Hz. This is another reason that should encourage the reception of CW signals with a beat frequency not greater than 700Hz and preferably between 300 – 500Hz rather than the usually-recommended 700 – 1000Hz. *Selectivity* of the human ear peaks around 300 to 400Hz where it is some 10 times better than at 1000Hz. The Fletcher-Munson curves, well known to hi-fi enthusiasts, show that the *sensitivity* of the ear peaks around 1000Hz. On a crowded band selectivity should take precedence over sensitivity.

### LOW-COST HUFF & PUFF STABILISED VFO

The frequency stabiliser for VFO designed by the late Klaas Spargaren, PA0KSB, in 1973 and to which I gave

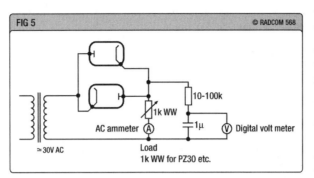

**Fig 5**
Method for testing the balance of thermionic diodes. Note that most diode rectifier valves have a voltage drop of about 15V at full load.

**Fig 6**
EB91 product detector designed by GM0HMR for use in an R1155 receiver.

the name "huff & puff", continues to attract experimenters seeking a means of stabilising free-running LC oscillators. Several variations are described in *Sprat* (No 122, Spring, 2005). Hans Summers, G0UPL, provides two simple designs, including a two-chip huff & puff stabilised 3.5MHz VFO that can be built for about £2: **Fig 4**.

G0UPL writes: "By using an inexpensive and common 32,768Hz watch crystal, the number of division stages required fits into the capabilities of the 74HC4060 oscillator/divider (IC1). The Q10 output is about 32Hz and determines the frequency step between lock points. A D-type flip-flop, IC2a (half of a 74HC74), latches the VFO signal and effectively behaves as a 1-bit frequency counter. The output is integrated via a simple RC network. By using two capacitors, the integrator voltage at switch-on is automatically initialised at 2.5V, leading to very rapid settling of the circuit. The integrator voltage is fed to a varicap diode in the VFO tank circuit. For the varicap diode, I just use an ordinary 5mm red LED and it works well! The 50 – 500k resistor depends on the inherent stability of the VFO.

"But that's not all! By studying the internal logic diagram of the 74HC74, I realised that if the clear (MR) input was held low, the path between the preset (PR) input and the Q-output would function as an inverter. As soon as you have an inverter, you have a potential oscillator! Using a KANK3333 inductor (but others would of course suit) and the circuit values shown, I got 170kHz coverage of the 80m band without further effort. Tuning uses the same 5mm red LED and a 47k potentiometer. The KANK3333 core is adjusted to obtain the desired coverage. By removing the 22pF capacitor from one side of the VFO tank circuit, I got complete coverage of 40m."

In his *Sprat* article, G0UPL also provides details of an extremely miniaturised huff & puff stabiliser using just one HC4060, but with locking steps of 64Hz. And, thirdly, it shows how a frequency counter and huff & puff stabilised two-chip VFO can be built provided that a 32.000kHz crystal is used. Another article in the same issue of *Sprat*, 'Huff & Puff Revisited Again', by John Beech, G8SEQ, also shows how the use of watch crystals can simplify an earlier design by Stefan Niewiadomski (*Sprat* No 63).

### VALVE DIODE HARMONIC MIXERS & PRODUCT DETECTORS

Robert B Kerr, GM4FDT, has followed for many years the discussions in 'TT' on mixers, product detectors etc and also thermionic valves. He was particularly interested in the technique of AC-coupling diode rings and the RA3AAE HF harmonic-mixer circuits (first reproduced in 'TT' from the Russian journal *Radio* way back in July 1977). He writes: "I can confirm that it works, the diodes developing a DC voltage to 'back off' the difference in forward voltage. I have even simulated the effect using a program called 'Croclips' by putting three diodes in series on one half of an RA3AAE product detector and one anti-parallel to them. **Fig 5** suggests a quick method of selecting diodes using 50Hz AC.

"I also noticed the comment in the July 'TT' from André Jamet, F9HX, 'Why not use vacuum diodes?'. As a retired TV service engineer with some *big* (power rectifier) diodes (PZ30, PY500, PY88, U808) – been there – done that!

"*Advantages:* They work well, are virtually indestructible, overload gently, do not seem to produce so many high order harmonics as semiconductors (no varactor effects).

"*Disadvantages:* (1) Heater supply required at odd voltages, eg 52V or 26V for PZ30, 42V for PY500, and a drive of 5W or so. The signal handling is excellent however, several volts input before overload. (2) The conversion loss is high, 14dB for **Fig 6**. (3) Small-signal diodes such as the EB91 (6AL5) have a limited current capability and need to work between high impedances. Fig 6 is a good example of their use: a product detector for a WWII R1155 receiver

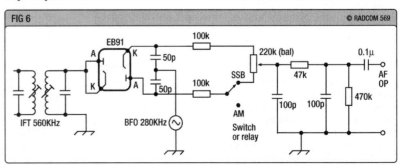

# TECHNICAL TOPICS

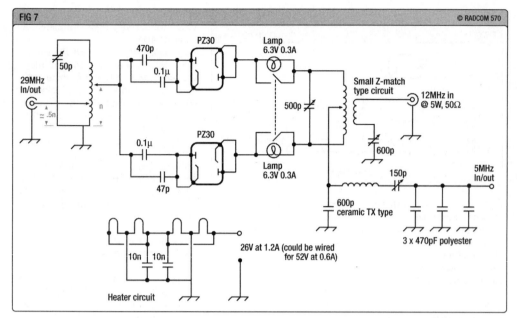

using an EB91 as designed by the late Stuart Martin, GM0HMR. In the R1155, the BFO runs at half the IF. For AM, the BFO is turned off and one diode disconnected by a switch or small reed-relay.

"As a transvert mixer, I use the circuit of **Fig 7** as part of my 5MHz station. It bilaterally transverts 29MHz from an HTX10 rig to 5MHz. To compensate for the conversion loss of around 14dB, a preselector is used for receive, and a 20dB parallel feedback cascode transistor amplifier running at 24V and 50mA is used in the transmit chain. This feeds via a half-wave filter to another cascode PA, a transistor in the cathode of a Class-A PL509 valve. The output is 5W. A 20P4 is used as an electronic T/R switch, This has a 38V heater, dropped from the 40V used for the PL509 from which an HT of about 50V is derived to operate the switch. The system uses negative grid-block T/R control. For an output of 20dBW, I can feed the 5W output from the PL509 to a passive-grid linear amplifier using, at the moment, three 829B double-tetrode valves. The heaters and the screen grids can be switched to allow the use of three QQZO6-40 2.1V quick-heat valves or three QQVO6-40 valves. At 100W, the amplifier is underrun. It can produce about 300W output on the other HF bands with about 10W of drive and 800V HT."

### G3DXZ's CERAMIC RESONATOR VXO

'TT' December 1985 summarised a long article by Albert D Helfrick, K2LBA, in *Ham Radio* (June, 1985, pp18 – 26) that showed how ceramic filter resonators, combined with variable mechanical or electronic-diode capacitors, could form very useful stable variable-frequency oscillators, stable over a significantly wider frequency range than VXOs using quartz crystals. K2LBA used 10.7MHz resonators (from FM receivers) and showed that stability could easily be maintained over a pulling range of some 2% compared with the 0.1% of quartz crystals, although the temperature drift is higher. The Q of a crystal can be as high as 500,000. For a ceramic resonator, the series resistance can be *lower*, the equivalent inductance is much lower and the Q is typically about 600 although this is much higher than a typical LC combination which is typically under 60.

Since then, considerable use has been made of ceramic resonators in VXOs, particularly for QRP transmitters. Even so, Chas Fletcher, G3DXZ writes to say that he has been "utterly amazed" at the stability-range of the VXO that he uses in 'An 80/40-metre, 6V, 5W CW Transmitter with Full Band Cover, Single Knob Tuning and QSK' (*Sprat*, No 122, Spring 2005, pp10 – 16).

He writes: "My reason for writing is that I would like to pursue the source of the stability in the VFO, and wonder if a brief airing in 'TT' might do the trick. Basically, I use a 3.56MHz ceramic resonator closely linked to a parallel resonant LC circuit. The oscillator runs from a 3V supply and I used a low Zener-diode-tuned circuit. As expected, changing the resonance of the tuned circuit pulled the oscillator frequency, but to my utter amazement, I managed to tune the oscillator down to 3500kHz without any serious change in stability. Having achieved 3.5MHz, the way was open to doubling – as in days of old – and on 7MHz the transmitter was still stable."

**Fig 8** shows G3DXZ's VFO and buffer/doubler circuit as presented in *Sprat*. The rest of the transmitter was also a little unconventional, as it used an un-tuned doubler and push-pull MOSFETs in the PA running at 6V.

G3DXZ continues: "I personally have not seen such an oscillator design published… My question remains, how does a pulled oscillator based on a ceramic resonator, retain such a degree of stability when so far from its natural resonant frequency?"

While the reported frequency range is within the 2% predicted by K2BLA in 1985 for a 10.7MHz resonator, my guess, and it is only a guess, is that the oscillator is basically an LC oscillator, (not a ceramic resonator oscillator), stabilised by the resonator in what is virtually a modern form of the Goyder Lock of the 1920s and 1930s. This was developed by Cecil Goyder who operated 2SZ, the licence held by Mill Hill School to make the first-ever UK – New Zealand contact in 1924 and became a professional radio engineer. Using quartz crystals, Goyder showed that LC oscillators could be locked by a crystal over a significant frequency range, although there was a tendency for the 'lock' to fail occasionally. One method by which G3DXZ could test this hypothesis would be to remove the resonator and replace it with a low-value capacitor – and see if the circuit continues to oscillate (unstabilised). It would also be interesting to know if the range of stable oscillation extends up to, say, 3600kHz as well as down to 3500kHz.

### HERE & THERE

John Walker, ZL3IB (Editor, *Break-In*), comments on the recent 'TT' items on cleaning components: "There is a much simpler way of cleaning ceramic variable capacitors, coil formers, rotary switches, antenna insulators, etc. I just include them in the family dish-washing machine and the rather alkaline dish-washer cleaner plus 'rinse-aid' leaves them all sparkling clean! Old Eddystone ceramic coil formers now look like new!" ♦

**Fig 7** GM4FDT's bilateral transverter mixer used for 5MHz with 29MHz drive.

**Fig 8** The VXO with ceramic resonator and the buffer / doubler used by G3DXZ in his 6V, 5W 3.5 and 7MHz CW transmitter with QSK, fully described in *Sprat*, Spring 2005. L1 Toko 10E core (from 10.7MHz IF transformer wound with 15-turn primary (4µH) 3-0-3-turn secondary).

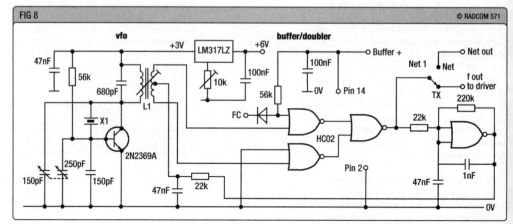

**Pat Hawker,** G3VA

37 Dovercourt Road, London SE22 8SS.

# TECHNICAL TOPICS

## Oscillators – valve or FET? ♦ VHF oscillators with inverted-mesa quartz resonators ♦ Aluminium - basics & care of ♦ Soldering to aluminium ♦ Lead-free solders

### OSCILLATORS – VALVE OR FET?

'Oscillator Stability with Valves and FETs' ('TT' December 1991, see also *TTS, 1990-94*, p122) reported the experiences of Ray Cracknell, G2AHU, who had found that LC oscillators based on valves were less sensitive to changes in local ambient temperature than those using FETs and could provide better long-term stability. I noted then that: "One of the very first widespread application of solid-state technology in amateur radio was to provide an LC oscillator that reached thermal stability from switch-on in a matter of seconds rather than the 20 minutes or so often required for valve oscillators. But this did not – does not – mean that a bipolar or FET oscillator is immune to changes in ambient temperature, either changes in shack temperature or changes resulting from the heat generated within high-power stages whether thermionic or solid-state.

"FETs with their high internal capacitance (Miller effect) and silicon construction can be difficult to compensate effectively over the temperature cycling that occurs in practice. A valve, with its much greater self-generated heat, tends to settle down after its (very long) warm up period and then becomes much more resistant to changes in ambient temperature. FETs have high input capacitances, sensitive to temperature changes; also to the varactor effect."

G2AHU was prompted to investigate a comparison of valve and solid-state oscillator stability when he found that, in practice, there was considerable discrepancy between the stability quoted for FET VFOs and their performance outside laboratories or temperature-controlled buildings. At the time, his shack was in a loft and suffered from wide temperature swings. He investigated temperature effects on FET oscillators by constructing an 11MHz FET test oscillator using a 2N3819 inverted-Hartley oscillator and 2N3819 source follower and buffer (Fig 4 of the December 1991 'TT') in a double-screened enclosure with heat insulation and including a 6V heater and thermostat. He showed that the output frequency swung from plus 3kHz to minus 3kHz when the unit was slowly heated from 20° to 60°C over a period of 140 minutes (including a change of 4kHz in the first 20 minutes).

This could not match the stability of his 18 – 20MHz VFO that used a 6J6 double-triode valve as a Kalitron oscillator, with 6CW4 (Nuvistor) cathode-followers. He wrote: "I found that the long-term stability of the FET oscillators was inferior to the 6J6 oscillator which can also produce a T9 note on considerably higher frequencies than 18 – 20MHz. But then we can't turn back the clock, can we?"

The main source of frequency/temperature drift in a valve VFO is usually the effect of the heat generated by the valves on the inductor and (to a less extent) on the variable and fixed capacitors forming the resonant tank circuit. The effect of heat (internally-generated or otherwise) on the small inter-electrode capacitances of the valve is relatively insignificant and can be made more so by a high-C tank circuit. Temperature compensation of the tuned circuit is reasonably straightforward. An alternative approach, that has been used in the past, was to separate the oscillator tank-circuit entirely, or by effective heat screening, from the oscillator/transmitter enclosure containing the valves, connecting the tuned circuit via short coaxial cables. The valve itself is thus a stable component although the heat it generates will effect adjacent components.

I was reminded of the 15-year-old experiments by G2AHU by an article 'XFY VFO', by Dr Andrew Smith, G4OEP (*Sprat*, No 123, Summer 2005, pp18 – 20). To quote selectively: "Conventional wisdom has it that valves get hot, they heat up other components, they create warm-up drift, they are 'old hat'; FETs and bipolars dissipate only milliwatts and avoid all that; they are 'the way forward'. I have never seen this dogma challenged [but see above – *G3VA*]. So when I made a valve VFO, I was amazed to find that a circuit with an EF95 was no more unstable than an ordinary bipolar design. This set me thinking. Temperature variations in a valve are extreme compared with a transistor – 2 or 3 watts total dissipation, and parts at red heat, as against 20mW or so, and not even barely warm... Valves must be basically much more thermally-stable than semiconductors, in which everything involves exponential functions of temperature. So reduce the heating effect and you must have a winner. Next stop – the junk box; out popped a XFY43 sub-miniature wire-ended pentode with a 1.4V, 10mA filament and 23V, 0.6mA anode rating, probably intended for hearing-aid applications. It will oscillate on an 'HT' of 8V, 0.4mA giving less than 20mW of total dissipation – no more than an average bipolar.

The result is a very simple VFO with spectacular stability. There is simply no warm-up drift. On a receiver there is an initial 'whoop' followed by an absolutely steady tone, and a frequency counter tells the same story. It reaches a steady condition in less than a second – the nearest thing to a crystal you can hope to find. What surprises me most is that I have never seen the use of miniature low-power valves recommended for VFOs, so perhaps the idea is 'new'. The thermal coefficients of the inductor and capacitors in the tuned circuit have the usual effect if you attack the circuit with a hair-drier, but these can be juggled or a simple compensating capacitor can be made using items found around the house... Don't be tempted to try using varactor tuning unless you want to ruin the VFO, and observe all the usual rules – solid construction, rigid screening, stable variable capacitor with proper bearings, etc. Although relatively insensitive to voltage it is best to stabilise the supply if intended for use in a transmitter. Resist the temptation to derive the heater supply from a pair of forward-biased diodes, use a dry cell for best results.. I was lucky to find the XFY43, other possibilities include the 6088, XFW10, XFW20 and XFW30, the last three having 0.625V filaments. The DF96 and DF97 with 1.4V, 25mA filaments are common B7G types designed for a higher anode voltages but might be worth trying at lower voltages."

G4OEP's 5 – 5.5MHz VFO is shown in **Fig 1**. It comprises an electron-coupled Colpitts oscillator. The directly-heated filament makes it necessary to have two filament chokes (L1, L2) or alternatively a choke with 10 bifilar turns on a 8mm ferrite bobbin. Further details, including the con-

**Fig 1**
The 5.0 to 5.5MHz VFO using a miniature, wire-ended pentode hearing-aid valve generating very little heat and found by G4OEP to reach thermal stability in a second or two with the VFO offering better stability than when using an FET or bipolar transistor. (Source: *Sprat*, Summer 2005)

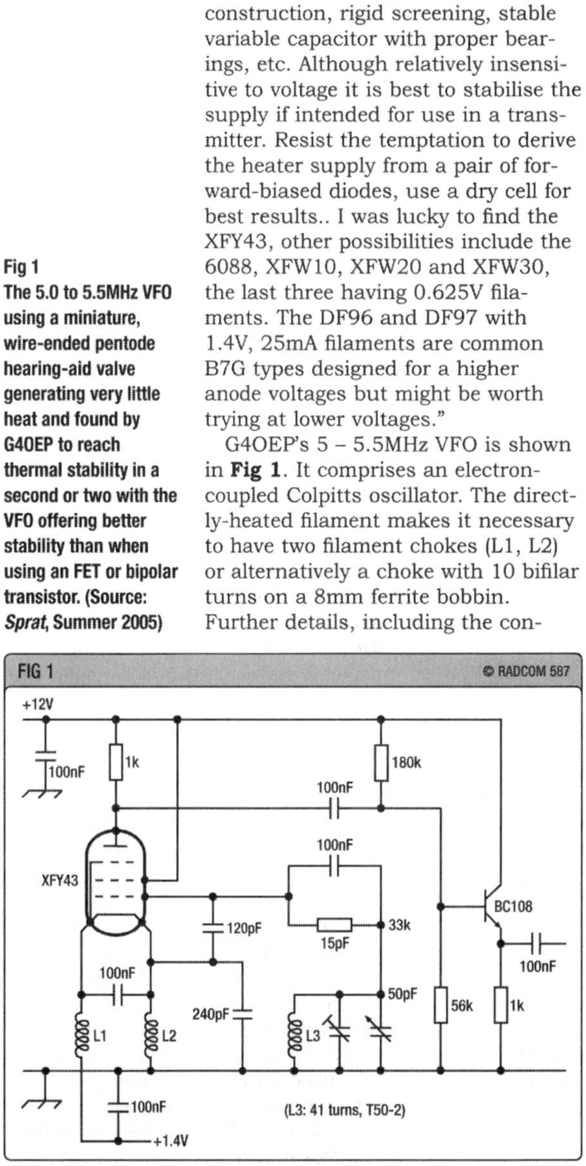

# TECHNICAL TOPICS

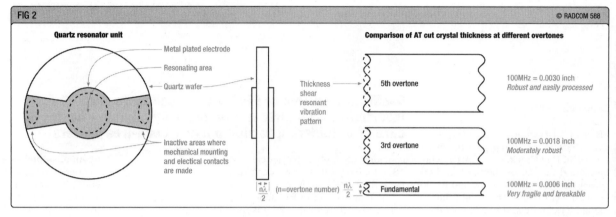

**Fig 2** Basic form of conventional AT-cut quartz crystal resonator showing the extreme thinness and hence fragility of a fundamental-mode 100MHz crystal. (Source: *RF Design*, June 2005)

struction of a simple bimetal thermal compensator are given in the *Sprat* article. Just one word of warning (applicable to both low-power valves and solid-state oscillators): higher-power oscillators can result in reduced noise sidebands.

**VHF OSCILLATORS WITH INVERTED-MESA QUARTZ RESONATORS**
A feature article 'Novel High-Frequency Crystal Oscillator Cuts Jitter and Noise', by Dan Nehring (*RF Design*, June 2005, pp32, 34, 36, 40, 42) carries an editorial note: "The inverted Mesa resonator offers a high-performing alternative to SAW-based oscillators and other bulk resonator types which use a noisy PLL or other such parametric multipliers that also multipliy noise. Oscillators have been developed to demonstrate the effectiveness of this design technique."

To quote from Nehring's introductory notes: "The fast data rates of today's digital systems continue to demand faster and better clock signals with low jitter and phase noise attributes – not to mention small form factors. Crystal oscillators are typically the devices that are providing the originating timing signals for these systems. Oscillators with internal phase-lock loops (PLLs) that generate a high-frequency output from an easily-produced low-frequency crystal are often used. However, the degraded phase jitter of this approach is prohibitive for many new applications. Other techniques, such as the use of surface acoustic wave (SAW) devices, have cost, availability and frequency-stability issues. We take a look at the design process, trade offs and the resultant benefits of oscillators built with pure analogue circuits and no frequency multiplication for providing the clock signals for jitter-intolerant systems."

**Fig 2**, adapted from a more detailed illustration in Nehring's article, provides an overview of quartz crystal resonators. Basically, the frequency of an AT-cut quartz resonator is determined by the thickness of the quartz wafer, with the frequency increasing as the thickness decreases. This is achieved by mechanically sawing and then lapping the wafer to the desired thickness. This results in fragile resonators, with the processing becoming more and more difficult as the resonant frequency increases. To achieve higher frequencies without reducing thickness, crystals can be resonated at an overtone (approximately odd-harmonic) frequency. Conventional flat wafers cannot be produced economically at fundamental frequencies higher than about 50 or 60MHz, limiting third overtone crystals to 150 to 180MHz at the most. Above this, higher overtones must be used. Nehring points out that fifth, seventh etc overtones are available and can be used in appropriate designs but, as the overtone number goes up, the resistance of the crystal inevitably increases and the number of spurious modes increases. For high performance, the lowest possible overtone available economically should be used.

I recall that pre-WWII, very few amateurs possessed 14MHz crystals, mostly using 7MHz crystals with frequency-doubling stages. The general use of an overtone resonance (in which there is no output in the fundamental mode) seems to have derived largely from US military equipment. Overtone techniques became widely used, post-WWII, by amateurs for VHF equipment.

Over the years, there has been a long struggle to produce higher frequency crystals at an economical cost that can utilise the lowest possible overtone at VHF. Dan Nehring claims that the major solution to this problem has been the development of the inverted Mesa-type resonator. 'Mesa' is derived from the Spanish word for 'table' - thus an inverted-Mesa implies a resonator with a central thin part of the wafer that determines the frequency supported by a thicker outer ring: **Fig 3**. He writes: "This type is effective in driving down the overtone number that can be used to hit any particular frequency. The inverted-Mesa is ideal for creating fundamental mode and third overtone-mode resonators up to 600MHz or higher." His firm, Valpey Fisher Corporation, is producing a tiny VF266 surface-mounted oscillator package (footprint 5mm by 7mm). He concludes, *inter alia*, "The development of better-performing (higher-frequency, lower-noise, smaller package) oscillators continues... The development of the inverted-Mesa resonator has produced a higher-performing alternative to SAW-based oscillators and other bulk resonator types which use a PLL or other such parametric multiplier that also multiplies noise..."

It is shown that the phase noise is generally 20dB lower than for competing PLL clock solutions. While inverted-Mesa resonators may not have an early impact on amateur radio, it is clearly a development worth watching.

**ALUMINIUM - BASICS & CARE OF**
John Rosindale, G0GUO, noted with interest the recent 'TT' items on cleaning silver-plated variable capacitors by immersing them in a solution of warm baking powder with the plates in contact with some aluminium foil. He writes:

"I can confirm that the reduction of the brown silver sulphide to a shiny silver surface requires an electron flow - so a good electrical contact is needed between the aluminium cooking foil and the object being cleaned. The electrons are provided by the aluminium as it dissolves. It is useful, from time to time, to turn the capacitor around on the foil; this ensures a more even cleaning operation. Current flow is akin to problems of 'throw' in the electroplating of irregular objects

"Aluminium is one of only a few metals to be attacked by both acids and alkalis. Baking soda (sodium bicarbonate) is very slightly alkaline, especially if partly changed into sodium carbonate by very hot water. Washing soda (sodium carbonate) is cheaper than baking soda and quite a bit more alkaline, although still acceptable for this cleaning job. 'Caustic soda' (sodium hydroxide) is too alkaline; the aluminium foil is attacked too quickly and there is a release of hydrogen gas to the detriment of the electrolytic action required to reduce the brown silver sulphide.

"Aluminium is strange in other ways - although plentiful in the earth's crust, it is a relatively reactive metal and is always extracted from bauxite (its oxide) by electrolysis. At the time of its discovery it was a curiosity, more expensive than gold; this was because it was then extracted from bauxite by using even more reactive metals to 'grab the oxygen'.

"Those of us who have spent hours 'chassis bashing' appreciate the soft-

# TECHNICAL TOPICS

ness of aluminium, even though it can have a bad effect on files, drill bits, etc. This softness presented a 'creep' problem when electricity boards used steel-cored aluminium cables many years ago when there was a world shortage of copper and it was necessary to re-design clamps for such cables.

"Chemically speaking, aluminium is quite reactive once the protective oxide layer is breached. Aluminium dust is used in 'thermite mixture' and reduces iron oxide to molten iron in a very spectacular way. Aluminium powder, with a good oxidising agent, has even been used in bombs!

"This chemical reactivity makes problems for radio amateurs when aluminium antenna elements are fixed with steel hose-clips or when brass or copper bits are used to connect coaxial cables. I live in a coastal region and the salt spray hastens the demise of VHF dipoles and 'Slim Jims'."

The problems encountered in the use of aluminium alloys for antenna elements, masts etc have been discussed on a number of occasions over the years in 'TT', but bear some repetition. In April 1991, Dick Biddulph, (M0CGN, formerly G8DPS) drew attention to the need to avoid contact between copper and aluminium in antenna installations, in addition to other precautions including the application of grease, lanolin or RTV silicone rubber. In June 1991, Steve Henderson, ZL1AOC, drew on a publication of the New Zealand Building Research Association concerned with combatting corrosion in TV receiving arrays. This stressed the effects of marine atmospheres in a country where most populated areas have high concentrations of chloride-containing sea-salt aerosol (which, like the sulphur-dioxide of industrial atmospheres, promotes corrosion). Incidentally, there is nowhere in England more than 72 miles from the sea, and industrial pollution affects most populated areas.

ZL1AOC pointed out: "All unprotected metal surfaces, except the few 'noble metals' such as gold and platinum, corrode or oxidise to some degree. How long this takes before it becomes a problem very much depends on the working environment. All too often one hears of an antenna where the telescopic tubes of an expensive Yagi array can no longer be adjusted or a trap in an antenna has deteriorated... Typical examples include: tinned-copper braided pigtail connections from a balun to a wire dipole completely disintegrated; alloy-aluminium bolts terminating the wire connections to traps in a wire dipole corroded to the extent that some had fractured; telescoping tubes of a Yagi corroded and seized, offering high resistance between sections; element mounting bolts rusted with corrosion to the extent that a VHF element may fall off after only a few months use; two-piece element clamps of diecast metal corroded such that the element sections no longer provide a continuous electrical path."

ZL1AOC continued: "Amateur radio antenna arrays can have a large number of tubular sections, many of them being required to have telescoping adjustable sections. If these are not protected when they are assembled, it will be impossible to dismantle them at some later date. Hardware supplied with some arrays is electro-plated; with others, stainless steel is provided. The preference is always to use stainless steel hardware – a point worth exploring when contemplating the purchase of a new antenna. Rust not only weakens elements but can result in the formation of electrical diodes and their 'rusty bolt' effects. Surplus compound should be cleaned off and the joint wrapped to seal it completely with a self-amalgamating tape

"In assembling an antenna for the first time, or after repairs, care should be taken to prepare all sections to prevent the entry of water. If a telescoping section is involved, all signs of corrosion should be cleaned off the metal. The sections should be liberally coated with grease or better still with one of the anticorrosive preparations. When the position of the sliding joint is finally determined [remember that element resonance lengths will be affected by the operational height above ground – *G3VA*] the surplus compound should be cleaned off and the joint wrapped to seal it completely with a self-amalgamating tape. If you have access to a hot-air gun, the joint could be covered with heat-shrink tubing. In assembling antenna elements to traps or to a boom, all the nuts, bolts, washers and clamps should be completely coated with a suitable compound... Always ensure that drain-holes in traps and other components are clear and face the ground, so that any water that may have penetrated the trap will drain away. Element tubes with open ends should be plugged to prevent water from gaining entry... Corrosion is frequently found on die-cast components, including clamps and die-cast casings of rotators..."

'TT', July 1993, included notes

### Table 1
**Relative galvanic series in sea water**
**ANODIC END**
Magnesium
Zinc
Aluminium
Mild steel
Iron
50 / 50 lead / tin solder
Stainless steel (US type)
Tin
Nickel (active)
Brass
Aluminium bronze
Copper
Nickel (passive)
Silver
Gold
**CATHODIC END**

based on a *QST* article by Scot Roleson, KC7CJ, who had experienced a problem with "a trusty Butternut antenna" just six months after installing this in the dry Californian climate. "One day it worked fine; the next it wouldn't load properly. As I dismantled the vertical, I noticed a fine white powder at each joint, corrosion had crept into every connection although, on installation, I had tightened each clamp and bolt securely."

KC7CJ explained how the use of different metals in a situation where an electrolyte is present at the junction results in bi-metallic corrosion, particularly where the metals are well separated in the galvanic series, see **Table 1**. He wrote: "Tin and gold are metals that illustrate how troublesome bimetallic corrosion can be. Both metals are commonly used to coat electrical connectors yet are galvanically remote. Sometimes connectors with these pins coated with these metals are inadvertently attached to each other. If the contact pressure is insufficient to keep out moisture or if in an environment where electrolyte forms easily, the tin surface oxidises. I've seen this happen in PCs, where plug-in cards with gold-plated edge connectors are plugged into tin-plated motherboard connectors. The resulting problems are usually intermittent and difficult to locate. Simply removing a card and reinserting it may remove enough oxide that the problem disappears.

"Aluminium and copper are not very galvanically compatible but connection of copper wires to aluminium antenna elements is often necessary. This problem can be minimised by tinning or solder-plating the copper wire, forming a gas-tight seal between the copper and plating. Then use stainless-steel hardware to secure the connection... use a stainless-steel washer between an aluminium surface and a tinned wire, or lug, connected to it... Another way to form a gas-tight seal with hardware is to use star washers which break through oxides and cut into mating surfaces [although] repeated assembly can damage a surface, possibly providing a path for moisture to

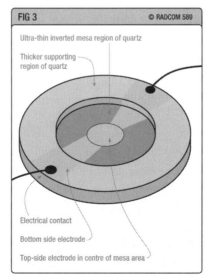

FIG 3
Ultra-thin inverted mesa region of quartz
Thicker supporting region of quartz
Electrical contact
Bottom side electrode
Top-side electrode in centre of mesa area

Fig 3
The inverted-Mesa quartz resonator. The central portion is chemically etched to make it thinner than the outer ring. The central part forms a thin VHF resonator, supported and made more robust by the thick outer ring of quartz. The technique provides a method of producing fundamental and low-overtone crystals in the VHF range. (Source: *RF Design*, June 2005)

# TECHNICAL TOPICS

enter the joint.

"Making and keeping good electrical connections in antennas is really simple – as long as you pay attention to the basics. The best way is to start with galvanically-compatible materials, then clean all connections well before assembly. To make sure these connections stay good, seal all contact points so moisture can't enter the joint. Electrical contacts occur between microscopic bumps and points where the metals meet... A smooth, clean surface ensures that there will be lots of these points and little between them to get in the way. For antennas, I've found it's best to first buff all joining parts with steel wool, emery cloth or a wire brush, then with a nylon scouring pad... For metal tubing, its important not to forget to clean the inside surfaces of telescoping parts. I wrap steel wool around a pencil or form it into a pencil-like shape so I can get to the tubing's inside surface. Finally, I use a clean rag to wipe off any powered metal and oxide. I do my best to refrain from touching the mating surfaces and contaminating them with body oils (use cotton gloves during antenna assembly is a good idea).

"Joint compounds, unfortunately, don't last forever. They harden and crack, or with time and temperature, simply flow away from joints. For this reason some sort of finishing barrier or overcoat is needed, such as plastic tape, paint, or silicon rubber sealant (bathtub caulking). Choose a material that is flexible and resistant to ultraviolet light. Many paints and plastic tapes eventually harden and become brittle from exposure to UV light... This argues for regular maintenance. Plastic tape is easy to remove and replace."

In 1978, Dick Biddulph, in discussing the protection of external metalwork had found that bolts coated with lanolin on his 144MHz array were easy to undo even after some five years in service without maintenance. The lanolin was dissolved in natural turpentine (*not* white spirit) and applied to the threads. After tightening the nuts, further solid lanolin was applied to the exposed threads.

A further tip is shown in **Fig 4** as a means of reducing metal fatigue in tubular sections due to wind flutter.

## SOLDERING TO ALUMINIUM

Back in 1978, a series of items in 'TT' noted the problems that are encountered when trying to make soldered connections to aluminium. For example, I reported a QST a tip by K0JFN: "An old trick for soldering to aluminium is to place a drop of oil on the aluminium and then scratch the metal with a knife or other sharp instrument until the area to be soldered is shiny. Apply a soldering iron and resin-cored solder. After the solder has taken, wipe the oil from the surface. A very neat solder base should then appear."

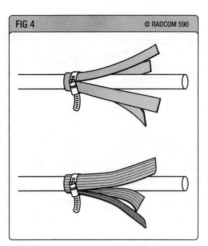

Fig 4
A method of reducing metal fatigue in tubular beam elements caused by flutter-type oscillation brought about by vortex shedding at specific, relatively-low windspeeds. Energy absorbers can be made from split rubber hose pipes or flat rubber or pliable plastic material. An alternative technique is to fill the inside of the tubes with the type of foam intended for sealing and insulating cracks and holes in buildings. Metal fatigue can cause the elements to fracture if the oscillation continues over months or years. (Further information in 'TT' December 1993)

However, in July, I warned that Harold J Reed, a professor in metallurgy, had pointed out in QST that the real problem is not that of making a good bond to aluminium but of making one that will last: "There are dozens of ways of making a good bond to aluminium, but no inventor has solved the corrosion problem that causes failure of the bond. Aluminium is an extremely active metal and, were it not for a self-repairing, protective oxide coating the metal and its alloys, could not be used for any commercial purpose. When aluminium is bonded to a metal or alloy (eg a solder) of considerably lower activity, the protective oxide film either cannot form or is imperfect. The aluminium becomes strongly anodic to the solder and is converted at the interface to corrosion products having little or no strength, and the bond fails. This may take a few hours, weeks or months, depending on the system and environment, You can be sure failure will occur sooner or later." What was required, he suggested, was a solder whose activity is the same, or nearly the same, as that in aluminium or its commerciql alloys.

Dick Biddulph (September 1978) agreed that an exposed soldered joint made to aluminium is subject to corrosion that will eventually weaken the bond, nevertheless such soldered joints are used widely in the electrical power industry. He commented: "The corrosion can be slowed up if, after soldering and cleaning off the flux, the joint is covered with adherent grease such as anhydrous lanolin and applied to the warm metal. An alternative, better for indoors, would be to heat the joint (again after cleaning) and coat with an epoxy resin such as Araldite."

## LEAD-FREE SOLDERS

Meanwhile, the date is coming ever closer (July 1, 2006) when the use of lead-free solder will become compulsory throughout the European Union – a directive that fills many electronics service engineers as well as amateurs with considerable foreboding It was noted in 'TT' October 2003 (see also *Technical Topics Scrapbook 2000-2004*) that, in most solders, the lead will be replaced by copper (or silver) with its appreciably higher melting point of tin-copper solders typically some 40 – 50°C higher than tin-lead solders, depending on the percentage of copper with its melting point of over 1000°C. Some television manufacturers including Sony and Panasonic are already using lead-free solders.

Adrian Gardiner in 'Bench Notes' (*Television,* June 2005, p490) notes that the removal of lead from solder has introduced a number of complications mainly arising from the increase in melting point. This can result in quality and stability problems, but also reduces processability, because of lower wetting and changes in flow behaviour.

He writes: "The new lead-free alloys are more temperature-sensitive than conventional solder. It's important therefore to avoid increasing the process temperature as the longer cooling time can cause microcracks. In addition, higher soldering temperatures can result in black layers on the soldering tip. This makes it unwettable, causing early fallout (charred flux, oxidised tin, tin-copper-iron fusion)."

A problem with using very high bit temperatures is that the flux burns off immediately and cannot work effectively. Flux is essential to soldering to deoxidise the metal surfaces and it facilitates diffusion of the metals into each other. Gardiner adds: "The way to overcome this is to be able to transfer increased amounts of thermal energy rapidly at lower temperatures. And this calls for hand-held soldering tools that are more efficient technically". He uses a Weller WSD81 soldering station, a more expensive tool than the conventional hand soldering iron.

In the August 2005 issue, a letter from Geoff Darby, a respected *Television* contributor, expresses his concern at the change to lead-free solders: "It has yet to be explained to me satisfactorily exactly what the issue is with leaded solder, particularly if a proper and responsible approach to the recycling of end-of-life electronic equipment is adopted. I fail to see how the lead in solder could have an appreciable effect on the public. He asked a friendly plumber of his experiences only to receive an unprintable reply (plumbers have for some time been banned from using leaded solder with open pipework systems for hot and cold water). He claimed that lead-free solder is a much more difficult material to work with even with a blowtorch. He said that it has completely different flow and resetting characteristics and that it is much more difficult to make a completely watertight join. The solder tends to melt very suddenly and rolls around in a less-controllable fashion. For electronics servicing there are warnings against mixing leaded and unleaded solders, and this would seem to include using leaded solder to repair a PCB where the manufacturer has used lead-free solder see Fig 4 of July 'TT'. I cannot help feeling that, for radio amateurs, lead-free solder is bad news. ♦

Pat Hawker, G3VA

37 Dovercourt Road, London SE22 8SS.

TECHNICAL TOPICS

**Facts sacred – comment free ♦ Small loops – final, final words ♦ Mobile phone risks ♦ Passive squarer for H-mode mixer**

### FACTS SACRED - COMMENT FREE

A great editor once wrote in his newspaper that "facts are sacred, comment is free" – advice that is sometimes lost in modern journalism. Too often most of us tend to take the easy course and accept that much of technical journalism is "scribbling on the back of advertisements." This reflects the fact that even a hint of adverse criticism can raise the hackles of firms and of some readers. It is easier, but in my view wrong, to take the line of praising new ideas and products unreservedly, anything for a quiet life. But there are times when even the most complacent of we hacks accept that we have a duty to our readers to attempt to tell it as we believe it to be, come what may.

When, back in 1958, I penned the first 'Technical Topics', I set out my stall as follows: "To keep abreast of current technical progress and practice in the amateur radio field has never been an easy task. New ideas and circuits are constantly being introduced, and old ones revived. Some have a short life, others are absorbed into the main stream of amateur practice...

"All we can hope to do is to survey a few ideas from the technical press, a few hints and tips that have come to our notice, with perhaps an occasional comment thrown in for good measure."

Of these rather pious aims, the one that has from time to time caused problems is "the occasional comment thrown in for good measure." Such a case occurred recently with my comments on the SD-1000 software-defined receiver.

In a letter to the Editor (forwarded to me) about 'Buying Overseas – *Caveat Emptor*' (in August), Gerald Youngblood, K5SDR (ex AC5OG), owner of FlexRadio Systems that markets the pioneering SDR-1000, writes *inter alia*: "I must say I am shocked and speechless that such baseless and harmful comments would be published in any amateur radio magazine, anywhere in the world. I have never seen anything even remotely similar in *QST* in my almost 40 years as a ham. The article contains only pure speculation with regard to FlexRadio and the SDR-1000. To my knowledge and his admission, writer has no experience whatsoever either with the company or the radio. The facts are that the SDR-1000 has been shipping for over two years, is in the hands of almost 700 customers worldwide, and *every* customer has received the support they requested. There has not been a single radio returned to the company that has not been repaired (other than for lightning damage)..."

He refers only to the short item in the August 'TT' with no reference to the much longer piece (nearly a page and a half) in the June 'TT', based solidly on the detailed six-page *QST* (April) product review by Steve Ford, WB8IMY, quoting his belief that the marketing of the SDR-1000 "opens a new chapter in amateur radio" and that "For the first time in ham history you can purchase 'off the shelf' an HF and 6m transceiver that uses software to define its functionality – a *software-defined radio*". But WB8IMY made it clear in a generally favourable review that there were important caveats, for example: "The SDR-1000 may mark the beginning of a new generation of amateur radio, but the pioneers who take it up may need a bit of a frontier spirit." I wonder what K5SDR made of the *QST* review and whether WB8IMY's caveats have been addressed?

WB8IMY put the performance as at least comparable to a traditional transceiver in its $1300 price class. The $1300 price range assumes that the potential user already possesses a high grade PC and suitable sound card and facilities, for example, to download the 100-page on-line instruction manual, etc. If, as a frequent user might find desirable, a dedicated PC is used then the total cost of installing and operating an SDR-1000 would be much higher than the cost of the kit alone. It was clear that WB8IMY had not found the SDR-1000 easy to install and get working and he stressed that a user would require considerable knowledge of computing. One might add that only a tiny minority of amateurs would have the ability to repair or possibly modify the circuit boards.

Following the publication of the June 'TT', I became aware that not every one of the 25 or so UK amateurs who had already purchased this transceiver was happy with the design of the RF firmware with its mixture of highly-sensitive analogue devices in close proximity to digital switching units etc. There can be (have been) problems with sending boards with SMD devices back to the States for repair. For example, if the board is lost in transport or is non-repairable (eg lightning damage), as far as I can tell there appear to be no facilities for replacing a single board, thus requiring the purchase of the complete set of three boards. Hence it may be advisable to take out insurance to cover the value of the three boards rather than the single board returned to FlexRadio. Only the very experienced would be capable of repairing the boards themselves or getting it done locally or finding a local source of the necessary components.

I had no knowledge of how most early UK users of the SDR-1000 felt about their purchases, although aware of problems that had arisen for one purchaser whose technical expertise I respect. So, in the August item, I sought further information without pin-pointing the actual problems of which I had learned.

Apart from the missive from K5SDR, the Editor has subsequently received e-mails from Peter Buck, G3LWT, and Klaus Lohmann, DK7XL. Both write enthusiastically

# TECHNICAL TOPICS

**Fig 1** Arrangement of the loop antenna reported by G3NOQ.

**Fig 2** Predicted and measured field strengths (5W RF power) of the loop antenna.

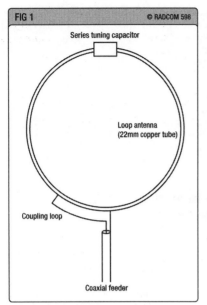

in favour of their SDR-1000 units, although both admit to having experienced faults that they had overcome themselves. G3LWT writes (abridged): "I am wholeheartedly in favour of this unit which has rekindled my enthusiasm for amateur radio construction and operation... I agree with you about *Caveat Emptor*, particularly for items not sold as complete working units. The reason these units are not fully assembled when shipped to the EU is simple – avoidance of the swingeing import duty costs. The current version vies with the very best transceivers at *vastly less cost*. Hardware has been frozen for some time, and the owner is not forced constantly to change the software unless he wishes to do. Gerald Youngblood has provided excellent backup... I would hate to see your comments misinterpreted – let's not discourage experimentation. Perhaps you should have added that such ventures are really for experienced constructors."

DK7XL, who wrote an article on it for *CQ-DL*, is another SDR-1000 enthusiast. He claims *inter alia* "Since working with the SDR-1000, ie experimenting with the hardware and dozens of upgrades in software, I discovered again the excitement of amateur radio and I enjoy being part of a community that is paving new roads to wireless community... The radio is not off-the shelf equipment that can be operated without going into technical details. FlexRadio is offering hardware from a bare system up to a complete set of additional equipment (100W PA, ATU, PC, etc). I have not heard from anyone who complained not to receive spare parts or getting warranty service if necessary... Your comments regarding the SDR-1000 cannot be justified nor are they fair... It is true that when FlexRadio advertises 'The radio that keeps getting better', FlexRadio is establishing a totally different relationship with the users/customers, exceeding everything that could be addressed when asking for 'after-sales-service' or warranty."

So there you are. Regrettably, today a divide exists between the digital and RF engineer, whereas the latter has been ignored by the former's ascendancy with all its serious repercussion on equipment design *vis-à-vis* RF parameters. I still believe my comments were justified and that the SDR-1000 is not a 'fit and forget' purchase for the majority of amateurs, nor does the basic firmware, in the form of the mixture of highly sensitive analogue and digital switching circuits without isolation/screening, represent the final word in high-performance software-defined radio. My judgment may be wrong, of course. All credit to FlexRadio for getting it on the market – and good luck to those experimentally-minded, analogue-RF expert and computer-wise amateurs who remain happy with their purchases. For others, I can only repeat *caveat emptor* – advice that extends to all serious purchases of new equipment, whatever the source.

## SMALL LOOPS – FINAL, FINAL WORDS?

A difficult task faces any columnist not prepared to accept uncritically claims made for novel ideas even when made by those for whom he has great respect. I have made myself unpopular in some quarters by consistently showing scepticism of the radiation efficiency claims made on behalf of very small loop transmitting antennas, including the CFL and more recently the small loops described by Professor Mike Underhill, G3LHZ and his student Marc Harper.

The small transmitting loop has, for several decades, established its place as a very useful transportable and space-limited antenna in amateur radio and in military and professional communications, particularly in regard to near-vertical incidence systems (NVIS) over distances up to a few hundred miles. But even with extremely low loss tubing, vacuum-variable or other very high-$Q$ tuning capacitors and effective matching to 50$\Omega$ feeders, the radiation efficiency (as shown by traditional methods such as Chu-Wheeler or professional *NEC* simulation) has long been considered low, especially for small loops at the low-frequency end of their ranges.

Some six years ago, G3LHZ, in a letter to the IEE's *Electronic Letters* and followed by comments in 'TT', and then at a professional IEE Conference on Antennas & Propagation in 2000, claimed that Chu-Wheeler had got it wrong, that *NEC* could not be applied to small loops and that the radiation efficiency of a small loop can exceed 90% throughout its frequency range. Despite the scepticism shown in a series of comments in 'TT' and expressed at the 2000 professional conference, G3LHZ has stuck to his claims, on the basis that they are based on real measurements rather than traditional formulæ, or computer simulations. He repeated his claims most recently in a two-part article 'New truths about small tuned loops in a real environment' (*RadCom*, August & September 2004), in which he concluded *inter alia:* "The small tuned loop continues to be seriously underestimated by those who prefer old theory and 'simulation' to 'real measurements'. Typical (intrinsic) efficiencies of transmitting loops of 80% to 90% or more are confirmed by proposed extensions to old EM theory".

Subsequently, a number of members (including VE2CV, G3UUR, G0GSF, etc) again disputed these claims, advancing a number of reasons why the measurements made by G3LHZ greatly exaggerate the radiated efficiency of his loops. Subsequent correspondence makes it clear that G3LHZ remains convinced that he is right and established theory is wrong.

One of the professional antenna engineers who has been concerned throughout the controversy is Alan Boswell, G3NOQ, a professional antenna engineer who chaired the session at which Professor Underhill presented his ideas (IEE 2000 International Conference on Antennas & Propagation). Together with colleagues Andrew J Tyler and Adam White (BAe Systems Advanced Technology Centre) he has investigated under rigorous conditions the 'Performance of a Small Loop Antenna in the 3 – 10MHz Band', to check once again whether there are any serious flaws in established the-

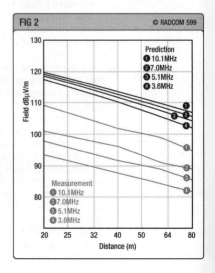

# TECHNICAL TOPICS

ory or in the use of *NEC* simulations.

The resulting paper has now appeared in *IEEE Antennas and Propagation* magazine, Vol 47, No 2, dated April 2005 (published July 2005); G3NOQ wrote the paper, did the theory and suggested what needed to be done. His colleague Andy Tyler, G1GKN, made the loop antenna, got it working, arranged the equipment, and supervised the apprentice Adam White, who made the measurements.

No reference is made in the paper to the work or theories of G3LHZ, presumably to avoid any professional disputation. Nevertheless it represents a further, and surely entirely convincing endorsement of the classical theory of small transmitting loops. The paper includes reference to the two-part article 'Performance of Electrically-Small Transmitting Loop Antennas', by J S Belrose, VE2CV (*RadCom*, June/July *2004*), and to a 1983 Racal Antennas publication (6559-1) reporting the performance of a square loop of 1.2m sides.

The tested antenna (**Fig 1**) was a circular loop of 1m diameter fabricated from 22mm- diameter copper tubing with a gamma coupling loop formed from a length of insulated stranded power wire of 1.5mm square cross-section, close to the loop, and connected to it by a clip at a point approximately one-eighth to one-quarter of the way round the loop. The position of the connection, together with the spacing between the loop and the coupling wire, gave the adjustments necessary to match the antenna at any frequency in its operating range after the series tuning capacitor had been adjusted to the required frequency. Inductance of the main loop was calculated to be 2.45μH, requiring a capacitance of 800pF for resonance at 3.6MHz, the reactance then being 55Ω.

The input impedance to the coupling loop was measured with a Hewlett-Packard network analyser type 8753C. The field measurements were made with the transmitting and receiving antennas placed 1.5m above ground, and orientated to lie in the same vertical plane. CW carrier power was generated with an Icom transceiver, with the power input to the antenna monitored by a Bird 'Thruline' wattmeter, Model 43, adjusted to show 5W forward power and negligible reverse power. Measurements were made at distances between 20m and 80m from the antenna at the four frequencies, see **Fig 2**.

The new paper once again confirms that the classical theory for loops is accurate when properly used. It also confirms that their 1m loop did not breach the Chu bandwidth limit. The results also lend support to the accuracy of modelling by computer codes such as *NEC* adding: "Computer analysis of small loops over lossy ground has been carried out extensively by Belrose" (see above).

**Table 1** shows the measured radiation efficiency, and the measured and predicted radiation resistance for the 1m loop antenna. The final paragraph in the paper's conclusions reads as follows: "The radiation-efficiency results described here are believed to be typical of loop antennas of similar dimensions in the HF band. As an example, a Racal Antennas HF loop – a square loop of 1.2m on the side – has a quoted gain of -16dBi at 4MHz and -9dBi at 7MHz. Although loop antennas of 1 – 2m² area do not appear, on present evidence, to possess good radiation efficiency at the low end of the HF band, this does not necessarily rule them out as effective components in practical applications. Successful communications links are often maintained with a radiated power of 1W or less at frequencies of 3 – 6MHz, especially for links of up to 300km where NVIS is used. Loops have to be judged against the performance of the available alternatives, taking into account the constraints imposed by mobile use on small vehicles. That is why small HF loops and half-loops enjoy a steady level of usage in the land-mobile, and other categories of radio systems, and this is likely to continue."

And that, as far as 'TT' is concerned, must surely be the final word on small transmitting loop efficiencies. There is still room for experimental work on small loops having two or more turns, on fabricating capacitors as well as elements that offer the lowest possible RF ohmic resistance etc.

### MOBILE PHONE RISKS

As one of the very small minority that neither possesses nor drives any form of motor vehicle, I hesitate to return to the controversy surrounding the use of mobile phones as recently debated in *RadCom* by David Taylor, G4EBT (June 2005), and Mike Grierson, G3TSO (August 2005). However, there is a duty to readers to report, if only briefly, the latest authoritative research findings brought to my notice by Dr G L Manning, G4GLM.

These are published in a four-page feature article 'Don't Phone and Drive – it Quadruples Your Risk of Crashing', by Suzanne P McEvoy (leading a team of six) in *BMJ* [British Medical Journal], 20 – 27 August 2005. Those involved represent: The George Institute for International Health, University of Sydney, Australia; the Insurance Institute for Highway Safety, Arlington, VA, USA; and the Injury Research Centre, University of Western Australia.

The article describes in detail the methodology used and results based on interviews with 456 drivers aged over 17 years who owned or used mobile phones and had been involved in road crashes necessitating hospital attendance between April 2002 and July 2004, plus company records of mobile phone use.

A summary of results indicates that "The driver's use of a mobile phone up to 10 minutes before a crash was associated with a four-fold increased likelihood of crashing (odds ratio 4.1, 95% confidence interval 2.2 to 7.7, P < 0.001). Risk was caused irrespective of whether or not a hands-free device was used (hands-free – 3.8, 1.8 to 8.0, P < 0.001); hand-held – 4.9, 1.6 to 15.5, P < 0.003). Increased risk was similar in men and women, and drivers aged over or under 30 years. A third (n - 21) of calls before crashes and on trips during the previous week were reportedly on hand-held phones. The conclusions: When drivers use a mobile phone there is an increased likelihood of a crash resulting in injury. Using a hands-free phone is not any safer".

The introduction notes that, because of concerns about risks of potential crashes, use of hand-held phones while driving is illegal in most countries in the European Union, all Australian states, and parts of Canada and the United States. "Most research on the safety of drivers' use of mobile phones has been experimental in design, involving volunteers, and has found that phone use affects reaction time, variability of lane position and speed, following distance, and situational awareness in simulated or instrumented driving tasks. Distractions are associated with conversations using both hands-free and hand-held phones. Studies have also reported effects of physical distraction from handling phones... Important questions remained about whether phone use affects the risk of more serious crashes involving personal injuries and whether the risk differs for hands-free versus hand-

### Table 1
**Measured radiation efficiency and the measured and predicted radiation resistance for the 1m loop antenna.**

| Frequency (MHz) | Radiation Efficiency (%) Measured | Radiation Measured | Resistance in mΩ From Equation* |
|---|---|---|---|
| 3.6 | 0.25 | 0.42 | 0.36 |
| 5.1 | 0.84 | 2.4 | 1.6 |
| 7.0 | 2.3 | 6.0 | 5.7 |
| 10.1 | 18 | 40 | 25 |

*$R_r \approx 20k_0^4 A^2$ Ω, where $A$ is the loop area and $k_0 = 2\pi/\lambda$ where $\lambda$ is the free-space wavelength.

# TECHNICAL TOPICS

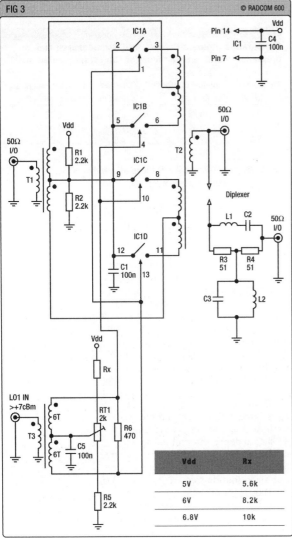

**Fig 3**
The I7SWX two-transformer H-mode mixer as used with a passive transformer squarer. Balance adjustment is done checking for the highest null of the LO signal at the mixer IF output. When using a 1:1:1 (1:4 CT) CT transformer the 470Ω resistor (R6) should be 200 to 220Ω. The basic two-transformer H-Mode mixer circuit was shown and discussed in 'TT' April 2003 and in Technical Topics Scrapbook 2000-2004, p168.

### Table 2
**I7SWX's Measurements on the mixer shown in Fig 3.**

| RF in at 0dBm | 2MHz | 7MHz | 14MHz | 21MHz | 30MHz | 50MHz |
|---|---|---|---|---|---|---|
| Conversion loss dB for downconversion at 9MHz | -5.5 | -5 | -5 | -5.5 | -6 | -7 |
| Conversion loss dB for upconversion (RF + LO) | -5.5 | -6 | -7 | -7.5 | -8 | -15 |
| LO null dB at IF | -53 | -50 | -53 | -56 | -51 | -26 |
| RF in null dB @ IF | -41 | -44 | -30 | -35 | -37 | -35 |

Data shown above are referred to an IF output at 9MHz (IF = LO − RF)
Upconversion data are IFup = RF + LO (eg 14MHz + 23MHz = 34MHz)

held phones. We studied drivers involved in injury crashes in Perth; since 1 July 2001, it has been illegal to use a hand-held phone when driving in Western Australia."

The authors point out that their findings reflect those already reported, providing a list of 11 references to other investigations into the hazards of using mobile phones while driving.

The possibility of another, possibly greater, driving hazard appears to be opened up by the success of tests in Helsinki of Digital Video Broadcast – Handset (DVB-H). The trial, in which 500 people were involved, used modified Nokia 7710 multimedia handsets, showed they could receive sharp, steady Finnish broadcast TV pictures while on the move: see 'Mobile TV Gets Good Reception', by Barry Fox (*New Scientist*, 3 September, 2005, pp22 / 23). Battery life at present limits viewing time to a maximum of around three hours. Clearly, the system, using the digital COFDM broadcast mode, is intended for passengers in cars or public service vehicles, but one can foresee the great temptation for some drivers to take quick looks at the high spots of sports programmes, such as the culminating moments of the recent Test matches between England and Australia, etc.

## PASSIVE SQUARER FOR H-MODE MIXER

GW4HBZ in 'Unnecessarily Good' in the August 'The Last Word', suggested that too much attention has been given to high-performance mixers and oscillator phase-noise. This was on the grounds that he found old receivers perfectly adequate, even on 7MHz. G3RZP responded in the September issue, but there is a further point that I would make. GW4HBZ wrote: "I have operated on 7MHz, unaware of the massive broadcast signals slightly higher in frequency. There is so much noise and interference on that crowded band that any receiver imperfections don't show themselves unless the equipment is particularly poor." GW4HBZ should ask himself why the band appears to have so much noise and interference – and does the noise disappear at those times of the day when the massive broadcast signals are not there? 7MHz is not inherently a noisy part of the spectrum compared to 3.5 or 1.8MHz. As we have stressed, the effects of intermodulation products and transmitted noise sidebands, etc is to produce an artificially-high noise floor and make the band *sound* noisy and full of spurious signals, obliterating weak DX signals.

One of the most important contributions made by 'TT' to high-performance mixers has been the first publication of the various forms of the H-mode mixer as developed originally by Colin Horabin, G3SBI, with subsequent contributions by Gian Moda, I7SWX (F5GVU). The H-mode mixer with fast-bus switches, together with G3SBI's development of the twin-tank low-phase-noise oscillator, truly represents the state-of-the-art. Both G3SBI and I7SWX continue to develop variations. I7SWX has also described, in 'TT' and elsewhere, simpler versions that can be incorporated in existing transceivers.

To summarise his most recent contribution: "During a 2004 visit to my friend Bill Carver, W7AAZ, I learned that Paul Kiciak, N2PK (of N2PK Vector Network Analyser fame), was working on an idea to simplify the squarer of the H-Mode Mixer using only a transformer in place of the 74AC86 or LVDS squarer to gate the FST3125. As I have still not seen anything published by N2PK on this, one weekend recently I spent a productive five minutes unable to resist checking out the idea of driving directly the FST3125 gates directly from the oscillator using a simple passive component such as a transformer. **Fig 3** shows my final tested solution. As always nothing is perfect and we have to compromise, but the results still provide a lot.

"The squarer is formed by transformer T3 and an adjustable bias circuit, formed by two resistors and a trimpot. T3 secondary is centre-tapped and drives directly the two groups of two FST3125 switches. The switches of the FST3125 do not switch at exactly Vdd/2 but somewhere around +1.5V. The use of adjustable bias helps to find the most suitable voltage for both groups of gates. I used a 4T:6T:6T transformer on a balun ferrite core No 43 (2402) to be sure that there is enough drive for the gates. A minimum of around +7dBm seems valid although I tested it up to +10dBm. A 50Ω local oscillator input termination is provided by the 470Ω loading resistor. Similar results, with a little higher drive, are given by a home-brew transformer with 1:1:1 ratios (three 4-turn windings). Commercial transformers should be better but costly.

"With a spectrum analyser or general coverage receiver, look for the LO signal at the mixer IF output and adjust the bias for the deepest possible null. I used a 1-turn trimmer (RT1) and found the adjustment a little tricky; a multi-turn trimpot might be better, but would require more space and would be more expensive if purchased new.

"**Table 2** shows my findings."

I7SWX has provided further details of his measurement set-up and a number of possible variations to Fig 3, but I hope that these notes provide at least an outline of this recent development. ◆

### CORRECTION
'TT' September, Fig 8. Earth connection to the HC02 IC should be Pin 7 not Pin 2 and there should be a 'chassis connection' to the 0V line.

Pat Hawker, G3VA

37 Dovercourt Road, London SE22 8SS.

# TECHNICAL TOPICS

**Supermodes – chordal hop, grey-line & TEP ♦ New steps in variable-width crystal filters ♦ Keeping the PA working**

### SUPERMODES - CHORDAL HOP, GREY-LINE & TEP

It might seem that, by now, just about everything must be known and understood about the vagaries of HF/VHF long-distance propagation of our radio signals. Yet, as any account of the past 60 years would underline, we continue to be surprised at what can sometimes occur. We tend to forget that classic propagation lore tends to be tied largely to high-power broadcasting and HF point-to-point professional or military links. Professional communicators, unlike radio physicists and radio amateurs, are seldom concerned with anomalous propagation that occurs only fleetingly and rarely and cannot be relied upon for traffic handling. As alternative forms of global communications - satellites, wideband cables, Internet etc - take over the role once occupied by HF point-to-point radio, the physicists research microwave rather than HF propagation. Investigation into HF antennas and anomalous propagation is increasingly being left to radio amateurs.

It seems a pity that so many of the introductory texts and articles in our journals on HF propagation still suggest that HF transmission at distances over 2500 miles depends on multiple hops between the earth and the F2 ionospheric layer. The significant attenuation of one or more ground reflections is largely ignored. Attenuation of reflections from the oceans is much lower, but there is reasonable evidence that most lowish-power amateur DX contacts over paths exceeding about 4000 miles are actually single-hop paths at close to or above the MUF, using what are generically called supermodes.

Chordal hop, long path (LP), grey-line, round-the-world (RTW) echoes and transequatorial propagation (TE or TEP) are all good examples of supermode propagation.

HF RTW echoes were first observed in the late 1920s. During WWII, German scientists developed a form of long-distance direction-finding from a single site by measuring the difference in times of arrival of direct and long-path (LP) signals, noting the consistency of LP signals. Soon after the war, the German radio-physicist and radio-amateur, Hans Albrecht, DJ2JR, operating as VK2AOU in Australia, noted the reliability of signals from Europe. He became convinced that this was due to single-hop transmission, and coined the term 'chordal hop'. This work was noted by the late Les Moxon, G6XN, publicised by 'TT' in the UK, and exploited by Les in connection with his own work on radiation at extremely low vertical angles. G6XN showed how, from suitable locations, reliable contacts could be made with VK when operating portable with powers of around 3W. Broadcaster investigations soon showed that optimum signals in Europe from Australian HF broadcasts were better and more reliable at around 10MHz than on the higher broadcast bands predicted by conventional theory. Again, this was found to be primarily due to single-hop, chordal hop propagation.

It also soon became clear that long-distance contacts could be made at modest power and with modest antennas on the 1.8 and or 3.5MHz bands by utilising the so-called 'grey-line' that exists, if only briefly, between locations situated along the boundary between day and night and where sun-rise and sun-set coincide. Interestingly, it has been shown that, at these times, relatively low-dipoles can sometimes perform as well as or better than vertical antennas designed for low-angle radiation.

In the early post-war period, amateurs soon found that 50MHz DX contacts could be covered in a north-south direction at times when even the extremely high sunspot maximum of 1947 could not readily account for them. These contacts seemed to be largely confined to stations located on opposite sides of the Equator, provided that the stations were neither too far North nor too far South. This soon became known as transequatorial propagation (TE or TEP).

'TT', January 1971, carried the following notes, based on a letter received from Oliver P Ferrell, the then editor of the American *Popular Electronics*, putting the historical significance of TE in perspective: "The first scientific notice of this radio amateur discovery appeared in the British publication, *Nature*, Vol 167, p811 of 19 May 1951, in my letter 'Enhanced Trans-Equatorial Propagation Following Geomagnetic Storms'. Several weeks earlier, I had given a paper at the Washington, DC, meeting of URSI titled 'Very High Frequency Propagation in the Equatorial Region' (abstracted in *Proc IRE*, June 1951, p719). The Washington paper gave the scientific community a chance to comment on this radio amateur discovery and, from that point, you will find a gradual building of intense interest."

'TT' included from Oliver Ferrell's Washington paper a diagram ('TT' January 1971, p30, Fig 8) providing a chart of the original amateur observations on 50MHz transequatorial propagation, based on the path between Buenos Aires and Mexico City (a path length of 4450 miles). It showed the concentration of openings around the equinoxes. The data was obtained from participating radio amateurs in a three-year 50MHz data gathering project, for which Ferrell was project supervisor, subsidised by the US Air Force.

It took a few years for amateurs to convince the 'professionals' that, in TEP, they had discovered something quite remarkable, not readily explainable by conventional theories of radio propagation. Ray Cracknell, G2AHU, has recently reminded me that in the late 1950s, while teaching in Southern Rhodesia (now Zimbabwe) and operating as ZE2JV, about 18° South, he had an article 'Transequatorial Propagation of VHF signals', published in *QST*, December 1959). He writes: "It seems a simple enough article, but a Japanese scientist published a paper suggesting that 28MHz signals to Australia did so in a single hop... At the same time the Russian Sputniks started sending 20 and 40MHz signals around the world. Meanwhile the Americans were preparing for greater advances.

"When my article was published, a senior member of the American project arrived on my doorstep. He sat and talked, inspected my gear in a friendly manner. The next day, I worked Chalky Whiting, ZC4IP (about 35° North) cross-band 50/28MHz as he was not permitted to transmit on 50MHz. We decided to investigate the propagation mode by measuring the time delays, with ZC4IP receiving my 50MHz pulses and then transmitting them back to me on 28MHz, where I displayed the outgoing and incoming signals on a CRT. The results were sent to ARRL

# TECHNICAL TOPICS

and they forwarded them to the US National Academy of Sciences who later wrote: 'Your unselfish co-operation with the Propagation Research Project has contributed significantly to the world's scientific knowledge in the field of VHF ionospheric radio propagation'."

The following years saw rapid progress in the exploitation of TEP. In a single day, ZE2JV worked all parts of the USA, Canada and one station in Central America. *QST*, November & December 1981, published his two-part article, compiled together with Fred Anderson, ZS6PW, and Costas Finneralis, SV1DH, 'The Europe-Asia to Africa VHF Transequatorial Circuit During Solar Cycle 21'. An earlier article (*QST*, December 1959) described the work of F9BG, ZC4IP, ZC4WR, and the St Helena 28MHz beacon, ZD7WR.

Partly as the result of a classic paper by Southworth ('Night-Time Equatorial Propagation at 50MHz' (*J Geophys Res* Vol 65 (1960), pp601 – 6 7), TE began to be investigated by several professional researchers, largely based on observations made by radio amateurs. A Japanese/ Australian study formed part of the IQSY research programme and showed that TE propagation, even in years of low sunspot activity, could extend to above 70MHz. Observations were made between 1965 – 68 in southern Japan on three 500W beacon transmitters (32.8, 48.5 and 72.65MHz) located near Darwin in northern Australia, a path length of 4850km (3000 miles). On 32MHz, TE-mode propagation regularly occurred during a large part of the time (except for a few hours in the mornings) despite low sunspot numbers. On 48 and 72MHz, reception was much less frequent, but signals were heard at good strength on many occasions, mostly evenings/nights (about 2000 – 0200 local time on 48MHz, about 2000 – 2400 local time on 72MHz).

The Japanese believed that the night propagation differed from that noted in the afternoons, showing significantly less fading margins, and followed what became known as 'Equatorial Spread F', often correlating with local Sporadic-E. Equatorial Spread F was well known to amateurs and HF listeners in the tropics as a cause of violent disturbances and distortion on long-distance HF signals:

'TT', February 1972, reported that Dr L F McNamara of the Commonwealth Centre, based on work carried out by the Australian Ionospheric Prediction Service in New Guinea and southern Papua, had thrown further useful light on the occurrence of evening-type TEP on the circuits between Japan/Okinawa to Townsville, Queensland. These observations strongly supported the view, already put forward several times in 'TT', that long-distance TEP paths do not always depend on intermediate ground reflection and are often a special form of chordal hop or supermode: **Fig 1**.

Dr McNamara in 1971 also reviewed the various forms of TEP as, by then, identified. The first, the afternoon type, characterised by steady signals and occurring most frequently between 1700 and 1900 local mean time (LMT) at the equator. The evening type seemed most frequent between 2000 and 2200 with deep flutter fading of the order of 5 to 15Hz, both types most frequent around the equinoxes. The optimum path is symmetrical about and normal to the magnetic equator and about 6000km in length; longer circuits tend to see only the steady afternoon mode, while places with magnetic latitudes around 30° or less, usually see only the evening fluttering mode.

That there are still TEP puzzles remaining to be solved is clear from the first of a two-part article 'Non-Traditional Mechanisms of Transequatorial Propagation: Part 1', by Dave Craig, N3DB, in W3ZZ's 'The World Above 50MHz' column in *QST*, October 2005, pp86 – 88.

**Fig 1**
Showing the difference between the conventional F2 double-hop path between Okinawa (about 23° north) and Townsville, Queensland (about 18° south) and the suggested supermode (chordal hop) path coinciding with range-spreading conditions above the ionospheric sound station at Vanimo (V) at 2100 hours local mean time. (Original source Dr McNamara's 1971 letter to *Nature*)

**Fig 2**
Showing how Sporadic-E refraction ('invisible $E_s$') is thought to form a mechanism to support TEP from stations at higher latitudes. (Source *QST*, October 2005)

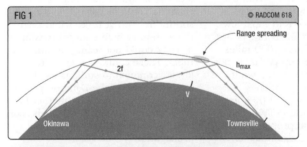

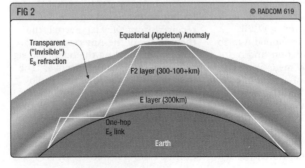

Conventional TEP theory suggests that TEP at high latitudes must include the presence of an intervening propagation mode, or modes. Only stations within about 30° of the geomagnetic equator have no difficulty in tapping into TEP propagation. To quote: "Barring an intervening propagation mode such as Sporadic-E, stations outside this zone cannot participate in these openings because their transmissions arrive at too steep an angle to allow refraction and waves continue through the ionosphere out into space: **Fig 2**."

N3DB is located about 48° North yet, on a significant number of evenings in Autumn 2000 he found, almost by accident while listening for 50MHz TEP, that he could occasionally hear stations in PY, LU and ZP working or calling Central American stations, most often between 0145 – 0245Z (local evening). Signals were fully audible but never strong. Beginning in August 2001 and up to the present, he has made 201 contacts (or beacon reception) on 50MHz that crossed the equator, sometimes when there was no evidence of Sporadic-E or scatter, particularly during the peak years of the solar cycle. Analysis of his contacts has revealed at least six in which TEP openings are apparently unlinked to Sporadic-E events. The most plausible explanation appears to be a chordal hop across the equator – a direct TEP link despite his northerly location. In Part 2 we are promised other reasonable explanations for these unusual TEP contacts.

## NEW STEPS IN VARIABLE-WIDTH CRYSTAL FILTERS

The emergence during recent years of high-performance programmable DSP filters to provide variable bandwidth selectivity has tended to push the classic MF crystal filter further down the scale of concern of many amateurs. The process had already begun with the emergence of the low-cost HF ladder filters based on low-cost crystal or ceramic resonators. There is also competition at MF from mechanical filters and at LF from the 'sliding doors' analogue twin low-pass triple-mix filters as developed by Rohde & Swartz and exploited recently by Dick Rollema, PA0SE (see 'TT' June 2002 and October 2004). Yet there are still possibilities in further development of symmetrical variable-bandwidth MF filters as used in the German wartime E.52 receiver (see the notes

## TECHNICAL TOPICS

by DJ6EV in 'TT', January 2000 pp50 – 57). There has also been the investigation of the often-overlooked Bridged-T crystal-filter by Jack Hardcastle, G3JIR (see 'TT' May 1998, pp58 – 60). All the above references can also be found in volumes of *Technical Topics Scrapbook*.

A substantial packet has come through my letterbox from Horst Steder, DJ6EV, containing, *inter alia* 'The Classic Single-Ended Two-Pole Xtal Filter – a New (?) Method of Bandwidth Control'. To quote his accompanying letter: "If you are surprised to receive this heap of paper, please blame Jack Hardcastle, G3JIR. I have been in contact with him for years on the topic of classic crystal filters, which led to the exchange of very many e-mails and numerous conversations on the 40m band. So I decided to consolidate all the findings and discussions into one paper [2003, updated June 2005] as a reference for both of us.

"Recently, I started to conduct some experiments with the 2-pole filter to verify the equations and simulations in the consolidated paper, and to try out some new ideas of bandwidth control. We both think that some of the experimental results may be of interest to others, although the whole topic is certainly only of nostalgic value [surely not! – *G3VA*].

"However, the method of controlling the bandwidth very elegantly by variation of the quality factor (*Q*) of the terminating LC circuit through positive feedback *(Q-multiplier effect)* has never, to the best of my knowledge, been mentioned in any publication. The possibility of continuous bandwidth control by a DC voltage allows a multiple-stage crystal filter to be implemented using just two FETs per stage. This method could possibly be extended to the MHz range, but there are two limiting factors which would then have to be addressed:
- With AT-cut crystals and their low inductances, the terminating L becomes very small – in the order of 20 to 50nH. This could be addressed with an appropriate capacitive divider in the LC circuit.
- Because the relative bandwidth becomes very small, and the filter is very sensitive to detuning of the LC circuit, the frequency stability of the LC circuit should approach that of an oscillator stage.

"The necessary *Q* of 1500 to 2500 to give a flat top, or with some pass-band ripple for a typical SSB bandwidth, can easily be achieved with a good design of the feedback parameters.

"I have included part 2 of the original 1937 nine-page article on crystal filters with continuously variable bandwidth control [German text] by W Kautter of Telefunken, because most of the references and equations in my basic 24-page (English text) compilation are based on this paper.

"If there is anybody interested in this whole topic, all the papers are available in the Adobe PDF format and can be requested via e-mail [h-g.steder@freenet.de]. Of course all the programs (*DOS* and the new ones for *Windows*) are also available."

There is space here only for some brief notes on DJ6EV's recent experimental work as reported in his eight-page 'Experimental Results with the two-pole Crystal Filter 464.2 kHz'.

**Fig 3** shows the basic filter used to verify the responses for two methods of bandwidth control, using either a 1MΩ log potentiometer as a variable resistor in parallel with the terminating tuned circuit or a 10kΩ log trimpot in series with the inductor. For convenience, both pots were soldered into the basic test filter. The responses for variation of the 10kΩ trimpot are shown in **Fig 4**.

DJ6JV's next step was to investigate the result of increasing the *Q* of the terminating inductor by means of positive feedback. The objective was to achieve flat-top responses at the narrower bandwidths, and to answer the question of whether it is possible to vary the filter bandwidth solely by a variable feedback control.

He modified the basic filter as shown in **Fig 5**. To obtain a narrower 3dB maximum bandwidth (about 2.4kHz instead of about 4.7kHz) without changing the inductor, a capacitive divider (two 325pF capacitors) replaces the single 160pF capacitor. The crystal now looks into an LC circuit with an effective L of about 180μH. Using this capacitive divider technique, it is possible to select any bandwidth less than the maximum defined by the inductor. The required *Q* to maintain the same pass-band ripple of 0.3dB is about 210, much higher than the actual *Q* of the inductance. DL6JV achieves

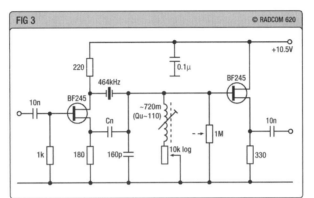

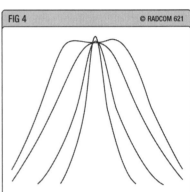

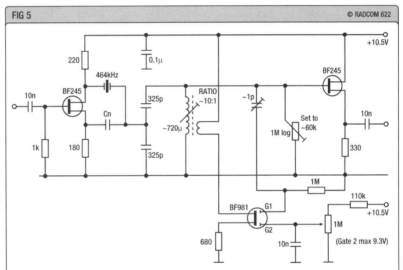

Fig 3
DJ6EV's basic test two-pole crystal filter (462.4kHz) with flat-top -3dB bandwidth of about 4.7kHz, but peaked responses at narrower bandwidths.

Fig 4
Measured filter responses for different values of the 10kΩ log pot. The responses match the simulated (calculated) responses very closely. Note the slight increase of amplitude at the narrowest setting, ie with highest series resistance.

Fig 5
Test oscillator modified to provide positive feedback (Q-multiplier effect). With feedback the maximum -3dB bandwidth is about 3.4kHz.

# TECHNICAL TOPICS

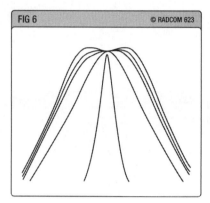

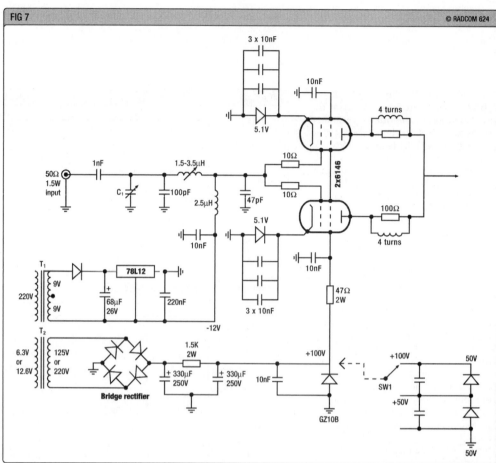

this by introducing some positive feedback (Q-multiplier technique) by means of an additional BF981 dual-gate FET as shown in Fig 5, although he believes the feedback circuit is capable of further refinement.

The turns ratio of about 10:1 is an estimate. The feedback trimmer capacitor (about 1pF) comprises a piece of insulated wire with a thin secondary winding enabling a very fine adjustment by adding or removing turns. The 680Ω resistor in the source of the BF981 was added to 'tame' the device, which has a high transconductance of about 15mS [$S$ = Siemens = $\Omega^{-1}$ – M5ACN].

Bandwidth setting with the 10kΩ series pot is no longer possible, since the bandwidth becomes very narrow even with very low resistance. This needs further investigation.

DJ6JV stresses that no attempt should be made to adjust the filter without a swept-frequency setup: "The interdependence of feedback setting, LC tuning, and load resistance is extremely sensitive and can be handled only with such a setup."

His early results showed: (a) It is possible, with positive feedback, to achieve the required narrow bandwidth with a flat top or even passband ripple. A flat-top SSB bandwidth with an initial inductance Q of <80 with the proper feedback level; (b) with a fixed feedback setting, bandwidth control can be effected with a variable load resistor parallel to the LC circuit, but the method using a series resistor does not work satisfactorily; (c) *It is possible to control bandwidth by varying the feedback!*

To achieve this requires a very delicate balance of setting the load resistance such that, at maximum feedback, the maximum bandwidth with the desired pass-band ripple is achieved while, at minimum setting, the desired minimum bandwidth is reached with tolerable increase of insertion loss.

A variation in bandwidth of nearly 10:1 requires a feedback-gain change of at least 40dB. A small, but noticeable, distortion of the response curve top (loss) in the intermediate settings is seen.

**Fig 6** shows the responses of his experimental 464kHz at various feedback settings.

### KEEPING THE PA WORKING

One of the continuing advantages of valve linear amplifiers, particularly the home-brew variety, is the relative ease of fault-finding and then getting them back into action. There is, however, one problem that can be time-consuming, if not impossible, to overcome – the present difficulty of locating a local source of high-voltage, high-wattage components. It is, however, often possible to overcome such problems with a little ingenuity and a firm grasp of valve-amplifier technology, as shown by Jorge Dorvier, EA4EO.

His 14MHz linear PA using two 6146 valves in parallel class-AB1 failed. He soon traced the fault to a failed grid-bias potentiometer (22kΩ 5W). Unfortunately, even in Madrid, a suitable replacement component was hard to come by. As in the UK and elsewhere, the majority of retailers who used to supply a range of components are now in the business of selling PCs and computer accessories.

EA4JO set about repairing his linear by devising a modified biasing arrangement using parts that are relatively easy to find, yet would give him the same biasing as the original classic -50V to the control grids to achieve a linear 100W PEP output.

His first thought was to have only two 5V Zener diodes in the cathode line, as shown in 'TT' August 2000, pp53 / 54. However, because his home-built amplifier used a special type of tuned-grid input (see his description in 'TT' September 2002, pp61 / 62) and his desire to maintain the same screen supply rather than the popular G2DAF system with four switching diodes, a double-bias system was adopted, adding a -12V to the grids using a cheap 78L12 IC regulator.

With this arrangement, the 6146s were operating in the AB1 linear region but, with the previous classic 195V on the screen, the 6146 anodes run dangerously red. The 195V supply used two GZ10B power Zener diodes in series. This gave the option of switching between 200V or 100V for full or reduced power for local contacts. A new, lower, screen voltage of 100V was provided using just one GZ10B power Zener diode. This permits the 6146 valves to run without problem. If the reduced-power option of the original is required, it would be better to use two 50V power Zener diodes in series, with similar switching. Both arrangements are shown in **Fig 7**. ♦

**Fig 6** DJ6EV's simulation of the CRT responses of the modified filter responses with bandwidth control by varying the positive feedback. Note the near flat-top responses at all bandwidth settings except the minimum.

**Fig 7** How EA4EO has modified the bias circuit of his 100W linear amplifier to overcome the difficulty of replacing a 5W resistor. C1 is a 420pF variable capacitor from transistor broadcast receiver; T1 a miniature transformer providing 300mA or less; T2 an inverted filament transformer. Optional SW1 power reduction switch for local contacts.

# Technical Topics

*More about software-defined radios, and tracing unwanted signals in SMD boards ♦ The facts about non-traditional TEP unfold…*

### SOFTWARE-DEFINED RADIOS
**– PROS & CONS.** It is clear that industry is becoming increasingly interested in the question of the future role of software-defined radio (SDR). At present, the pace is still being set by the large involvement of the military in the USA. According to a feature article by David March 'Software-Defined Radio Breaks Cover', in *EDN-Europe* (March 2005, pp19 – 22, 24, 26, 28), the US military has earmarked as much as $25 billion for SDR development through the US *Joint Tactical Radio System* initiative. The US Department of Defense is working with agencies in Canada, Japan, Sweden and the UK to foster SDR development.

As an editorial note explains: "Software-defined radio has long been the Holy Grail of communications engineers working mainly with military projects. Although all-digital implementations are still some way off, hybrid architectures now appearing may pave the way for strategic development within the commercial field". The FCC defines SDR simply as "In a software-defined radio, functions that were formerly carried out solely in hardware, such as the generation of the transmitted signal and the tuning and detection of the received signal, are performed by software that controls high-speed signal processors". The SDR Forum defines an SDR device as one that functions independently of carrier frequencies and can operate within a range of transmission-protocol environments. The prime attraction of SDR to the military is clearly its ability to cope with current and future digital and analogue transmission modes on a wide range of frequencies without requiring any change in the basic hardware. **Fig 1** provides a block schematic of an ideal software-defined two-way radio intended to carry out carrier-wave data conversion and processing entirely in the digital domain, other than the initial low-noise amplifier for VHF / UHF. Currently, however, it is still impracticable to carry out the initial analogue-to-digital conversion (ADC) directly on an HF or VHF signal, with the result that, at present, it is necessary to incorporate a traditional RF front-end to convert the signal to a low- or zero-IF (direct-conversion) with an image-rejection (I, Q) mixer.

The thrust towards SDR is thus primarily the capability of being able to handle virtually any transmission protocol, combined with ease of manufacture, at relatively low-cost with performance in terms of sensitivity and dynamic range still governed largely by the RF engineering of the analogue front-end. Not all potential commercial users are convinced that SDR is likely to be widely adopted beyond the military field in the near future. A news item 'SDR is Good for the Military But Not Commercial Mobile Phones' (*Electronics World*, August 2005, p8), quotes Hans Otto Scheck, principal engineer for Nokia Networks, as saying: "We are in the second cycle of industry hype in SDR. When all the fog disappears, we will get an improvement. But SDR is part of a trend, not a revolution. It is not a disruptive technology, as some people like to think, but it's an evolutionary one, and we have to take things step by step."

So what about amateur radio? I have already ('TT' July 2005) expressed the view that "for the computer-cum-radio enthusiast, [SDR] opens up interesting new possibilities, although stand-alone HF transceivers with front-panel controls are likely to be still with us for the foreseeable future."

My comments on user-experience of the FlexRadio SDR-1000 have been considered controversial and I do not wish to add to them. An entirely non-critical description of the early SDR-1000 was given in G3RJV's 'QRP' column in *RadCom*. September, 2003, p83, and seemingly induced a number of RSGB members to purchase the original kit. The RF extension board was added later, and has apparently been recently modified. The majority seems to have been happy with its purchase, but I still feel that my *caveat emptor* was a justified service to readers.

Surprisingly, *QST* (October 2005) has published a second, very long 'Product Review' of the SDR-1000, that no longer includes the various *caveats* that appeared in its first Product Review in the April 2005 issue, summarised in the July 'TT' and mentioned in the November 'TT'. However, an unsolicited letter from Paul Widger, G0HNW, provides a further balanced report on user-experience. He writes: "In my view, your comments are fully justified and not by any means baseless. I have the basic SDR-1000 kit of three boards. The addition of the RF expansion board would reduce, but not eliminate, some of the effects I have noted. It makes me wonder if those praising the unit have much 'RF experience'.

"*Bad points:* (1) Large number of DDS-related 'birdies', confirmed as not due to my computer or external signals. Reduced but by no means eliminated by extra decoupling on the boards, the method of construction does not help in this respect. (2) Other spurii due to use of switching regulators – cured with a linear power source.

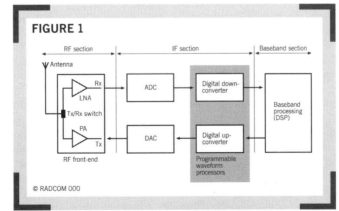

FIG 1: THE FUTURE IDEAL SOFTWARE-DEFINED RADIO WILL PERFORM ANALOGUE / DIGITAL CONVERSION DIRECTLY AT THE SIGNAL FREQUENCY AND THEN OPERATE ON SIGNALS EXCLUSIVELY IN THE DIGITAL DOMAIN. THIS IS STILL SOME WAY OFF, WITH THE RESULT THAT THE SIGNALS ARE DOWN-CONVERTED TO A LOW- OR ZERO-IF FREQUENCY BY MEANS OF AN ANALOGUE-TYPE FRONT-END, ALTHOUGH TUNING CONTROL ETC IS UNDER THE CONTROL OF MULTIPLE DSP, FPGA DEVICES OR BY MEANS OF A SOUND CARD IN AN EXTERNAL PC. (SOURCE: *EDN EUROPE*)

## TECHNICAL TOPICS

(3) Severe 'ghost' reception of broadcast and amateur signals due to DDS harmonics, plus very poor RF preselection ahead of Tayloe mixer. (4) Bad breakthrough of FM radio due to DDS harmonic mixing. Reduced, but not eliminated, by using a low-pass filter in the antenna lead and by housing the unit in a heavily-filtered diecast box. It would need this box putting inside another to eliminate the breakthrough, and would allow all the audio and computer cables to be fully filtered. (5) Sensitivity adequate up to 14MHz, then it gradually deteriorates to abysmal at 50MHz. (Note the RF Extension board would no doubt improve this as it has RF gain – if I understand correctly a 26dB gain amplifier with 10dB attenuator on input?). (6) I couldn't get on very well with any of the AGC arrangements in spite of trying the *Handbook* suggestions – it is tiring to listen to a 'blank' frequency – noise spikes upset the AGC much more than any other AGC I have ever heard. (7) Poor frequency stability – akin to the KW2000! – but OK with external oscillator.

"*Good Points:* (1) Quality of demodulated signal very good. (2) Versatile filtering etc. (3) Useful measurement and analogue facilities. (4) Used as a digitalised 'Q5er' gave best results, as the selectivity and modest gain of a properly-designed front-end allowed the gain of the op-amps after the Tayloe mixer to be reduced and thus reduce some of the adverse effects noted above. (5) Very good sideband suppression. (6) It is a tinkerer's delight – keeps you off the air for weeks!"

**TRACING UNWANTED SIGNALS IN SMD BOARDS.** A problem with closely packed boards carrying a mixture of digital and analogue devices is the appearance of odd parasitic-type RF signals at frequencies that bear no obvious relationships to those required by the design. Boards for SDR units represent only one example of where such unwanted signals can occur. Dimitrii Derevensky in 'Watching the Currents Flow', in *EDN Europe*, 1 September 2005, p28, provides a useful miniature RF current probe that he used to trace the source of a 50MHz signal fed into the network from an embedded six-layer DSL modem much stronger than the -40dB noise allowed. To quote briefly and selectively: "Building mixed-signal, quiet boards is an art, and textbook recommendations on layout for signal integrity can fail if you take them literally. We could have been dealing with inductive coupling from the currents flowing in

A REPLICA WHADDON MKVII BUILT BY PETER SIMPSON, G3GGK, IN A GENUINE MKVII METAL CASE, EMULATING THE GENUINE MINIATURE 'SPY-SET' AS CLOSELY AS POSSIBLE, ALTHOUGH USING MODERN RESISTORS AND CAPACITORS AND A BRITISH ARMY MINIATURE KEY WHICH IS NOT SILENT LIKE THE ORIGINAL KEY. THE SET WORKS ON 3.5 AND 7MHZ WITH AN OUTPUT OF ABOUT 4W INTO A 75Ω POWER METER. G3GGK ADDS: "THE RECEIVER IS USABLE ON 3.5MHZ, BUT I HAVE TO AGREE WITH YOUR COMMENT THAT ONE NEEDS SAFE-CRACKER'S FINGERS TO VENTURE ON TO 7MHZ. THE WHOLE PROJECT HAS GIVEN ME IMMENSE SATISFACTION AND I AM NOW COLLECTING PARTS TO BUILD A REPRODUCTION OF THE WHADDON MK33 TO SIT ALONGSIDE MY HRO-5T RECEIVER."

the ground plane, but my suggestion to watch the current in internal layers surprised the team… currents come only in loops… finding or predicting these loops is part of the art of mixed-signal design…

"We made a handy AC-current probe, a Magnepick. It is simply a toothpick with three to four turns of thin wire making a tiny coil on its tip. It picks up the magnetic field parallel to the toothpick. We also connected it to a spectrum analyser (don't forget the 50Ω termination) and, sniffing with it around the board, could – with maximum intensity – map the paths of selected frequency components. We also made an improved version from the head of a floppy drive… the revelation was striking. The [textbook] recommendations to allow one 1μF capacitor per digital IC, and to avoid breaking any planes were our biggest enemies… When we tinkered with bypass capacitors between the pairs – voila! – the noise currents flowing out of the CPU area shrank." [Thanks G3HJK]

**MORE ON NON-TRADITIONAL TEP.** The December 'TT' included an item 'Supermodes – Chordal-Hop, Grey-Line & TEP', outlining some of the ways in which long-distance HF/VHF contacts can be made without intermediate reflections from Earth, sometimes over paths that are apparently well above the classical 'Maximum Usable Frequency' due to rare anomalous short-lived propagation conditions. The item drew attention to Part 1 of a *QST* (October 2005) 'The World Above 50MHz' feature 'Non-Traditional Mechanisms of Trans-Equatorial Propagation', in which Dave Craig, N3DB, reported a number of unexpected 50MHz TEP contacts or beacon reception. These were from his home location which is at a northerly latitude conventionally considered beyond the area of direct TEP unless supported by an intervening propagation mode or modes such as Sporadic-E. In Part 2 *QST*, November 2005, he outlines further possible explanations.

N3DB stresses that ionospheric layers are usually depicted as smooth although, in practice, this may not be the case. There can be travelling clouds and patches of ionised plasma that can produce tilts supporting TEP from sites well beyond the 30° North and South latitudes implied for TEP supported only by Spread-F over the Equator.

My files include an article by Dr Bruno Beckmann (then Head of the Ionospheric Warning Centre at Darmstadt, Germany) 'Can Hams Make Contacts Above the *Classical* MUF?'. This appeared in a 1960s publication of the International Amateur Radio Club and emphasised that the ionosphere does not consist of an ideal and regularly-stratified ionised medium with concentric surfaces of equal ionisation level, but that considerable divergences are observed which are characterised as 'irregularities'. These consist particularly of cloudy structures and of stronger horizontal gradients of ionisation. Two detailed tables were included; the first listed six kinds of irregularity: Cloudy structures in the E-Region (Sporadic-E); cloudy structures in the F-Region (Spread-F); cloudy structures in the D-Region; ionisation pillars along the earth's magnetic lines of force, strengthened by auroras; ionisation in meteor trails; greater slopes and horizontal gradients of layers (tilts).

The second table listed propagation paths that can result from the irregularities noted in the first table. Eleven forms of propagation mode were designated A to L together with the approximate highest frequency; path range; type of irregularity concerned; approximate signal strength, and general remarks.

In effect. Bruno Heckmann showed there exist several long-distance propagation modes above the classical MUF that can be exploited, if infrequently, by amateurs including: (a) scattering by F-region reflections; (b) super-modes of normal propagation (tilts) including TEP; (c) super-modes of the Pedersen ray (up to 7000km (tilts)), including East-West circuits at frequencies rather above the MUF; (d) duct mode due to signals propagating between the F- and E-regions; duct mode in the F-region (chordal-hop). At 50 – 80MHz there can be reflections or scattering from Sporadic-E layers; auroral reflections; and scattering from the F-region.

To revert to the 50MHz TEP work of N3DB, he gives the following advice: "Successful operation at 6m is often a case of catching an *abnormal* occurrence and taking advantage of it. With the exception of widespread events like 30 March, 2003, and *some* evening events during cycle peak years, the propagation I believe to be non-E related is virtually always marginal, at the very edge of practicable usage. Excellent antenna systems with low-loss feed-line and a quiet noise environment are *absolute necessities,* at least on the northern end of such paths. As a new generation of bigger and better antenna systems goes on-line in mid-latitude regions, I suspect that more and more operators will discover that they too can participate in these openings, whatever the propagation mechanism(s) at play."

**HYBRID CASCODE AMPLIFIER.** In the 'Circuit Ideas' feature of *Electronics World,* November 2005, p51, P F Gascoyne contributes an item 'Cascode Buffer Gives Op-Amp 600V Output Swing' that with modifications might well form the basis of a linear 100W amplifier or of a Class-C 200W transmitter amplifier provided the 10kΩ load resistor is replaced by a tank circuit [and output filtering – G3VA], the 10kΩ feedback resistor disconnected and the biasing suitably adjusted for Class-C operation. It could then form the power amplifier for a grounded-grid power amplifier, although P F Gascoyne mentions that this has not been tried. To quote: "This circuit was designed to interface the output of an operational amplifier to the control grid of a high-voltage tetrode but it could be used wherever an output swing of several hundred volts is required." As shown (**Fig 2**), the circuit has a nominal voltage gain of 100. The transmitting valve type TT100 is no longer available, but the basic principle could be applied to other tetrode power valves. The item carries an editorial warning "Circuits such as this one should only be built and used by those proficient in the safe handling and operation of high-voltage devices."

**LEAD-FREE SOLDER AGAIN.** The October notes on lead-free solder ended on a rather pessimistic note. It seems worth reporting the somewhat more hopeful tone expressed in the October issue of *Television.* A letter from Michael Bennett points out that there appear to be significant exemptions from the European directive covering the use of lead in electronic equipment. He writes, "Here are a few points that might be of interest: Article 2,

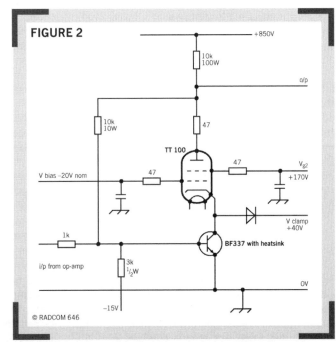

FIG 2: CASCODE AMPLIFIER PROVIDING A 600V OUTPUT SWING FROM A LOW-LEVEL OP-AMP DRIVE. COULD PROBABLY BE MODIFIED TO PROVIDE A LINEAR OR CLASS-C TRANSMITTER PA. (SOURCE: *ELECTRONICS WORLD*)

Paragraph 3: 'This directive does not apply to spares for the repair, or to the reuse, of electrical and electronic equipment put on the market before 1 July 2006'. Under the list of exemptions, one of them exempts 'lead in solders for networking infrastructure, equipment for switching, signalling, transmission, as well as network management for telecommunications'. The *Television* correspondent also states that his company has been told that components for military and medical equipment are exempt. This seems to suggest that there is a known reliability problem with joints made using lead-free solder. It appears to be mainly the consumer market that is being forced to use lead-free solder."

Another correspondent, Michael Maurice, in the November issue writes: "From what I hear, products have become less reliable as a result of the use of lead-free solder. Soldered joints are now failing in sections of [TV] sets where we never had this trouble before… Lead-free solder does produce unreliable joints. That is why its use is prohibited in the production and servicing of equipment in the medical and aviation fields…"

I do not know whether amateur radio equipment is classified as consumer equipment or is entitled to exemption from the directive, but it is clear that lead-tin solder will not disappear but continue to be used, at least in some fields.

In the October issue of *Television,* Adrian Gardiner in 'Bench Notes' reports his findings of testing three types of the lead-free solder now available, showing that there is considerable difference in how these perform. The differences seem to relate directly to the flux. All three have the same basic composition, 90.7% tin and 0.3% copper.

Type X39 no-clean flux has only a tiny flux content and in use was found to produce the typical dull appearance of lead-free joints made by manufacturers. Although the results with X39 were 'OK' when re-making lead-free joints, results were very poor when mixing into a leaded joint – as a result, the joint has to be cleaned of old solder before being remade. Type 502 no-clean flux with a mildly-activated flux was again

# TECHNICAL TOPICS

found to be OK for lead-free reworking, but unsuitable for remaking leaded joints. "A significant factor is a nasty smell that is given off in use, so bad that I would not be able to use this product on a daily basis."

His third type was "Resin-flux solder" with five cores of halide activated-resin flux: "I was surprised by the solder's behaviour in use and had to check that I had in fact ordered a lead-free product. The wetting characteristics are much better than those of the other two varieties tried and it flows easily into the joint. I had no problems when using it to remake lead-free joints but, more importantly, it mixes well with leaded solder used in older equipment. The only slight disadvantage is that a certain amount of flux is left after rework. This should ideally be cleaned off. Overall this solder gets 10 out of 10 for ease of use… A 1.2mm, 500g reel can be ordered using CPC code SD00521."

Adrian Gardiner also reports: "Lead-free solder requires a higher melting point than the traditional type. In practice, I have found that a working temperature of 360°C (10° higher than I would use with leaded solder) works well for most applications. I increase this to 410°C when working with large areas of copper… These temperatures ensure rapid melting of the solder so that the process can be completed quickly."

### THE OLD ORDER CHANGETH

In this increasingly digital world, the role of HF radio (broadcast, communications (amateur and professional), and television retailing and servicing are all changing at what, to some of us, appears to be an ever quickening pace. Inevitably, this is affecting the electronics publishing industry.

The latest manifestation of this process is the disappearance of *Short Wave Magazine*

AUSTIN FORSYTH, G6FO (THIRD FROM LEFT) LONG-SERVING EDITOR OF *SHORT WAVE MAGAZINE* WITH OTHER TECHNICAL JOURNALISTS AT ONE OF THE RSGB AMATEUR RADIO EXHIBITIONS IN THE 1960S. LEFT TO RIGHT: PAT HAWKER, G3VA – COMMUNICATIONS EDITOR *ELECTRONICS WEEKLY*; JOHN ROUSE, G2AHU – EDITOR OF THE *RSGB BULLETIN*; G6FO; JOHN WILSON, G3BGP – EDITOR OF *ELECTRONICS WEEKLY*; JOHN CLARRICOATS, G6CL – RSGB GENERAL SECRETARY AND FORMER EDITOR OF THE *RSGB BULLETIN*; AND F L DEVEREUX, FORMER 5FA, EDITOR OF *WIRELESS WORLD*. MOST NOW SILENT KEYS.

(*PWM*) now being merged with *Radio Active* to form *RadioUser*. For many of us, *SWM* in its heyday as a purely amateur radio publication, edited by Austin Forsyth, G6FO, evokes nostalgic and fond memories. In my case, it was the first commercial publication to carry in 1938 a schoolboy article I wrote and for which G6FO paid me the princely sum of 7s 6d (37.5p). G6FO was not the first editor of *SWM*, it had been launched in 1937 under the editorship of Basil Wardman, G5GQ, who, for professional reasons, soon handed it over. G6FO remained editor until his death in 1977, apart from the wartime break when he served in the RAF, as did his assistant editor, 2CUB, who became a wartime fatality.

Recent years have also seen changes in ownership of *Electronics World* (formerly *Wireless World*) and *Television and Consumer Electronics* (one-time *Practical Television*), both now published by Nexus Media Communications. A new in-house editor, Boris Sedacca, has been appointed for *Television*. He succeeds John Reddihough, who has been associated with the magazine for some 38 years and who, more than 50 years ago, was my close colleague in Newnes Technical Books working on such books as the many volumes of *Radio and Television Servicing* and *Radio & Television Engineers' Reference Book*.

It should not pass unnoticed that the death during 2005 of Ray Herbert, G2KU, has seemingly meant the closing of his *Baird Newletter*. Ray was a prolific writer on the Baird Company, for which he worked as a young engineer in the 1930s, participating in Baird's pioneering work for the French on airborne television using the Intermediate Film system. Ray contributed to many magazines including *RadCom* and, as the foremost authority on the Baird Company, was consulted by many authors, television and radio producers, researchers etc, often taking part in their programmes. During recent years he reactivated the original Baird call G2TV. He was a wartime Voluntary Interceptor (VI) for the Radio Security Service (Box 25). I first met him while he was Chairman of the EMC Committee of the British Radio Equipment Manufacturers Association.

Another old-timer who became a 2005 silent key was John Piggot, G2PT, whose interest in amateur radio dated back to the era of bright-emitter valves. As a professional engineer, he worked at the Post Office Research Station at Dollis Hill, a wartime member of Eric Flower's team that developed the Colossus code-breaking computer. He later worked on the new wideband ocean cables, becoming a key player in the design of ship-borne cable-laying equipment. In his submission for his pre-war licence (an era when one had to provide technical reasons for requiring a radiating licence) he proposed experimental work to establish whether or not the lunar cycle had any effect on the propagation of radio waves. [*Thanks G4HMC*] I still recall an excellent article on the Morse code, Morse keys etc he wrote in the 1950s for the *RSGB Bulletin*.

LF aficionados should also mourn the passing during 2005 of Kenneth Budden, FRS, for long associated with the Cavendish Laboratory, Cambridge. In 1966, he was elected a Fellow of the Royal Society in recognition of his experimental and theoretical work on the propagation of very long radio waves. His research showed *inter alia* that very long waves respond in a characteristic way to solar flares etc that produce the sudden ionospheric disturbances of a different kind at HF. He was one of the first to use the power of the EDSAC 1 computer to advance radio physics. [*Thanks G8MOB*]

# TECHNICAL TOPICS

PAT HAWKER
37 DOVERCOURT ROAD,
DULWICH, LONDON SE22 8SS.

G3VA

# Technical Topics

*Automotive electronics & mobile QRO ♦ Small loops, NEC, capacitor & ground losses ♦ Electrically-small self-resonant antennas ♦ Throwaway or serviceable? ♦ Optimum radial earth systems*

**AUTOMOTIVE ELECTRONICS & MOBILE QRO.** To quote an editorial in *Electronics World* (December 2006, p3): "One area that has been quietly chugging along in the background – without too much fuss, as in wireless communications, and without the glitter of consumer electronics – is automotive electronics. Even though the growth rate of new cars produced is almost flat, the electronics content in them is exploding."

*New Electronics* (13 September, 2005, p31) puts it thus: "Today's cars typically contain more than 100kg of electronics, close to 2km of wiring and 50 or more embedded processors. To operate these processors, cars contain more than one million lines of computer code per vehicle".

The danger of RFI to microprocessor-controlled automobiles, avionics and industrial machines has been recognised since the 1980s, and EMC and safety recommendations have been tightened up. But most airlines still prohibit the use by passengers of lap-top computers for fear of interference with aircraft navigational and control systems.

EMC regulations normally define a limit to the protection that must be provided based on the strongest signal levels likely to be encountered plus a further safety margin. As amateurs, we now appear to be increasingly contemplating the use of high-power transmitters in vehicles, well above the powers of other civilian mobile transmitters. I wonder if we can be sure that these pose no potential RFI (or exposure) hazards? In the December, 2005, *RadCom*, Peter Hart, G3SJX, reviewed two mobile HF linear amplifiers capable of providing some 500W output. In the States, 1kW mobiles are not unknown. Is there a paradox in these trends? Are all automotive electronics RFI-proof to the extent that they are unaffected by high-power transmitters with their antennas mounted directly on the vehicles? I wonder if any reader can answer the question with conviction?

ARRL's recently published *Amateur Radio on the Move* (reviewed in *RadCom* October, 2005, p40) devotes several pages to 'Interference From Automobile Systems to Amateur Radio', (pp1.40 – 1.44) but dismisses 'Interference From Amateur Radio to Automobile Systems' with the sub-heading 'Immunity is Ensured by OEMs' and a single paragraph (p1.44): "Just as vehicle electronic systems may interfere with mobile receivers, mobile transmitters may also affect the vehicle electronics if appropriate immunity measures are not in place. Electronic modules that radiate energy may also receive energy. This energy may come from on-board transmitters, nearby radio and TV stations, or any other device or event that generates an electric or magnetic field. Manufacturers use proven design techniques and run extensive tests to ensure RF immunity in their vehicles." So there is no need to worry? Can amateurs be sure that running 500W or so will never affect their own or a passing car that may have less-immune automotive electronics?

In the early days of automotive electronics, a 'TT' item 'Vehicle EMC Affects Reliability and Safety' (May 1988) reported an IEE colloquium on 'vehicle electromagnetic compatibility'. This noted: "Two trends have transformed vehicle EMC from being largely a question of suppressing ignition and accessory interference to in-car entertainment systems to one with important safety and reliability overtones: (a) the proliferation of car electronics systems based on CMOS microprocessors; and (b) the increasing use of 'composites' such as glass-reinforced plastics (GRP or fibreglass) rather than metals in vehicle bodies, with RFI/EMI and ESD (electrostatic discharge) now recognised as 'probably the most effective killers of electronic modules in cars'. While car manufacturers have begun to take seriously the problem of immunity against strong RF fields, including those from nearby radio, television, radar and carphones, it should not be forgotten that few manufacturers contemplate the possibility that a radio amateur may wish to install a high-power transmitter in the vehicle."

In 1988, only a few manufacturers such as BMW, Jaguar and Opel had models using up to about 10 microprocessors, a far cry from the 50 or so now reported by *New Electronics* as being fitted by many manufacturers together with almost 2km of wiring. The magazine claims: "Vehicles are increasingly controlled by complex electronics containing large numbers of 8-, 16- or 32-bit microcontrollers. High-end vehicles such as the BMW7 Series and Mercedes S-Class – use 70 to 100 microcontrollers each, and all new vehicles have upwards of 20 processors. Increasingly, automotive electronics are made up of subsystems connected through a shared, 'safety critical' communication network, separated by gateway controllers."

Interference could presumably compromise safety. And while undoubtedly much has been learned about screening and filtering etc, and more stringent EMC regulations introduced, can we be sure that these will always be effective against high-power mobile transmitters with antennas mounted directly on the (possibly glassfibre) vehicle? Even if your own vehicle is RFI-proof to this level, there could remain the possibility of affecting passing vehicles some of which may have been manufactured in the early days of automotive electronics.

**SMALL LOOPS, NEC, CAPACITOR & GROUND LOSSES.** What hope is there of ever writing a "last, last word" on any engineering controversy? Mike Underwood, G3LHZ, appeals for fairness and balance, and believes that his critics are beset by dogma. He tells me: " I do not expect you to change your mind. Everyone will have to choose between the Chu-Wheeler criterion and the First Law of Thermodynamics. There is no other choice in this case. You have clearly made your choice. Do not expect everyone else to follow you lead… I do not expect you to change your mind. That is what is so sad about dogma. …." G3LHZ puts his views in 'The Last Word' (December 2005).

One of G3LHZ's critics is Jack Belrose, VE2CV, VY9CRC. He has been concerned during the past 15 years with trying to clarify controversial views about unusual antenna systems, about the fundamentals of electrically-small antennas, and with power transfer to antenna systems (CFA and EH antennas, electrically-small loops, and conjugate match). He now thinks that "time spent [in such debates] can sometimes be a waste of time. Debates and rebuttals can go around-and-around,

"An interesting book detailing how scientists and engineers think is *The World Treasury of Physics, Astronomy and Mathematics*, edited by Timothy Ferris (Little Brown and Company, 1991). A quotation that caught my attention: "Isaac Asimov in *The Nature of Science* (p783) has written 'Scientists (and engineers) share with all human beings the great and inalienable privilege of being, on occasion wrong, of being egregiously – sometimes, even monumentally – wrong. What is worse still, they are sometimes purposely and persistently wrongheaded. And since that is true, science can be wrong in this respect or that."

It is not for me to say which side of the small loop argument is being put perhaps perversely and persistently wrongheaded. Who can be sure that the fundamentals of electromagnetic or even quantum theory may not be wrong? An article in *The Guardian* (November, 2005, p9) reports that Randall Mills, a Harvard University medic claims to have built a prototype power source that generates up to 1000 times more heat than conventional fuel. Independent scientists claim to have verified the experiments... Dr Mills says that his company, Blacklight Power, has tens of millions of dollars in investment lined up to bring the idea to market and claims to be just months away from unveiling his creation. The problem is that, according to the rules of quantum mechanics, the idea is theoretically impossible... A recent economic forecast calculates that hydrino energy would cost around 1.2 (US) cents per kilowatt-hour compared to 5 cents per kWh for coal and 6 cents for nuclear energy". *The Guardian* explains that hydrinos are theoretical, and to many scientists impossible, particles made by forcing hydrogen atoms to shrink.

The full-page article ends: "If it's wrong it will be proven wrong, said Kert Davies, research director of Greenpeace USA. But if it's right, it is so important that all else falls away. It has the potential to solve our dependence on oil. Our stance is of cautious optimism."

By comparison, G3LHZ's challenge to Chu-Wheeler and *NEC* applied to small loops must seem relatively small beer. From his letter in 'The Last Word' (December 2005) it is clear that he sticks to his radical views and dismisses G3NOQ's *IEEE Ant & Prop Magazine* paper (digested rather inadequately in 'TT'. November 2004) that I had hoped would be the final word.

It is true that some of those who strongly oppose G3LHZ's views are unhappy at the apparent differences between G3NOQ's *NEC*-predicted and his measured results. As explained below, these differences are far smaller than might be gathered from Fig 1 of the November 'TT'. VE2CV suggests that the prime cause for the relatively small differences are the losses incurred by G3NOQ's use of a broadcast-type tuning capacitor (combined with a fixed capacitor at lower frequencies) rather than an ideal vacuum variable capacitor free of the resistance associated with a sliding contact to a rotor. He points out that his own measurements, showing close affinity with *NEC-4* predictions, were made on AMA loops which overcome the contact resistance by employing a specially-fabricated tuning capacitor, a disc welded to the rotor, and a spring-loaded wiper to make contact with this disc. He believes that capacitor resistance plays a large part in the total series resistance of the loop, with *NEC-4* taking ground-induced resistance into account.

Dave Gordon-Smith, G3UUR, another confirmed critic of G3LHZ's efficiency claims, was also decidedly uneasy at the large discrepancies (between 12 and 20dB) between the figures given in Table 1 (p75) and Fig 2, (p74) of my digest of G3NOQ's paper and his endorsement of *NEC* for small loops. The answer, I must confess, was almost entirely due to my over-simplified digest of G3NOQ's paper and this may have confused other readers. What I should have made clear was that the 'predicted' curves of Fig 2 were calculated for a theoretical antenna radiation efficiency of 100% and that the large discrepancy between 'predicted' and 'measured' curves form a striking endorsement of what the debate has been all about, ie the relatively low radiation efficiency of small loops, The wide differences between the predicted and measured curves represent, in effect, a practical confirmation of the relatively low radiation efficiency of small transmitting loops.

G3UUR was clearly misled by my digest and wrote "These discrepancies are extremely large and have to be explained. [See above – *G3VA*.] I know that VE2CV has absolute faith in *NEC-4D* simulating ground loss accurately. But having seen two sets of comparative data from small, low loop experiments, my scepticism about its ability to handle real earth situations at such low loop heights is increasing, not diminishing!"

G3UUR adds (abridged in light of my failure to explain the basis of Fig 1): "I would be interested to know if either VE2CV or G3NOQ monitored the ground moisture content during their experiments. All [the professional experts] need to do to clear this up is to do some experiments with small loops at various heights (I would suggest 0.5, 1.5 and 3m in the first instance). The moisture content of the ground beneath the loop can have an enormous influence on the conductivity, and hence ground loss. But whatever the outcome of my comments and queries, the G3LHZ small transmitting loop has been shown to have no better efficiency than a conventionally-fed, small loop."

As a non-user of *NEC* programs, my own deep scepticism is confined to G3LHZ's claim of 90%-plus radiation efficiency of small loops at the low-frequency end of their range! Clearly both capacitor and induced ground losses contribute to the loss of efficiency, although the major cause stems from the extremely low radiation resistance. No matter what G3LHZ may say about the First Law of Thermodynamics – an argument that falls by the way if, as G3UUR has suggested previously, there are serious flaws in the way in which G3LHZ measures his RF power loss (see also G3UUR's letter in 'The Last Word', January 2006).

G3UUR has provided details of how the influence of induced ground loss can be checked by amateurs on their own installations. He notes that many of the loops used on 7, 3.5 and 1.8MHz are larger than those investigated by VE2CV and G3NOQ and that, at any given height, the induced ground loss will be lower. G3UUR writes (abridged): "Plotting the bandwidth between the 2:1 VSWR points at several different heights can identify the lowest

## TECHNICAL TOPICS

height at which the loop can be operated with negligible induced ground loss [**Fig 1** shows the relationship between bandwidth (ie Q) and height]... A halving of the loop bandwidth indicates a 3dB improvement in efficiency and hence radiation. Those using 1m diameter loops for transmitting on 10MHz and above, should definitely check out their loops. It's obvious from G3NOQ's experimental results that the magnetic induction ground losses can be considerable with such a small diameter loop at these frequencies, even at a height of 1.5m over certain types of soil. I've been trying [unsuccessfully] to convince VE2CV and G3NOQ that there is a ground loss problem with low, small loops. It was unfortunate that my failure to explain adequately Fig 2 in the November 'TT' added to G3UUR's belief that *NEC-4D* does not take ground losses fully into account. Seasonal variations in moisture content could, however, affect losses.

**ELECTRICALLY-SMALL SELF-RESONANT ANTENNAS.** For those who wish to delve further into the performance of electrically-small wire antennas (but not small loops), a long 15-page paper 'A Discussion on the Properties of Electrically-Small Self-Resonant Wire Antennas', by Dr Steven R Best (formerly President and earlier Director of Engineering of Cushcraft Corporation) appears in *IEEE Antennas and Propagation Magazine*, Vol 46, No 6, December 2004. The paper covers the normal-mode helix, the meander-line antenna and several arbitrarily shaped monopole wire antennas. Examples are based on UHF (408MHz) antennas but the results are generally applicable from MF to UHF.

There is space here only to quote briefly. The Abstract reads: "The performance properties of several electrically-small, self-resonant antennas are compared as a function of their total wire length, geometry and effective volume. The radiation properties considered include resonant frequency, radiation resistance, and quality factor (Q). It is shown that the resonant properties of these antennas are directly a function of the antenna's effective height and effective volume, which are established by both total

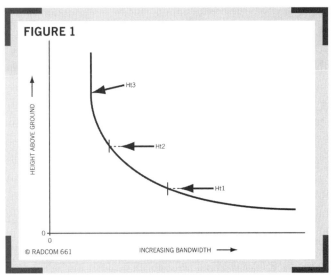

FIG 1: BANDWIDTH OF SMALL LOOP EXPERIENCING GROUND LOSS VERSUS HEIGHT ABOVE GROUND. ALL BANDWIDTHS MEASURED AT THE SAME VSWR (2.0:1 IS SUGGESTED BY G3UUR, BUT OTHERS CAN BE USED)

wire length and geometry. When the total wire length and geometry of these antennas are configured such that the antennas exhibit the same effective height and volume, their resonant properties are essentially identical, independent of any difference in their total wire length and geometry. Both computed and measured data are presented to support the comparison of the antennas' resonant properties."

The paper does not present 'conclusions' but has a useful concluding 'discussion' section that provides useful general guidance to 'vertical' antennas that depend on self-resonance (quarter-wave) rather than capacitance or inductive loading. To quote selectively: "In the previous sections, the radiation properties of several electrically-small, self-resonant wire antennas were considered. In each case the total wire lengths and geometries of the antennas were adjusted to make all of the antennas self-resonant at the same frequency. The resonant radiation properties of these antennas were essentially the same, independent of the significant differences in their physical properties. Optimisation of the radiation resistance and quality factor were not, significantly, functions of antenna's geometry or total wire length. It is evident that improving the antenna's performance properties requires that the antenna's effective height and volume be increased to the greatest extent possible... It is evident that the narrower-diameter antenna structures require significantly more total wire length to achieve resonance than their wider-diameter counterparts. To achieve self-resonance with the least amount of wire, the wire arrangement should be as open as possible, and should occupy the greatest possible physical volume... Optimising the antenna's geometry becomes a function of how much total wire length is required to achieve self-resonance. Geometries requiring less total wire length to achieve resonance will be more efficient, as the total loss in the structure will be minimised. These general conclusions apply to the simple linear antenna structures. It was shown that other self-resonant antennas, such as top-loaded monopoles and folded monopoles, exhibit different resonant performance characteristics as a function of antenna height. Additionally, the configurations were shown to exhibit improved performance characteristics. This indicates that the performance-optimisation process requires an evaluation of numerous antenna types."

**THROWAWAY OR SERVICEABLE?** The days when it was common practice for most amateurs to repair or at least bodge their own equipment faults are long gone, at least for the large majority using equipment manufactured within the past decade or so. The first sign of difficulties to come began with the change from hard wiring to printed circuit boards, then integrated circuitry and the end for many of us came with surface mounted devices and multi-layer boards.

Fortunately, modern equipment is reasonably reliable, although the problems of dry soldered joints and vulnerability to lightning or accidental abuse, such as the spilling of coffee etc, remain. It has become the practice to return faulty equipment to the original retailer who can usually be expected to offer after-sales service, at least for products for which the firm is an accredited agent. There has also be the growth of small firms or individuals offering specialist repair of communications equipment. The downside is that outside of warranty any repair work can be expensive and involve quite lengthy periods with the equipment out of use.

However, with SMD, multilayer boards,

microscopic components etc, the situation is again deteriorating. The latest circuit boards are now so difficult to service even in well-equipped professional servicing workshops that the boards are often having to be returned, even by the accredited retail agents, to the manufacturers, usually overseas, introducing further delays and adding to costs.

Servicing problems are not confined to amateur radio equipment. The general field of consumer electronics is now well into the era of throw-away electronics where it is cheaper to purchase a complete replacement than have an obscure (or even a simple) fault traced and repaired. My long-established local television dealer/service shop closed down last year. A letter from Larry Brown in *Television* (October 2005, p747) reflected the pessimistic view of many service engineers: "With most skilled trades – for example, plumbing, joinery, decorating – the longer you have been doing the job the easier it becomes. The exact opposite applies with the electronics repair business. Many electronic products are now just too complicated for the ordinary engineer to repair. Who would be happy to try fault-finding to component level with LCD and plasma TV sets? As for panel-swapping etc, it would be sickening to pay £100 or more for a board then find that the cause of the fault lies elsewhere. If you were clever enough to be able to fix a DVD recorder, would Joe Public want to pay for the repair when supermarkets are selling these products for under £70?... Some engineers I know who work in the repair trade are not earning much more than I did 17 years ago when I worked in the service department of a well-known electrical retailer… A job stacking shelves in a supermarket sounds like a nice, stress-free alternative."

It is still possible to maintain or repair at an economic cost old but potentially still useful equipment, particularly the older hard-wired valve and early solid-state units. Some 50 years ago, I pointed out in *Radio Servicing Pocket Book* that the routine

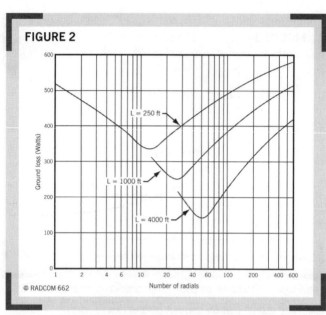

FIG 2: GROUND LOSS VERSUS NUMBER OF RADIALS FOR SEVERAL TOTAL WIRE LENGTHS WITH A QUARTER-WAVE VERTICAL OPERATING AT 3.60MHz OVER AVERAGE SOIL (CONDUCTIVITY, $\sigma$ = 0.0045S/m). (SOURCE: N4UU (QST))

repair of broadcast radio receivers required a relatively limited amount of test equipment and tools, although the time taken to repair or restore early equipment could be reduced by gradually adding extra facilities. I listed as "absolutely essential" a universal testmeter (volt-ohm-milliamp, VOM, meter) preferably with a sensitivity of at least 1000 ohms / volt on DC ranges, with 20,000 ohms / volt or more sensitivity preferred for solid-state receivers. Experience suggests that a classic Avo Model 7 or preferably Model 8, both having an effective meter-overload trip is likely to prove longer-lasting than most low-cost analogue or digital units. I also listed as essential a general-coverage signal generator or at least some local RF source. Yet, it still remains possible to trace and cure many faults on old valve receiver simply with the aid of a multimeter (VOM), soldering iron and a few hand tools.

**OPTIMUM RADIAL EARTH SYSYEMS.** During the 1930s, Dr George Brown (RCA) carried out the definitive investigation into ground systems for MF vertical broadcast radiators, leading to the adoption by the FCC of a standard of 120 radials. However George Brown's paper covered a mass of information that can be and has been drawn upon by radio amateurs using electrically-short monopole antennas on 1.8 and 3.5MHz.

Robert Sommer, N4UU, in 'Optimum Radial Ground Systems' (*QST,* August 2003, pp39 – 43), digs into George Brown's paper and comes up with some useful suggestions for radial earth systems on 1.8, 3.5 and 7MHz. These are based on the use of a fixed length of wire split into an optimum number of buried radials. For this work he used some 100 hours of computer time, whereas Dr Brown must have depended on a slide rule!

In brief, **Fig 2** shows ground loss versus number of radials for several total wire lengths with a quarter-wave vertical operating at 3.60MHz over average soil; (conductivity $\sigma$ = 0.0045S/m) and 1kW output power. N4UU suggests that the radial wire need be of the order of 20-gauge rather than the 8-gauge recommended by Dr Brown for MF broadcast stations, since the current is divided between the radials. N4UU provides a number of other calculated diagrams and tables covering the bands from 1.8 to14MHz including 10.1MHz, for optimum and low-power losses. His calculations for small losses, based on a total wire length of 11,570ft (73 radials each 158.5ft) for 1.8MHz; 6290ft of wire (66 radials each 99.8ft) for 3.50MHz; 2850ft of wire (50 radials each 57.0ft) for 7.15MHz; and 1870ft of wire (44 radials, each 42.5ft) for 10.1MHz, etc.

For other considerations, N4UU writes: "In addition to using the optimum number of radials, several other aspects should be considered to ensure their maximum effectiveness. Copper wire is the best choice [aluminium wire is now often used by broadcasters for economic reasons – G3VA] and it can be bare or insulated. Radial wires should be equal in length, evenly spaced and run radially from the antenna without meandering. The wires can be laid on the surface of the soil or buried shallowly; deep burial reduces their effectiveness. Wire as thin as 20-gauge (AWG) can be used at amateur power levels since the total current is divided among *n* radials. It just needs to be rugged enough to avoid breakage. Lastly, be aware that no benefit is obtained at radio frequencies by incorporating ground rods as part of the radial system."

However, when considering radial systems, we should not overlook the detailed investigations carried out in the 1980s by

# TECHNICAL TOPICS

Arch Doty, K8CFU, and friends into the question of the relative efficiency of radials, ground-screens (ground mats) both buried and on the surface, and elevated counterpoises. This elaborate project (first noted in 'TT' February 1983) involving over 10,000 measurements went a long way towards resurrecting the value of elevated counterpoise systems. This was later made the subject of computer calculations and trials by Al Christman, KB8I, that confirmed the value, even for MF broadcasting, of using a few elevated radials instead of 120 buried radials (see 'TT' August 1988, February 1992 etc). Amateurs have long used elevated radials inherent in the HF/VHF ground-plane antenna, which incidentally was originated by Dr George Brown.

HERE & THERE. An Australian reader, Michael Ong, contributes a useful tip in the 'Circuit Ideas' feature of *Electronics World*, December, 2005, p54. He shows that it is possible to use a power transistor, such as the 2N3055, as a high power Zener diode. Such Zener diodes are expensive and can be hard to locate, particularly for powers of 10W or above. For a bipolar power transistor to function as a Zener diode, the base is connected to the collector: **Fig 3(a)**. To ascertain the specific reference voltage for a given power transistor, the test circuit of **Fig 3(b)** can be used fed from a 15V or higher DC power source with R1 about 1k$\Omega$, and M1 a DC voltmeter. Michael Ong notes that the actual reference voltage of a specific type of power transistor may vary between different manufacturers and different batches from the same manufacturer. He notes that as a benchmark, a Motorola TIP31c (40W) was found to have a reference voltage of 9.0V, while a Motorola 2N3055 (115W) with adequate heat sink was found to be 11.60V.

Harry Leeming, G3LLL, writes: "Quite a few of the older rigs have some component parts, especially in such areas as the VFO and PLL, but also in other odd corners, secured against vibration by a kind of rubbery adhesive. This is extremely difficult to remove during servicing. When necessary, I have attacked it with freezer to try to make it brittle, or a hair dryer to try to soften it. I have then had to pick it off with a pair of long-nosed pliers. I wonder if any readers have found a better way of dealing with it? In recent years I have encountered a few strange and curious intermittent faults that seem to have been caused by this glue. PLLs which would not lock until the set had warmed up for an hour or so. Several VFOs with frequency jumps. An FT-290 with a 'dead' receiver but with its S-meter reading full scale for the first half-hour after switching on. All these faults disappeared while I was trying to get rid of the glue to get at suspected components. At first, I thought that I had somehow disturbed a bad connection or moved two short-circuiting parts. But the 'cure' has occurred so often that I am now convinced that something is happening to this type of glue, after being in place for many years. Does it become conductive? I wonder if other readers have had similar experiences and whether there is much 'beyond economical repair' equipment sitting around in odd corners, with nothing wrong except that some of this rubbery glue needs removing."

Ian Braithwaite, G4COL, writes: "I normally monitor my antenna system SWR using a classic Stockton wattmeter as adored by QRPers. This is a dual directional coupler based on ferrite toroidal transformers, but the important point here is that it monitors forward and return *transmit* power. My antennas are a little long for their intended 10.1 and 14MHz bands, and a tuner is used to provide a match to the transceiver. I recently wanted to find the resonant frequency reasonably accurately and did so using a dip (GDO) meter. As a cross-check, I also built a noise bridge which, by tuning for minimum signal on my receiver confirmed the frequencies. However, I was then surprised to find that the tuner itself was not apparently presenting a good match. I realised that the roller-coaster inductor in the T-match tuner was not making good enough contact at low signal levels. Visually, it looked quite reasonable, but cleaning it produced the expected good match to the noise bridge. Presumably, the transmit power was sufficient to punch through the microscopic layer of insulation. In summary, the noise bridge, operating at very low signal levels, produced an unexpected benefit in diagnosing a tuner problem that otherwise would have gone undetected."

'TT', January 2006, p73, in commenting on the future role of software-defined radio (SDR) noted that "At present, the pace is still being set by the large involvement of the military in the USA... [which]... has earmarked as much as $25-bilion for SDR development through the US *Joint Tactical Radio System* initiative. The US Department of Defense is working with agencies in Canada, Japan, Sweden and the UK to foster SDR development." However, it is becoming clear that SDR development is not going according to plan. A news report in *New Scientist*, November 19, 2005) headlined 'US Military Hit by Jitters Bug', continues: "A $5-billion programme to build a universal radio to replace the 200-plus types now used by the US military has run into trouble. The Pentagon's ambitious 'jitters' programme, begun in 1997, aims to create a universal communicator that will connect its personnel with one another no matter what frequencies or modulation system they use. But prototypes have fallen short of expectations No affordable antenna can work over the planned frequency range, which runs from 2MHz to 2GHz, without cumbersome power amplifiers, according to a source close to the project. And they need too much computer power to decode multiple formats. The Army has just ordered 100,000 non-Jitters radios, fuelling suggestions that the Defense Acquisition Board is set to curtail the project severely.

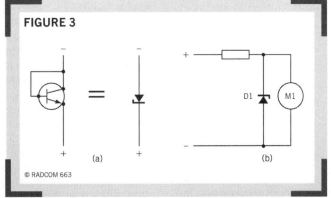

FIG 3: USING A POWER TRANSISTOR AS A HIGH POWER ZENER DIODE (a) CIRCUIT; (b) TEST CIRCUIT TO DETERMINE ZENER VOLTAGE. (SOURCE: *ELECTRONICS WORLD*)

# TECHNICAL TOPICS

PAT HAWKER
37 DOVERCOURT ROAD,
DULWICH, LONDON SE22 8SS.

G3VA

# Technical Topics

*Wide-range tunable oscillator with stable output ♦ HF supermodes or multi-hop ♦ Temperature controller ♦ Improving earth-rod performance ♦ Thoughts about Zepp-fed antennas*

**WIDE-RANGE TUNABLE OSCILLATOR WITH STABLE OUTPUT.** Over the years, 'TT' has included a number of oscillator circuits featuring the use of two active devices in order to permit the use of two-terminal tuned circuits with one side-grounded. For example, as long ago as 'TT' September 1970, the late LA8AK sent along details of a simple source-coupled FET oscillator that could function with an appropriately-tuned circuit from about 1MHz right up to VHF: **Fig 1**.

A recent three-page item, 'LC Oscillator Has Stable Amplitude', by Julius Foit of the Czech Technical University, in the 'Design Ideas' column of *EDN* (October 17, 2005), begins: "Many applications call for wide-range-tunable LC oscillators that can deliver a nearly-constant frequency, nearly-harmonic-free output, even when the circuit's output load changes. From a design viewpoint, eliminating both inductive or capacitive LC circuit taps and transformer couplings within the frequency-determining circuit simplifies fabrication and production, as does the option of grounding one side of the tuned LC circuit. These requirements suggest a circuit that can automatically and efficiently adjust loop gain internally, the basic requirement for oscillation. In addition, the circuit must provide sufficient gain to oscillate with low-impedance LC circuits and regulate the oscillation's amplitude to improve frequency stability and minimise THD (total harmonic distortion)."

Julius Foit discusses in some detail the design processes leading him to come up with the complete oscillator circuit shown in **Fig 2**. In his application as a radar-marker generator, the oscillator operates at around 280kHz with an L of 147μH and C of 2200pF. But it is claimed that the circuit will provide stable amplitude oscillation as low as 5kHz and up to 50MHz with no adjustment of component values apart from the L and C of the tuned circuit. Variations in load impedance will affect frequency stability and for VFO applications it would be advisable to use a buffer stage. At the higher frequencies, stability will be influenced by mechanical and constructional factors, as for any

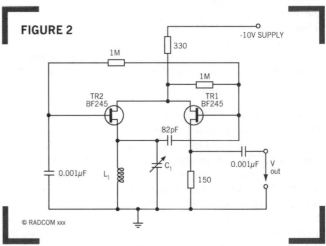

FIG 2: WIDE-RANGE LC-TUNED OSCILLATOR WITH STABLE AMPLITUDE AS DESCRIBED BY JULIUS FOIT IN EDN USES TWO N-CHANNEL BF245B JFETS. WITH SUITABLE LC VALUES IT IS CLAIMED TO WORK WELL FROM ABOUT 5KHZ TO 50MHZ.

other oscillator. Foit claims that the best active devices for this oscillator are the selected N-channel, medium-grade BF245B devices with a drain current of 5mA at a gate-to-source voltage of 0V and a drain-to-source voltage of 15V.

**HF SUPERMODES OR MULTI-HOP?** In the December, 2005 'TT', I pointed out: "It seems a pity that so many of the introductory textbook and articles on HF propagation still suggest that HF propagation at distances over 2500 miles depends on multiple hops between the earth and the F2 [or night-time F] ionospheric layers. The significant attenuation of one or more intermediate ground reflections is largely ignored."

As the item was primarily concerned with 50MHz transequatorial propagation (TEP), the question was barely touched on of the much greater path loss occurred in multihop signals compared with the various forms of HF supermode propagation in which there are no intermediate hop(s). Indeed, I may have left the impression that

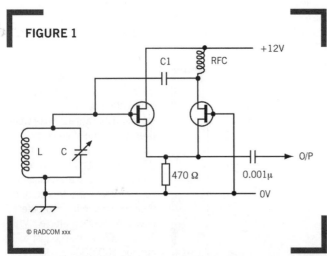

FIG 1: SOURCE-COUPLED FET LC-TUNED OSCILLATOR USING TWO MPF120, 2N3819 ETC, AS DESCRIBED BY THE LATE LA8AK IN 1970. THE VALUE OF C1 SHOULD BE CHOSEN TO SUIT THE FREQUENCY, BUT OFTEN OF THE ORDER OF 10PF. FOR USE FROM ABOUT 1MHZ TO VHF.

# TECHNICAL TOPICS

the prime attenuation of a multihop signal is caused by poor ground reflection(s), No mention was made of the daytime D-region that is the major factor in the attenuation of all daytime HF sky-waves. To quote *HF Communications – Science & Technology*, by John M Goodman (Van Nostrand Reinhold, 1992): "The attenuation of a radio wave corresponds to additional loss it suffers from factors other than the range spreading loss ($r^{-2}$). This attenuation could be the result of ionospheric absorption or the dissipative effects associated with imperfect ground, foliage, and atmospheric factors. *Ionospheric absorption occurs predominantly in the D-region where refractive bending is negligible at HF, and this form of attenuation is termed nondeviative*" [italics added].

The D-region is the lowest of the various ionospheric regions or layers, about 70 – 90km above Earth. It also has the lowest concentration of free electrons – about 10,000 per cubic centimetre during daylight, compared to some 100,000 in the E-region (about 110km above Earth) and one-million in the F2 layer (about 300km km above Earth).

To quote from a footnote on p107, "The D-region has insufficient ionisation for introducing a strong refractive interaction with radio waves in the HF band. At the low end of the HF band (3MHz)... refractive bending is negligible. On the other hand, absorption (which in the upper D-region is proportional to the product of the electron density and the collision frequency) is not at all negligible..."

The three upper layers, E, F1 and F2, all possess the property of reflecting (refracting) HF radio waves of different frequencies. But the D-region, as far as HF is concerned, acts as an absorbing region, and not as a reflector. F1 is observable only during the day; the E-layer ionisation also falls to a very low value during the night, but can become very intense during periods of Sporadic-E when it may reflect signals up to about 150MHz. The D-region virtually disappears between sunset and sunrise but, while present, can strongly absorb HF signals. In practice, the D-region determines the lowest usable frequency between any two points, attenuating HF signals to a rapidly-decreasing extent as the frequency increases. It has been shown that the absorption is inversely proportional to the square of the frequency. The ionisation of the D-region, however, can be increased by solar flares to the extent that the region virtually blocks the passage of all HF signals for periods of from a few minutes to several hours (Sudden Ionospheric Disturbance, or SID).

It is evident that, while a single-hop transmission passes twice through the attenuating D-region, once on the way up, once on the way down, a two-hop signal traverses the region four times, and a three hop signal six times. Classical theory would suggest that, for a transmission path exceeding about 10,000km, signals passing along an all- or mainly daylight-path, would have to pass at least eight times through the lossy D-region. A high-power broadcast transmission could survive such losses, but I find it difficult to believe that signals from an amateur 100W transmitter would often provide a receivable signal at this range – yet we know that contacts with Australia and New Zealand can often be made from the UK with a power as low as a few watts.

As the late Les Moxon, G6XN, put it in Chapter 1 of his *HF Antennas For All Locations*, (RSGB): "Whereas high angles [of radiation] present few problems and are optimum for the shorter ranges, much confusion exists in regard to very long ranges. It is almost universally supposed that these involve multiple reflections between the atmosphere and the earth's surface, and that there are optimum angles of radiation which are greater at lower frequencies... It is ironic that, whereas chordal hop (tilt-mode) theories of propagation, which were first put forward by Albrecht in the early 1950s, owed their origin to the experience of amateurs, the amateur literature has persistently followed the 'official line' [of multiple hops]... (a) there is convincing evidence that 'multihop' modes are abnormal, at least under those conditions of relatively low path loss which permit amateur or other low-power communication, and (b) for optimum DX performance it is best to aim always for the lowest possible angles of radiation at all frequencies."

**TEMPERATURE CONTROLLER.** Nyall Davies, G8IBR, writes: "I often want to control the temperature of a unit, but do not have any thermistors in my box of bits. The circuit shown in **Fig 3** does the job with a couple of diodes and an operational amplifier.

"As shown, it switched the fan on to fast with a rise of 7.5°C with the room temperature at 19°C. At 19°C, the voltage across the bridge was 29mV and the op-amp offset added another 1.5mV. With the voltage drop across each of the two diodes lessening by 2mV/°C (that is 4mV/° differential, this produced the 7.5°C switching-temperature rise. This drop can be adjusted by varying the 1.2kΩ resistor. The circuit was first designed for a 12V rail and worked nicely with a 1kΩ resistor in place of the 1.2kΩ resistor, but it needed to be increased to 1.2kΩ for 13.8V.

"The system was built to operate in a 20A switching-mode power supply. The supply was adequately cooled with 50Ω in series with the fan, but it proved rather too noisy. The reason for letting the circuit switch with a rise of only 7.5° is that the additional 39Ω resistor slowed the fan sufficiently to quieten it, but it still produces sufficient cooling so that, for normal intermittent operation even with only 7.5°,

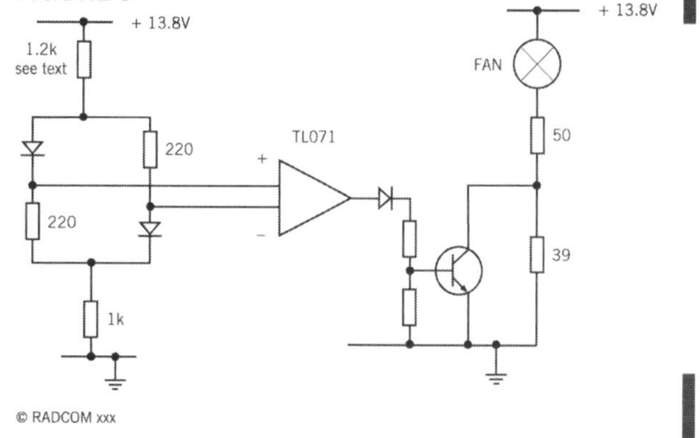

FIG 3: G8IER'S TEMPERATURE CONTROLLER BASED ON A COUPLE OF DIODES AND AN OP-AMP. ARRANGEMENT SHOWN SWITCHES A FAN TO 'FAST' WITH A RISE OF 7.5°C.

# TECHNICAL TOPICS

| Top resistor (Ω) | Differential (mV) | Approx temp rise (°C) |
|---|---|---|
| 1100 | 11 | 3 |
| 1200 | 37 | 9 |
| 1300 | 61 | 15 |
| 1400 | 83 | 21 |
| 1500 | 103 | 26 |

TABLE 1: THE OP-AMP DIFFERENTIAL INPUT VOLTAGE (FIG 3) AS DETERMINED BY THE TOP RESISTOR.

the circuit switches on only occasionally. See **Table 1** for some performance figures.

IMPROVING EARTH-ROD PERFORMANCE. Despite the increasing use of vertical dipoles, elevated counterpoises, radials and earth mats, particularly on the higher HF bands, the use of one or more metal rods (preferably copper or aluminium) driven into the ground, remains an important feature of many antennas on the lower HF bands. While the performance depends ultimately on the earth conductivity of the area – that can vary by a factor of some 20:1 (ie alluvial soil is 20 times more conductive than rock, with sea water about 1000 times more conductive than average soil) – soil resistivity can range from 500Ωcm to 1MΩcm. The aim should be to reduce this to a maximum of 10kΩcm although the lower the better. In seasons of low rainfall, some amateurs pour tap water (despite the almost infinite resistance of pure water) around the rod, in order that it may dissolve some of the salts present in the soil. However, it should be appreciated that an abundance of water does not necessarily result in good conductivity, particularly in areas of poor earth conductivity as found in urban and rocky areas.

A quarter of a century ago, in 'TT' May 1980. I wrote: "Many years ago one used to be able to buy a small copper pot containing crystals of copper sulphate for use as an earth 'rod' for domestic wireless sets. I have no idea how effective this was, but the idea of using certain salts to reduce the resistance of earth electrodes, particularly in areas of poor earth conductivity, is of long standing. The term 'chemical earth' is often used where deliberate action is taken, for instance in placing chemicals in trenches around and above earth electrodes [rods] in order to lower soil resistivity in their vicinity.

"The basic principles of 'chemical earths' are well explained in 'A Solution to Lower Earth-Electrode Resistance' by W Hymers of Marconi Communications Systems Ltd in *Electrical Review,* March 1980 (brought to my attention by Brian Castle, G4DYF). The author makes it clear that chemically-treated earthing systems can be spectacularly effective where soil resistivity is high, but tend to provide only a temporary respite, since the salt solution becomes progressively diluted by rain, leaching, gravity etc. Common salt (NaCl) for example, can result in extremely low-resistance earths, but its very high degree of solubility means that large amounts of salt are needed to prevent the soil solution from becoming progressively weakened.

"W Hymers indicates that a far more suitable substance is calcium sulphate ($CaSO4.2H_2O$), commonly known as gypsum and used for making Plaster of Paris, plaster board, cement, glass, fertilisers etc. Gypsum can be applied in particle form as a top dressing over earth rods, earth mats etc, and then left to percolate through the soil with the surface water. He points out that 'this also applies where retrospective action is taken to improve existing installations, or mixed with excavated soil when refilling. On new sites which have to be landscaped later it would not be impracticable to top dress entire areas of the site and plough the gypsum into the sub-soil.' Gypsum is much less soluble than common salt, and should remain effective over a much longer period [than common salt etc]. The particle size of the gypsum is important, as microcrystalline particles could impede the flow of moisture: 'a grading equivalent to coarse sand would be ideally suitable', he suggests."

To quote my science and technology encyclopædia: "Gypsum is the most common sulphate mineral, hydrated calcium sulphate. It is formed by precipitation from evaporating seawater. Huge beds of gypsum occur in sedimentary rocks, where it is associated with halite. It can be clear, white or tinted and it crystallises in the monoclinic system as prismatic or bladed crystals…"

A final quote from Hymer's 1980 article: "Any form of salting slightly increases the corrosion of the metal electrodes, but it is claimed that sulphates are not particularly aggressive in this respect; however, it is advisable *not* to put gypsum too close to concrete building foundations.

"In view of the importance of low-resistivity earths for vertically-polarised HF antennas, particularly monopoles, the use of a gypsum top-dressing seems a technique well worth trying out."

Surprisingly, in the intervening years since the item first appeared in 'TT', I cannot recall receiving any comments based on experience on the value of gypsum-enhanced earthing. It seems high time that the suggestion was given another airing!

THOUGHTS ABOUT ZEPP-FED ANTENNAS. The very first transmitting antenna that I used when G3VA was licensed in October 1938 (following two years with the artificial aerial licence 2BUH) was the then popular Zepp(elin) antenna with 67ft top and 33ft open-wire feeder for 7 and 14MHz; used also on the 1.7MHz band with the feeders joined at the base and fed against earth. It brought me contacts on these bands (the only ones covered by that new pre-war licence) despite the 10W DC input limit then imposed on new licensees. In the immediate post-war years, a Zepp with 66ft top and 17ft feeder brought me contacts on HF/MF bands and helped me achieve DXCC No 321 worldwide.

I have no idea whether my 1938 Zepp worked strictly according to the book, as I simply followed the published material in the ARRL's *Radio Amateurs Handbook,* the RSGB's *A Guide to Amateur Radio* and its then new *Amateur Radio Handbook*.

I assumed that the system had been developed for use in the German Zeppelin airships in the 1920s, and widely used by amateurs from the late 1920s up to the post-war years when it gradually fell out of favour as a system liable to cause interference to Band I television reception. It was also adversely criticised, in its classical form, by 'aerial wizards' Dud Charman, G6CJ, in 1955 and later by Les Moxon, G6XN. The Zepp-feed system still remains popular for use in the 144MHz J-pole antenna and the later 'Slim Jim' folded version first described by Fred Judd, G2BCX in *Practical Wireless* (April, 1978).

In the USA, the classical form for amateurs was, as far as I can trace, first described in detail in *QST* in an article 'The Zepp' by ARRL's James J Lamb, W1CEI (September, 1928), although in this he refers to "the familiar Zeppelin" so clearly it was in use before 1928. Lamb's article was sub-headed 'Facts and Figures for the Design of the Hertz Antenna with Twin-Wire Voltage Feed', and the design was soon included in the *ARRL Handbook*.

In the UK, the first detailed description was by A E Livesey, G6LI in *The T & R Bulletin,* December 1931, pp184 – 189) (**Fig 4**) and in a shorter review of antennas by Austin Forsyth, G6FO. These articles soon evoked critical comments from Dud

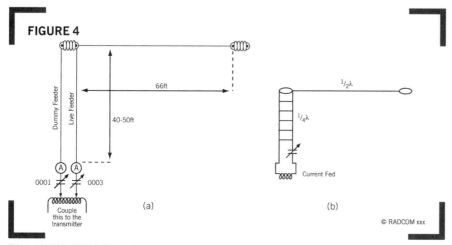

FIG 4: (A) 'ALL-WAVE ZEPP' ANTENNA AS DESCRIBED AND DISCUSSED BY G6LI IN THE T&R BULLETIN IN 1931. (B) ZEPP ANTENNA AS COMMONLY USED IN THE 1930S WITH A HALF-WAVE TOP (OR MULTIPLES THEREOF) AND WITH THE TWIN-WIRE 'FEED' LINE AN ODD MULTIPLE OF A QUARTER-WAVE (OFTEN WITH A VARIABLE CAPACITOR IN EACH 'FEED' LINE IN AN EFFORT TO ACHIEVE BALANCED CURRENTS).

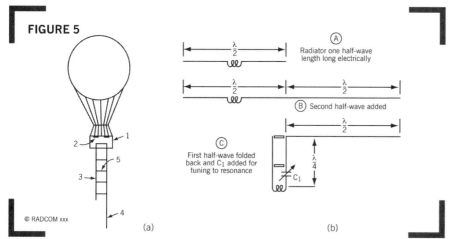

FIG 5 (A) ILLUSTRATION FROM ORIGINAL 1908 PATENT TAKEN OUT BY DR HANS BEGGEROW IN GERMANY FOR AN ANTENNA SUSPENDED FROM A BALLOON OR AIRSHIP REQUIRING NO DIRECT CONNECTION TO EARTH. (B) EVOLUTION OF THE ZEPP ANTENNA AS USED BY AMATEURS. (SOURCE: RADIO HANDBOOK)

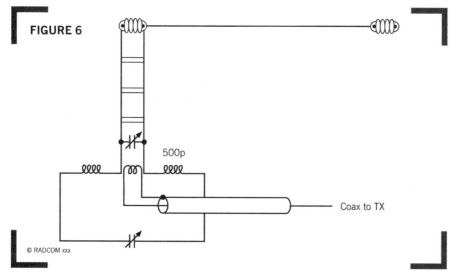

FIG 6: SUGGESTED 'ALL-WAVE ZEPP' IN WHICH THE COMPLETE 'TOP' AND THE 'LIVE' AND 'DUMMY' FEED-LINES ARE BROUGHT INTO SYSTEM RESONANCE BY THE USE OF A BALANCED P-TYPE TUNER WITH NO DIRECT EARTH CONNECTION. THE TWIN FEEDER SHOULD HAVE A SPACING OF FROM ABOUT 6IN TO 10IN.

Charman, G6CJ, who in February 1932, in a survey of 'SW Aerials and Feeders', wrote: "One other system requires comment, the type known as the 'Zeppelin'. This falls into the AOG [Act of God] class, and consists of an inverted-L, with twin-feeder for the lead in, the top being one half-wave, and one of the feeders being free. The idea is that the free wire suppresses the radiation from the other feeder but it is doubtful if it does, as no two people seem to get the same results, and the system is often very directional."

The previous issue (January 1932) had also carried a long letter from G6CJ and a reply from G6LI, with, in the February issue, a long letter from H A M Clark, G6OT, 'Cleaning up the 'Zepp' Question'. G6OT agreed with G6CJ that "we are forced to Mr Charman's conclusion that the ordinary Zepp feeder must be working unbalanced. Reference may be made to an article by Roosenstein in *Experimental Wireless*, June 1931 for further details." This was a theoretical article on open-wire balanced transmission lines.

In the UK, at least, the Zepp antenna was thus mired in controversy from the start of its use by amateurs. So why did it become firmly established and still 75 years later remains in most amateur radio handbooks, etc? Primarily, I would suggest because it is often working, not as a classical Zepp, but quite effectively as an inverted-L, multi-band antenna with the 'free' feeder wire acting as a vertically-oriented counterpoise/radial providing the 'return current' without direct connection to earth, the whole system being brought into resonance by the tuner. If fed from a balanced π–network tuner with the 'feeder' wires spaced by 6 – 10in, it seems entirely possible that the resulting antenna would radiate effectively. This may apply whether or not the dimensions of the top and feeders are themselves resonant and whether or not the feeders are accurately balanced to suppress most of the vertically-polarised radiation. It then becomes a variant of the end-fed inverted-L antenna analogous to the W3EDP antenna. In other words, a useful antenna that does not depend on an earth rod - capable of reasonable if not high performance.

In support of this suggestion, I would point out that in 1988, a letter by Alois Krischke, DT0TR/OE8AK (*Ham Radio*, November, 1985, reported in 'TT, January 1986), claimed that the antenna was originally patented in 1908 by Dr Hans Beggerow for use suspended from balloons and airships: **Fig 5(a)**. It seems to me that since 1908 was still firmly in the era of spark transmitters and LF/MF waves, it is highly unlikely that Beggerow's 'feeders' were a quarter wave long or the 'element' a full half-wave. No, surely the vital point was that it provided an antenna that did not require an earth connection, impossible in a balloon or airship! I have never seen a description of the antenna system actually used in German Zeppelin airships and from the 1920s this may well have been for HF transmission. Commercial frequencies were not harmonically-related and there must have been a requirement for an antenna that would function on various frequencies, at least for the later trans-Atlantic trips, etc. I rest my case, but suggest that it might be worth experimenting with the arrangement shown in **Fig 6** on 3.5, 5 and 7MHz.

# Technical Topics

**PAT HAWKER**
37 DOVERCOURT ROAD,
DULWICH, LONDON SE22 8SS.

**G3VA**

*Hybrid cascode power amplifier* ♦ *1.8MHz bandpass filters for the 'Buccaneer'* ♦ *Looking again at the classic 'Windom'* ♦ *New DDS chips*

**HYBRID CASCODE POWER AMPLIFIER.** Chas Fletcher, G3DXZ, admits that his grey cells were stirred by the suggestion from *Electronics World* (November, 2005) – reported in the January, 2006, 'TT' – that a hybrid (valve/transistor) cascode amplifier could form a useful HF power amplifier as well as a high-voltage amplifier, based on the circuit shown as Fig 2 in the January issue.

He writes: "The cascode amplifier has been one of my interests for many years and I have used them for pre-amplifiers in my home-brew receivers with some success. I also fancied it as a power amplifier and about eight years ago built one using two transistors. My best effort used a Motorola 15004 high-voltage device to handle the150V supply with a IR510 MOSFET tail. This produced 80W on 3.5MHz but, unfortunately, destroyed itself in a spasm of self-oscillation.

"The January 'TT' piece stirred up renewed action on this unfinished business. I found an old ex-TV PL500 in my junk box and enough old bits, some going back to 1950, for a 900V power supply and for a tank circuit. To keep it simple, I used link coupling to the tank circuit, with a sliding link coil – plastic drainpipe can be very useful! [Advisable to check first with the microwave oven test as described in 'TT' August 2003 or *TTS 2000 – 2004*, p185 – G3VA]. I stuck to the IRF510 MOSFET as a tail device as it is a little easier to use than HF power transistors: see **Fig 1**.

"The exercise has proved very interesting. After re-learning long-unused valve techniques and some nervousness when using a 900V HT line, the result is a very efficient and passive amplifier. Using my QRP transceiver running at 2W as a driver, the amplifier produced some 95W on 3.5MHz with an anode efficiency that appears to be well over 80%, with very similar results on 7MHz. The amplifier runs in Class-C with a duty cycle of about 30% and does not exceed the rating of the old PL500 or even appear to stress it unduly.

Using a reed relay to switch the QRP feed from transmitter to receiver results in excellent break-in (QSK) operation.

"Tuning is simple but needs a little understanding of what is happening. For good efficiency, it is essential that the PL500 anode voltage swings down to around 55V. This is mirrored by a sharp peak in the screen-grid current which is a much better indicator of the 'in-tune' condition than the anode current. Typically, anode current 125mA, screen-grid current 18mA. The final operating power is simply set by the DC bias of the MOSFET. All-in-all, I found tune-up a more satisfying process than with my all-solid-state units as one is 'in amongst the action', having to juggle tank tuning, link coupling and bias as well as the ATU. Yet, in practice, the whole operation takes only seconds."

It seems possible that by adjusting the MOSFET bias, the amplifier could be run in a linear mode for SSB, although with appreciably reduced output than as a Class-C CW amplifier as used by G3DXZ.

**1.8MHz BANDPASS FILTERS FOR 'THE BUCCANEER'** etc. Dave Gordon-Smith, G3UUR, was surprised to see a Cohn-type front-end filter used in "The Buccaneer" – the experimental high-performance binaural 1.8MHz receiver fully described by Phil Harman, VK6APH. and Steve Ireland, VK6VZ (*RadCom*, July 2005, pp78 – 83).

He writes: "I am surprised this filter design is still being perpetuated by anyone. It is an abysmal design from the KISS (keep it simple, stupid) point of view. It requires much effort to construct because of the two centre coils with inconveniently large values of inductance, and, when it is completed, it rewards the constructor with inferior performance compared with a simple, third-order, top-coupled Butterworth or 0.01dB-ripple Chebyshev filter. One could claim that the Cohn filter has high-performance compared with the usual second-order designs, but any third-order design would have! A top-coupled arrangement using capacitors between the three parallel-tuned circuits with identical coils (see **Fig 2**) gives improved attenuation on the low-side of the response, which is just where you want it on 1.8MHz to get rid of local medium-wave broadcast signals. Note the reduced coil inductance of 49$\mu$H, a more convenient figure than the 70$\mu$H used in 'The Buccaneer'. Additionally, although not important on 1.8MHz, the suggested filter of Fig 2 has lower insertion loss."

**LOOKING AGAIN AT THE CLASSIC 'WINDOM'.** Last month, March 2006, 'TT' looked at the origins and potential value of the classic 'Zeppelin' antenna. This month it is the turn of the 'Windom' off-centre-fed (Hertz) dipole – a widely used (and often abused) antenna in the 1930s – to come under scrutiny.

To quote from my 13th edition (1974) of The ARRL Antenna Book: "A multiband antenna that enjoyed considerable popularity in the 1930s is the 'off-centre feed' or 'Windom', named after the amateur who wrote a comprehensive article about it. Shown in **Fig 3**, it consists

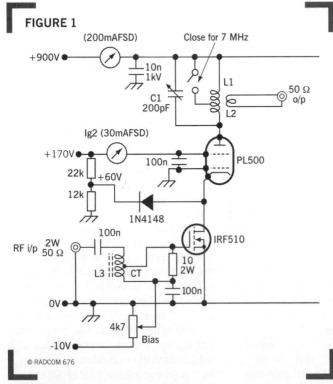

FIG 1: G3DXZ'S HYBRID CASCODE POWER AMPLIFIER USING PL500 PENTODE VALVE AND IRF510 POWER MOSFET. COMPONENT VALUES: L3 – LARGE TWIN-HOLE FERRITE 5-0-5 TURNS BIFILAR WOUND. L1 – 11.8$\mu$H, 16 TURNS 14SWG ON 2in-DIAMETER (50mm) FORMER, 1.75IN-LONG TAP AT 7 TURNS FORM COLD END. L2 – 4 TURNS ON 1.5IN (40mm) FORMER TO SLIDE INSIDE L1. C1 – SPACING TO WITHSTAND 2KV

of a half-wavelength antenna on the lowest-frequency band to be used, with a single-wire feeder connected off-centre as shown. The antenna will operate satisfactorily on the even-harmonic frequencies, and thus a single antenna can be made to serve on the 80, 40, 20 and 10m bands. The *single-wire* feeder shows an impedance of approximately 600ω to ground, and since the return circuit for the feed system is through the earth, a good ground connection is important to the effective operation of the antenna. Also, the system works best when installed over ground having high conductivity.

"Theoretically, the single-wire feeder can be any convenient length, since its characteristic impedance is matched by the antenna impedance at the point where the feeder is connected. However this type of feeder is susceptible to parallel-type currents just as much as the two-conductor type and some feeder lengths will lead to 'RF in the shack' troubles, especially when the feeder goes directly to a p-network in the transmitter. Adding or subtracting one-eighth wavelength or so of line usually will help cool things off in such cases."

The RSGB's *Radio Communication Handbook* is not enthusiastic about single-wire transmission line: "There are three main types of transmission line: (a) The single-wire feed arranged so that there is a true travelling wave on it, ie the line is terminated in its characteristic impedance... Single-wire feeders are usually connected to a point on a resonant antenna where the impedance formed by the left- and right-hand portions in parallel match the impedance of the wire... This is now rarely used, being basically inefficient due to losses in the return path, which is via the ground and to radiation from the feeder which acts also to some extent as a terminated long-wire antenna; against this, it offers the advantages of light weight and low visual impact, and for short lengths (up to about 0.5λ) the losses should normally be less than 1dB..."

The 'comprehensive article' by L G Windom referred to above was "Notes on Etheral Adornments", sub-headed "Practical Design Data for the Single-Wire-Fed Hertz Antenna", in the September 1929 issue of QST. This has an editorial note "The use of the linear Hertz radiator fed by a single-wire line has been restricted in amateur work because of lack of data on its design and adjustment. This article explains how these systems may be erected with the assurance that the voltage and current distribution on both the radiator and feeder will be correct."

L G Windom introduced the system as follows: "Sooner or later in the course of amateur development, one must have some sort of antenna, skyhook, or as you like it. In the earlier stages, it consists generally of merely 'a' antenna, then later after much deep (?) thought, it is 'the' antenna. These few notes concern themselves only with that much-cussed atrocity, the single-wire-fed (cross-breed, voltage-current) Hertz. This type has the advantages of simplicity, ease of erection, very high efficiency and, as will appear later, can be designed on paper and erected without the usual pruning operation."

He continued: "The information herein contained is due to the efforts of John Byrne of the Bell Telephone Laboratories, ex-8LT, W8GZ, W8ZG, W8DKJ; Ed Brooke, also of the Bell Telephone Laboratories, W2QV and ex-8DEM; and Jack Ryder, W8DQZ, under the direction of Prof W L Everitt of the Department of Electrical Engineering, Ohio State University. The writer acts solely as a reporter and all credit is due the above-named men.*

It is clear that the calls W8GZ and W8ZG were held by the university and it is uncertain whether Windom held his own amateur licence at the time. It is paradoxical that the antenna has borne his name, rather than that of Byrne, Brooke or Ryder for almost 80 years! What is clear is that the design data etc resulted from a supervised experimental project at the Ohio State University, although the concept of the single-wire feeder had originated even earlier. Byrne and Brooke erected a special antenna at the university, built so that it could be varied in every possible way. Ryder came into the frame later when he found that the formulæ derived by Byrne and Brooke held true for feeder efficiencies exceeding 85%

The 1929 article continues: "Interest in the single-wire-fed Hertz antenna for amateur work started mainly with an article by Williams, 9BXQ in the July, 1925 QST followed by several others including the re-hash in the July, 1926, issue. It is perhaps best to disregard all this previous material in relation to the single-wire feeder system and start from the beginning."

A system of movable trolleys was devised to allow measurement of the relative RF current at various points along the half-wave radiator. It was soon found that the then common practice of adjusting the radiator by maximising current at the centre of the element was unsuitable when applied to an off-centre feed: **Fig 4(a)**. Instead, it was found

FIGURE 2

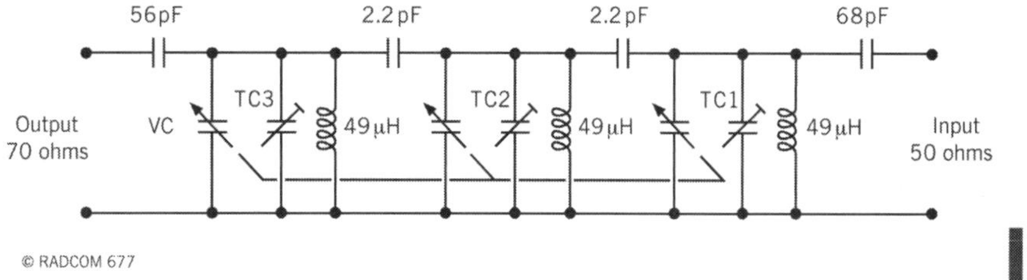

FIG 2: G3UUR'S THIRD-ORDER 1.8MHz TUNABLE BAND-PASS FRONT-END FILTER WOULD BE EASIER TO CONSTRUCT AND PROVIDE BETTER PERFORMANCE THAN THE COHN FILTER USED IN 'THE BUCCANEER' RECEIVER FULLY DESCRIBED IN *RadCom* JULY 2004. VC IS A TRIPLE-GANG 5-50pF VARIABLE CAPACITOR. TC1 39PF + 2-22pF TRIMMER. TC2 100pF + 2-22pF TRIMMER. TC3 47pF + 2-22pF TRIMMER.

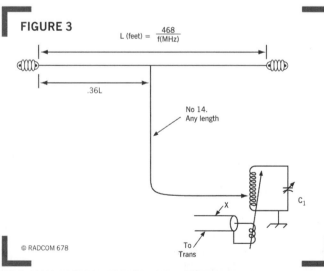

FIG 3: CLASSIC VERSION OF THE SINGLE-WIRE-FED WINDOM ANTENNA AS DESCRIBED IN *THE ARRL ANTENNA BOOK*. THE SINGLE-WIRE FEEDER CAN BE CONNECTED DIRECTLY TO THE 'HOT' RF OUTPUT TERMINAL OF A p-NETWORK IN THE TRANSMITTER. ALTERNATIVELY, THE LINK-COUPLER CAN BE USED WITH A SEPARATE GROUND TERMINAL AS SHOWN. THIS TYPE OF COUPLING HELPS REDUCE TROUBLES FROM RF CURRENTS ON THE STATION EQUIPMENT.

# TECHNICAL TOPICS

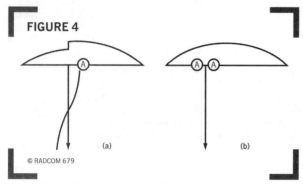

FIG 4: (A) WINDOM'S 1929 ARTICLE SHOWED THAT THE CORRECT TAPPING POINT CANNOT BE ACHIEVED BY MEASURING FOR MAXIMUM CURRENT AT THE CENTRE OF THE RADIATOR, BUT AS SHOWN IN (B) BY MEASURING AND EQUALISING THE CURRENT FLOWING IN THE ELEMENT CLOSELY SPACED EITHER SIDE OF THE FEEDER TAP. THIS METHOD WAS USED TO DERIVE THE DESIGN FORMULAE GIVEN IN THE TEXT.

necessary to use two meters closely spaced either side of the feeder tap and adjusting for equal current flowing into each section: (**Fig 4(b)**). After the radiator difficulties had been resolved, the position of the feeder was varied and the current distribution along it measured. When the correct position was located, there were no standing waves along the feeder, no matter its length, and the radiator still showed excellent current distribution.

As a result of all this work, two general formulae were derived:
Fundamental wavelength = 2.07 x length of element in metres.

An editorial footnote points out that his agrees quite closely with the usual formula:
Length in feet = desired fundamental wavelength x 1,56, or
Length in feet = 468,000 divided by desired frequency in kHz.

The correct feeder point from the centre of the element is equal to
Length of antenna in feet times 25/180.

Since the impedance of the feeder depends to some extent on its diameter, it is pointed out that the factor of 25 applies to No 14 (AWG) copper wire, and increases progressively to 30 for No 24 wire. It is emphasised that the position of the tap should have no effect on the fundamental resonant frequency of the radiator. The feeder efficiency for average feeder runs is well over 95%"

I assume this disregarded ground losses. For average or poor ground conductivity areas it would probably be an advantage to use a counterpoise wire a few feet above ground beneath the radiator fed from a non-earthed p-network ATU output as noted last month for use with a Zepp antenna. With such an arrangement, the antenna should also work, to a reasonable degree, as a top-loaded asymmetrical-T, or as an inverted-L on bands such as 10.1, 18, 21 and 24MHz.

Les Moxon, G6XN, in my RSGB's *Radio Communication Handbook* (6th edition), is not altogether enthusiastic about single-wire feeders. He recognises (p12.17) that this is one of the three main types of transmission line for HF antennas: "Single-wire feeders are usually connected to a point on a resonant antenna where the impedance formed by the left- and right-hand portions in parallel matches the impedance of the wire... This is now rarely used, being basically inefficient due to losses in the return path, which is via the ground, and to radiation from the feeder which acts also to some extent as a terminated long-wire antenna. Against this it offers the advantages of light weight and low visual impact, and for short lengths (up to about 0.5λ), the losses should normally be less than 0.1dB"

To check that the single-wire feeder is actually working as a transmission line rather than as part of the radiator, it was the practice in the 1930s to run a neon bulb along a section of the feeder wire to check that there were no pronounced standing-waves present. Today, it would probably be easier to use a clip-on RF current probe.

The name 'Windom' is now often used in connection with an off-centre-fed dipole using 300Ω balanced line feeder that has been described a number of times in 'TT'. To quote again *The ARRL Antenna Book*: "A newer version of the off-centre-fed antenna (mis-called 'Windom') uses 300Ω TV twin-line instead of a single-wire line (Fig 5). The claim has been made that the 300Ω line is matched by the antenna impedance at the connection point both on the antenna's fundamental frequency and on harmonics, but there is little theoretical justification for this. The system is particularly susceptible to parallel-line currents because of the unsymmetrical feeder connection and, probably in many cases, the line acts more like a single-wire feeder than a parallel-conductor one. The parallel currents on the line can be choked off by using balun coils, as shown in the diagram. The same balun can transform the impedance to 75Ω, in cases where the line actually shows a resistive input impedance of 300Ω.

"With either of the off-centre-fed systems, the feeder should be brought away from the antenna at right-angles for at least a quarter wavelength before any bends are made. Any necessary bends should be made gradually." [This implies ideally a height of at least 66ft for 3.5MHz, usually impractical, but can often be implemented for 7MHz and above – *G3VA*].

Popular in Continental Europe is the "TD4 Windom", a commercial four-band antenna. I do not have details of this product, but it seems to work satisfactorily, and may well be a twin-wire-fed off-centre-fed antenna with balun as discussed above.

The twin-hyphen wire "Windom" is discussed in detail by G3LDO in *RadCom*, March 2006, p68.

**NEW DDS CHIPS.** Colin Horrabin, G3SBI, draws attention to some interesting recent developments by Anolog Devices in chips for DDS oscillators. The AD9951 is a 14-bit part with 32-bit accumulator and 14-bit control of relative phase and amplitude output. The new device is specified at 400MHz, which means that you can produce an output directly up to about 150MHz. Fig 6 shows the functional diagram.

Colin mentioned this device to Harold Johnson, W4ZCB, and found that he was already using one (it is possible to get two free samples from analog.com via the

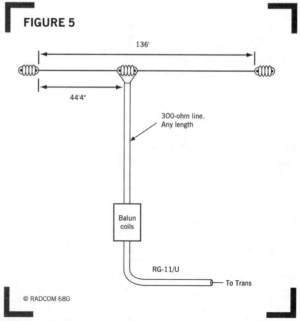

FIG 5: A NEWER VERSION USING 300Ω TV TWIN-LINE INSTEAD OF A SINGLE-WIRE LINE.

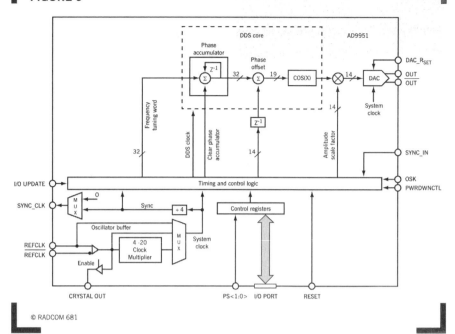

FIG 6: FUNCTIONAL BLOCK DIAGRAM OF THE NEW ANALOG DEVICES AD9951 DDS CHIP.

web). He reports that W4ZCB has made some useful observations: "For a start, it is worth generating your 400MHz clock externally to the chip, otherwise the use of the on-board clock multiplier makes the device run very hot despite the new chip consuming significantly less power than earlier DDS chips. In any event, you achieve the lowest phase noise when you don't use the on-board clock multiplier. Used in this way, W4ZCB has been able to operate the device with a 750MHz clock, meaning that it is possible to generate up to 250MHz output frequency."

G3SBI continues: "Chris Muriel, G3ZDM, a long-term member of the UK Analog Devices team, tells me that the firm is shortly to release another DDS device that will be a 14-bit, 1000MHz clocking part. It is possible to synchronise these chips on power-up so that multiple outputs of different phase can be produced. There are plenty of applications for this type of advanced technology.

"Clearly, these new components represent major steps forward in this technology, with Analog Devices still a leading-edge company in electronics."

HERE & THERE. Memory plays odd tricks, particularly with names and calls. In the January 'TT' in referring to Short Wave Magazine under the editorship of Austin Forsyth, G6FO, I wrongly gave, without checking, the call of the pre-war assistant editor as 2CUB. John Wightman, ZL1AH (ex-G3AH), writes: "I served [in the RAF] for some months in 1940 with another former CWR (Civilian Wireless Reserve) man who said he was the assistant editor of SWM – his name was Stan Clarke, 2AMW, and he died during the war of illness in Cairo".

In the February, 1994 'TT' attention was drawn to the change, as a result of a European agreement, that the UK AC power supply mains voltage would, from January 1, 1995 become 230V with tolerances of +10% and -6%, representing a minimum of 216V and maximum of 254V. At the time I wrote: "Since these tolerances cover the present 240V, I remain uncertain whether a voltage change will actually be introduced next January." In practice, there was little noticeable change. But it does appear that the companies are these days taking full advantage of the wide range of tolerances. This winter, using an Avo Model 7 multimeter, I have found my supply to be often in the region of 220V, and have measured it as low as about 216V and only very occasionally above 235V. As a result, I changed the mains adjustment plugs on my KW2000A PSU from 240 to 220V, giving an appreciably better performance. But care needs to be taken to keep this under review when the demand for electricity falls as warmer weather approaches (I now fit my 500W Variac to keep the KW2000A supply to approximately 240V.

John D Wightman, ZL1AH (G3AH, 1937 – 1950), earlier commented that the 'The Coming of Coax' (TT' August 2005 pp77 / 78) stirred memories of WWII days in the RAF. He wrote: "I spent the spring of 1941 at Lanivet, Cornwall [the village site of the Bodmin transmitter]. RAF West Drayton sent two of us down there to install some new gear at the Bodmin Beam Station under the supervision of one of Louis Varney's engineers. [G5RV was then still with Marconi]. RF at the Bodmin station was generated by Marconi SWB 8 transmitters and we added an SWB 25 amplifier. The beams were Sterba curtains on 400ft towers and we had to construct a new run of coax cable; 3in copper outer conductor and 1in inner with ceramic spider spacers. Matching was interesting; with the full 25kW input, we were getting an output of about 15 – 17kW. Matching was done at 2kW, and flash-overs due to any mismatch were spectacular. The SWB 25 used both air and water cooling, requiring us to acquire new (plumbing) skills. At that time we were controlled from RAF Birdlip, Gloucester. The station manager discovered I was proficient in Morse, and I had the privilege of doing the testing. This took place on 32MHz during the afternoons with VOAC in Newfoundland. I felt pretty chuffed, keying 15kW to a Sterba curtain! Billeted in the peaceful village of Roche, we lived the life of Riley, spending our evenings in the *Victoria Inn*."

Someone I do remember rather wryly is Dr C K Kao, who was praised in a *Guardian* leader column (January 26, 2006): "Forty years ago tomorrow, Dr C K Kao told an IEE meeting about research that he had been doing into 'guiding light energy along spherical types of optical conductors' at the Harlow laboratories of STC (now part of Canadian Nortel). Some of the optical waveguides exhibited an information-carrying capacity equivalent to 200 television channels or over 200,000 telephone channels… This demonstration in 1966 was almost completely ignored by the Press but we now know it was not just a milestone in the development of optical fibres, it was a day that changed the world because, without fibre-optic communications, it is difficult to see how there would have been an Internet, mobile phones or multi-channel TV… Every phone call we make, every e-mail we send and receive, every website we watch, is made possible by optical-fibre communication… To this day, Dr Kao is little-known outside his profession, though deeply respected within it." My own memory of meeting Dr Kao was in March 1968 when I visited STL in the course of an 'STC Profile' for *Electronics Weekly*. I was much impressed by meeting Alec Reeves who was not only the pre-war inventor of Pulse Code Modulation (the basis of digital communications) but had also been the first to develop HF SSB (with pilot-carrier). But I also talked in his laboratory to Dr Kao who explained what he was hoping to achieve with glass fibres. But he admitted that to achieve a practical system would still need the development of extremely low-loss fibres. I seem to recall that he said it needed fibres with 1/1000 of the attenuation so far achieved. This seemed pie-in-the-sky research and to my eternal shame I devoted much attention to Alec Reeves and other STL scientists but omitted to mention what proved to be the iconic work of Dr Kao. In mitigation, I can say only that it was several years before practical optical fibre systems appeared.

**TECHNICAL TOPICS**

PAT HAWKER
37 DOVERCOURT ROAD,
DULWICH, LONDON SE22 8SS.

G3VA

# Technical Topics

*Using the AD9951 DDS chip ♦ Power reducer using two transformers ♦ 'Ancient' series modulation ♦ Reduced radials for ground plane/monopole antennas ♦ More on ceramic resonators ♦ More on RFI to automotive electronics*

**USING THE AD9951 DDS CHIP.** In last month's 'TT' (April 2006), Colin Horrabin, G3SBI, drew attention to the new Analog Devices AD9951 chip with its potential for improved DDS performance. Even before the April issue appeared, a letter from Nyall Davies, G8IBR, further endorsed this new chip. He writes: "I recently made myself a DDS local oscillator using the AD9951. This is a 14-bit DDS chip able to clock at 400MHz and thus will give a considerable reduction in DDS spurs compared to the 10-bit chip used in the 'PIC 'n' Mix' Digital Injection System (see the series of articles *RadCom*, January to May, 1999) by Peter Rhodes, G3XJR. I built one and found I needed extra filtering to tame the spurs.

"The results with the AD9951 are very encouraging. Using a crystal oscillator at 100MHz multiplied by three, I now have a phase noise of -139dBc at 2kHz off carrier with the 14-bit DDS chip clocking at 300MHz. I now know why I have had to use so much filtering in the past." Another advantage not mentioned by G8IBR is the significant reduction in the power consumption of the AD9951.

**POWER REDUCER USING TWO TRANSFORMERS.** Dave Gordon-Smith, G3UUR, recalls that in 'TT' August 2003, G4LU criticised G3GKG's dynamic power reduction circuit (June 2003) on the grounds that he had not included in his calculation the coefficient of coupling of his ferrite transformer; G3GKG countered by pointing out that his power reducer worked out well in practice.

G3UUR writes: "The question of coupling coefficient in current transformers needs to be addressed since the figure (k = 0.3) given in the literature for primary windings consisting of a screened coaxial wire passing through the centre of a toroid seems to be wildly out and can cause quite serious errors if constructors use this figure as a guide". Experience with a two-transformer VSWR bridge described by Hank Perras, K1ZDI (*Ham Radio,* August 1979) has shown G3UUR that the coupling coefficient is nearer 1 than 0.3.

G3UUR continues: "The two-transformer hybrid circuit used by K1ZDI struck me as a good way of reducing the power of my 28MHz exciter to drive a 70cm transmit converter without the need for variable attenuators to get precisely the right level: see **Fig 1**. After completing the initial design, I was surprised to find the output power was 100mW or 10dB higher than the 10-turn winding with a k = 0.3 should have given… Since then I have characterised with crude instruments every current transformer I have made, and k has always been near to unity. I think the real coefficient of coupling for a screened conductor going through the centre of a toroidal ferrite core is between 0.9 and 1.

I've tried using cores with permeability between 70 and 5000, and it doesn't seem to affect the k. Certainly, G3GKG's 2003 results seem to confirm mine, and it would be interesting to know what other experimenters have found.

**'ANCIENT' SERIES MODULATION.** There is still a role for so-called 'Ancient Modulation'. There are currently several AM nets operating on the 3.5 and 7MHz bands. A return to AM on VHF might improve coverage compared with NBFM; AM would be fully acceptable on 1.8 and 28MHz, although the added spectrum requirement for its double sidebands rules out routine AM use on 14 and 21MHz or on the narrow WARC bands. The inherent simplicity and ease of adjustment should appeal to home-brewers.

In the heyday of AM for amateur radio, most amateurs used high-efficiency 'plate' or 'plate and screen' modulation of a high efficiency RF Class-C power amplifier with bulky iron-components. Audio chokes (Heising modulation) or 'modulation transformers' with high-power audio amplifiers made bulky, heavy transmitters and would today be relatively costly to implement. There were lower efficiency alternatives such as grid-, cathode-, suppressor-grid modulation requiring far less audio power, although one needed care to avoid 'downward' modulation and reasonable audio quality.

An early 'high-efficiency' system was 'series modulation'. By the late 1930s, this was little used by amateurs, although remaining in use for some radio and television broadcast transmitters. By the 1940s, series modulation was seldom covered in the amateur radio handbooks, although I have found it mentioned briefly in my 1942 edition of *The Radio Handbook*: **Fig 2**. To quote: "Another form of plate modulation is known as *series modulation* in which the RF tube and modulator are in series across the DC plate supply. Series modulation eliminates the modulation choke required in the usual form of Heising modulation. Although the system is capable of very good voice quality, the antenna coupling must be carefully adjusted simultaneously with the C bias on the modulator in order to maintain at least 20 per cent more plate voltage across the modulator than

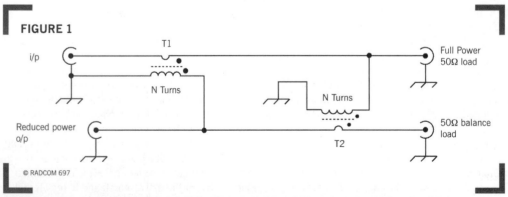

FIGURE 1: G3UUR'S POWER REDUCER USING TWO-TRANSFORMER HYBRID CIRCUIT. T1 AND T2 ARE IDENTICAL FERRITE TRANSFORMERS WITH A SINGLE-WIRE PRIMARY THROUGH THE CENTRE OF THE TOROIDAL CORE AND AN N-TURN SECONDARY. THE 50Ω BALANCE LOAD SHOULD HAVE A POWER RATING EQUAL TO THAT OF THE LOW-POWER OUTPUT LEVEL. POWER REDUCTION EQUALS 20 LOG N, WHERE N IS THE NUMBER OF TURNS ON THE TOROIDAL CORE. FOR EXAMPLE: 10 TURN WINDING 20dB; 14-TURN WINDING 23dB; 20-TURN WINDING 26dB AND 31-TURN WINDING 30dB. THE CHANGE IN OUTPUT POWER IS ABOUT 1dB/TURN AT 10 TURNS AND 0.3dB/TURN AT 30 TURNS. THE INDUCTANCE OF THE WINDING SHOULD BE AT LEAST SUFFICIENT TO GIVE A REACTANCE OF 500Ω AT THE LOWEST FREQUENCY BEING USED.

**FIGURE 2**

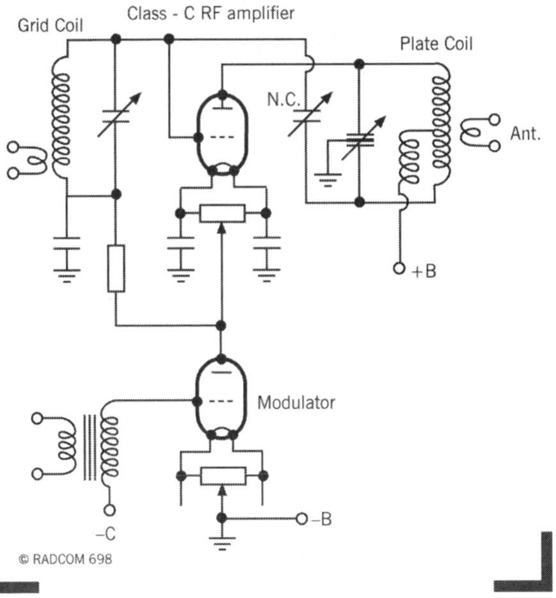

FIGURE 2: BASIC SERIES-PLATE MODULATION AS FOUND IN THE 1942 EDITION OF *THE RADIO HANDBOOK*.

that which is measured from positive B to RF tube filament".

My attention was drawn to the possibilities of a very simple low-power 1.8MHz AM valve transmitter using series modulation by an article 'Receiving FM Programmes in the Medium Wave Band', by Peter Lankshear in *Radio Bygones*, No 18, August/September, 1982, pp4 / 5. He described, with constructional details and a selection of suitable valve-types, a mini-transmitter operating in the MW band and intended to distribute programme material from an FM receiver to AM sets within the house: **Fig 3**. For this application, such a device breaches UK regulations. A few simple tests showed that the system worked well on MW and could almost certainly with a few modifications and a good antenna be shifted into the 1.8MHz band. It would then become, with microphone input, a simple 1.8MHz QRP AM transmitter (or used to provide grid-drive for a high gain grid-driven linear amplifier), although I never got round to trying this out in practice.

Peter Lankshear, in a section headed 'Series Modulation' wrote: "Various systems were considered and series modulation using valve technology seemed to be the best option. Many broadcast transmitters of the 1930s era used series modulation with its advantages of wide frequency response and not requiring a modulation choke or transformer. Solid-state transmitters often use a form of series modulation, but suitable amplifiers and their adjustment are more complex than is warranted for this project.

"Early radio was very fortunate that Class-C oscillators could be easily modulated with low distortion. In fact, many of the first generation of medium wave transmitters were simple modulated oscillators. Separate modulated RF amplifiers were eventually needed to meet the requirements of crystal control and reduction of frequency shift with modulation, factors that are largely academic in this project. With this in mind, some experimentation with a double triode as a combined oscillator and series modulator produced encouraging results and the addition of a simple voltage amplifier provided full modulation with an audio input of less than 1V. The addition of a simple power supply would make a compact, self-contained unit a viable proposition…"

I was again reminded of AM series modulation, this time based on semiconductors, by an item 'Simple Amplitude Modulator', contributed by Reza Golparvar Roozbahani to the 'Circuit Ideas' feature of *Electronics World*, December 2005, pp56 – 59. This points out that, while the simplest method of AM is by using a four-quadrant analogue multiplier chip, this can present problems with supply voltages and current consumption. To quote: "Thus, we need a simple and flexible amplitude modulating circuit with simply modifiable characteristics that may easily be designed for any specific application. **Fig 4** shows a simple AM circuit with a minimum number of elements."

In the example shown, a frequency of 1MHz from a 600Ω source is applied to the base of TR1 and 1kHz audio from a 600Ω source to the base of TR2, both type 2N3904 transistors, with a supply of 9V at about 1mA. Reference to the *EW* article shows how the modulation index depends on the component values. For fuller details see the original item in *EW*.

### REDUCED RADIALS FOR GROUND PLANE/MONOPOLE ANTENNAS.

The ground plane antenna (GPA) with four sloping quarter-wave radials remains one of the most popular HF antennas for use in restricted-space locations, often with a quarter-wave rod element mounted on a chimney piece and with three or four sloping radials to facilitate matching to 50Ω coaxial cable feed. However, this type of implementation is not easy for 7MHz or lower-frequency bands in small- or medium-sized gardens due to the length of the radials.

It should not be forgotten that the GPA was originally developed in the 1930s by Dr George Brown of RCA as a low-VHF antenna with just two horizontal radials forming the artificial ground plane. The other two radials were added primarily to meet market pressures when it was adopted by the American police forces then operating on about 7m. They were not convinced that substantially omnidirectional radiation could be achieved with just two radials.

The use of quarter-wave grounded and artificially-grounded monopole antennas with short length counterpoises (single radial) was investigated by the late Les Moxen, G6XN, in the 1970s and 1980s and a number of his ideas were published in 'TT'. For example in 'TT', March 1981 (see also *Antenna Topics*, p166) as 'A Basic No-Radial Monopole Element'. G6XN emphasised that vertical antennas near ground are inherently asymmetrical and

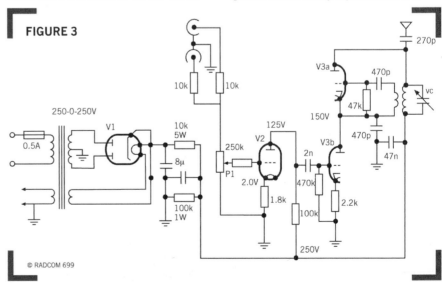

FIGURE 3: CIRCUIT DIAGRAM OF A LOW-POWER MF 'TRANSMITTER' WITH SERIES MODULATION AS DESCRIBED BY PETER LANKSHEAR IN RADIO BYGONES AS A MEANS OF DISTRIBUTING FM BROADCASTS TO IN-HOUSE AM MEDIUM-WAVE RECEIVERS BUT WHICH, IT IS SUGGESTED, COULD BE MODIFIED TO FORM THE BASIS OF A LOW-POWER 1.8MHz AM TRANSMITTER. A WIDE RANGE OF TWIN-TRIODE, TRIODE AND RECTIFIER RECEIVING-TYPE VALVES MAY BE USED.

# TECHNICAL TOPICS

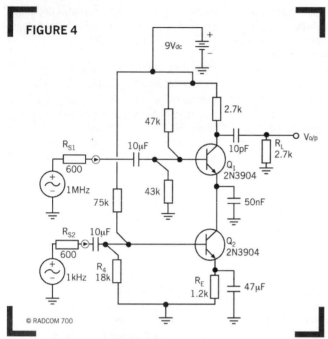

FIGURE 4: SIMPLE SOLID-STATE AM CIRCUIT USING SERIES MODULATION AS DESCRIBED IN *ELECTRONICS WORLD*.

that it is almost always difficult to ensure that no common-mode current flows on the outer screen of the coaxial feeder.

I was reminded of G6XN's work by a letter from Nick Brooks, G4BMI, who writes: "I have been working for some time with the object of substantially reducing the length of elevated radials for an HF ground plane antenna, while keeping losses to a minimum. I would like to encourage discussion amongst readers, especially lowband (1.8, 3.5 and 7MHz) DXers.

"I have devised a method of maintaining an omnidirectional radiation with little loss in bandwidth or efficiency, but with a reduction in radial length to only 34%, or even 30% of that of two quarter-wave radials.

"The two [conventional quarter-wave] radials are replaced by one [counterpoise] conductor which is folded back on itself and extends back past the centre point: **Fig 5**. The diagram shows a 3in spacing between the wires but, in practice, 450Ω ribbon line can be used with only slight adjustment in the length of the open end (I could not model ribbon feeder on *EZNEC* so used 3in spacing). Note the wire is a different length each side of the centre point – this is necessary to produce omnidirectional radiation.

"During my [*EZNEC*] investigation, in order not to fall foul of Occam's Law, I kept the size of the vertical constant throughout, only altering one component at a time. My findings are as follows:

"The bandwidth of this G4BMH system is slightly less than with the reference two-quarter-wave radials; however, it still comfortably covers the 7MHz band with low VSWR. There is some additional loss compared with my reference antenna (**Fig 6**): an additional 6Ω over poor ground (0.001, 5); 5Ω over average ground (0.003, 20); and 4Ω over good ground (0.03, 20). These figures are of little importance provided that the radiation resistance of the antenna is reasonably high (figures are for 7MHz, they will differ slightly on other bands).

"All the 'rules' regarding elevated radials still apply: the height above ground of the radial system should ideally be about 4ft on 7MHz, 8ft on 3.5MHz, 15ft on 1.8MHz. I have found that on 3.5MHz it works satisfactorily with the radial system on the top of a 6ft garden fence.

"An advantage of the system that may not be readily apparent is that as the currents in the earth system are in series, changes in earth conductivity or other conducting objects nearby will not alter the radiation pattern nearly as much as when two quarter-wave radials are used. It is well known that as the input impedance of a quarter-wave radial is extremely low, small unbalances in the radials can cause massive differences in radial currents, significantly affecting the radiation pattern.

"The system can be scaled for other bands. The saving of space on 3.5 and 1.8MHz is enormous. There is only one 'hot' end on the radial, making for convenient adjustment. The length of the free end can be altered quite a lot before the radiation pattern, according to *EZNEC*, changes to any great degree, but the ratio of 1:1.2 seems ideal.

"It would also seem possible to fold the free end back again, saving even more space: **Fig 7**. In this case, the ratio of the free-to-folded length needs to be 1:1.5 and space saving is some 80%, but my preference, unless space is really at a premium, is for the single-fold arrangement as this keeps the 'hot' end well away from the feed-point.

"An additional radial at right angles to the first was modelled; bandwidth increased slightly and there was about 1Ω less loss, giving only a minimal improvement overall.

"My conclusion is that an effective 3.5MHz DX antenna can be constructed on an average-size plot by mounting the earth system on top of a fence at the side of a house and, being of different lengths, makes them easier to accommodate in some gardens. The vertical element extends up to 40ft to a pole mounted on a chimney and along the apex of the house, and if necessary extending downwards again. Provided the horizontal portion is perpendicular to the radial system, it will radiate omnidirectionally, with low-angle radiation. Indeed, a useful all-round antenna!"

**MORE ON CERAMIC RESONATORS.** Dave Gordon-Smith, G3UUR, notes that the topic of HF ceramic resonators has turned up again ('TT' September, 2005). He writes: "Just after 'TT', December 1985 (see also *Technical Topics Scrapbook 1985 – 90*, p69) briefly reported the original long article on HF ceramic resonators (*Ham Radio*, June 1985), I found that Ambit International was selling a restricted range of them and bought a few. Since then I have found that many other standard frequencies have become available, and have used them in all sorts of applications.

"But first, the history of the development of materials for these HF resonators is interesting, and dates back to the 1980s. The modes and additives chosen to optimise the Q and frequency stability took longer to establish for HF than for lower frequency resonators, possibly because the emphasis at first was for LF/MF applications. In the mid-to-late 1970s, it became evident that there would be a market for HF ceramic resonators in digital applications and their development was given more impetus.

"HF ceramic resonators use the same parent binary solid solution of $PbTiO_3$ and $PbZrO_3$ as the LF resonators, but the modes of operation and additives necessary to achieve optimum Q and frequency stability are different. It is very difficult to get proprietary information from manufacturers about what they are using to achieve the best results. However, it is clear that they are varying the additives used for each standard frequency to achieve optimum performance. The slight variations in dielectric constant for each frequency within the Murata CSA HF series provide an indication that this is being done.

"The odd snippets of information I've gleaned from scientific papers on material development would suggest that the additives used are combinations of Mn, Fe, Cr, Co, Sb and Ni. A thickness shear mode

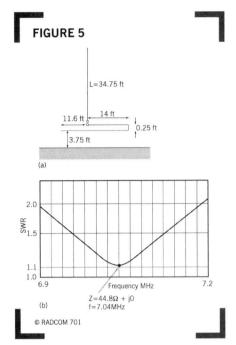

FIGURE 5: (a) 7MHz GROUND-PLANE ANTENNA WITH G4BMH'S ELEVATED RADIAL SYSTEM SUITABLE FOR LIMITED SPACES AND NOT REQUIRING THE USE OF A GROUND STAKE. (b) BANDWIDTH SIMULATION USING ARRL EZNEC (11 SEGMENTS EXCEPT SHORTEST 1 SEGMENT. GROUND 0.005, 13. WIRE 0.05in DIAMETER.

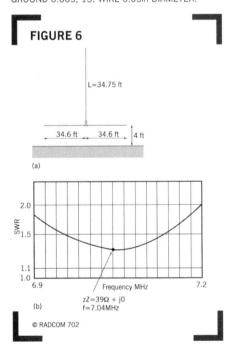

FIGURE 6: REFERENCE 7MHz GROUND-PLANE ANTENNA WITH TWO-QUARTER WAVE RADIALS AND ASSOCIATED EZNEC SIMULATION FOR COMPARISON TO FIG 5.

of vibration is used for HF resonators. The stability of resonant frequency requires changes of density, compliance and thickness of the material to compensate each other effectively as the temperature varies. Density, compliance and dimensions of the host material can be controlled to a certain extent by the careful substitution of additives on the T and Sr sites in the crystal structure. Murata have obviously done this fairly well in their CSA range of HF resonators, judging by Q values and frequency stability.

"I have used 2MHz resonators for 160m VXOs and 40m local oscillators, 3.58MHz resonators for 80m CW transmitters, and 2.45, 2.5, 3.27, 4 and 4.18MHz as LO sources for superhet receivers. The 4.43MHz resonators divided by 32 can give 136 – 138kHz VXOs. Q values of 3.58 and 3.69MHz resonators that I have measured have typically been in the range 1000 to 1200."

### MORE ON RFI TO AUTOMOTIVE ELECTRONICS. 

The item 'Automotive Electronics & Mobile QRO' ('TT' , February 2006, p72), drew attention to greatly increased use of automotive electronics, including some up to 100kg in weight with over 50 embedded microprocessors and some 2km of wiring. Linear amplifiers with outputs of some 500W are now being offered on the UK amateur radio market specifically for use in mobile stations. It was stressed that vehicle manufacturers now use proven design technologies and run exhaustive tests to ensure RF immunity in their vehicles, I posed the question "Can amateurs be sure that running 500W or so will never affect their own or a passing car that may have less-immune automotive electronics?

Nyall Davies, G8IBR writes: "I have often wondered as I go past Droitwich or any other big transmitting station how remarkable it is that modern motor cars keep going… I have two stories that have no other solution than RF getting into embedded microcontrollers.

"Some years ago, I was overtaking a lorry on a dual carriageway approaching Bury St Edmunds at 70mph or so when I glanced at the speedo and saw that it, the rev counter and all other gauges were reading zero. Attempting to accelerate indicated that that the engine had gone out. Fortunately, I had enough speed to pass the lorry and into a lay-by. Switching off, I thought 'what do I do now?' Against all logic, since it had gone out in gear, I tried to start the engine. It started and has never missed a beat since. This was in open country, the car a 2000 model Citreon 2.0 Hdi. The only cause I can think of was that the lorry driver was using CB or amateur radio.

"Yesterday, while approaching Witham on the A12 in a Ford Focus 1.8 Turbo Diesel, I was caught in a traffic jam resulting from an accident. At the scene of the accident, I passed quite closely to two police cars blocking the inside lane. Immediately afterwards I saw that the 'battery no-charge' indicator was on. I was 60 miles from home on a Saturday night in the dark. Hoping that the battery would last, I drove home as quickly as possible on dipped headlights. On arrival the lights were undimmed and I began to wonder whether I really had no charge. The logical

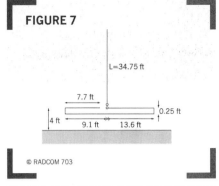

FIGURE 7: 7MHz GROUND-PLANE ANTENNA USING DOUBLE-FOLDED ELEVATED RADIAL SYSTEM.

thing was to recharge the battery, but it would not take a charge. A check with a meter showed 13.4V and on starting the engine it went to 14.4V showing that it was indeed charging. And now the battery light went out. Police cars have transmitters!

"Neither experience provides any proof. Perhaps other readers have other experiences: windows winding down unexpectedly or such like. I can say only that I am glad my bonnet catch is operated mechanically with a key."

Apparently, some manufacturers warn against using even low-power mobile phones with their antennas inside the car and recommend plugging in an externally mounted antenna.

*The Guardian* (February 22, 2006, p5) carried a long report headlined 'Countdown to Zero on the Road to Cromer – Secretive Radar Base Blamed for Car Breakdowns', by Patrick Barkham which began: "Modern motor cars rattle with fear when they take the winding road from Mundesley past RAF Trimingham to Cromer. Engines have stalled. Fuse boards and microchips have fried. Speedometers have roared up to 150mph or down to zero. Dashboards have gone black. Clocks have conked out… 'A microwave radiation hazard exists beyond this point' warns a danger sign on a barbed wire fence… Local mechanic Kevin Abbs has been a busy man in recent days after a spate of mysterious breakdowns outside a radar station on the north Norfolk coast… The mechanics at Crayford and Abbs say nearly 100 motorists have phoned them in recent days reporting… problems on everything from [a Nissan Almera] to a new Honda and a Renault Laguna… The local garage is not charging customers to fix cars zapped by the radar. Most cars can be repaired by disconnecting and then reconnecting the battery. But they have lodged a complaint with the Ministry of Defence."

Patrick Barkham concludes: "Your correspondent bravely subjected his utterly unreliable *10-year-old* [Italics added] car to the full force of Trimingham's radar eight times yesterday. Everything carried on winking and bleeping as before." Even allowing for a degree of journalistic licence, this report seems strong evidence that even modern electronically-replete cars are more vulnerable to strong local radiation than the older mechanically-orientated models.

# TECHNICAL TOPICS

PAT HAWKER
37 DOVERCOURT ROAD,
DULWICH, LONDON SE22 8SS.

G3VA

# Technical topics

*Non-mains power sources* ♦ *Petrol-electric-generator lore* ♦ *HF propagation – questions remain* ♦ *More on the classic Zepp*

**NON-MAINS POWER SOURCES.** The approach of summer implies that portable, holiday and field day operation increasingly occupies the thoughts of many amateurs. The development of very compact, lightweight – yet reasonably powerful – transceivers operating from 12VDC or with switch-mode PSUs from 230/110VAC supplies, plus the availability of fibre-glass GRP telescopic masts, have eased many of the difficulties. But there still remain valid question-marks over the supply of electric power while operating away from local supply mains.

Truly portable QRP gear can operate from conventional storage batteries, preferably using rechargeable cells. Mobile operation in vehicles can draw on the large-capacity lead-acid batteries. But the prime need for field operation of digital systems, including the use of lap-top computers, hand-held cellphones, DAB and, DRM receivers, and (potentially) SDR transceivers, is to operate over extended periods away from heavy vehicle batteries. This must be achieved over longer periods than is currently possible at economic cost with rechargeable cells such as NiH, Lithium-ion, Lithium-polymer, etc. A comparison of the basic characteristics of the main existing rechargeable cells is given in **Table 1,** reproduced from 'TT', September1995.

Another technique that has come into use in the past few years to provide temporary supplies for low-consumption battery-operated systems is the 'Super-capacitor', a compact electrolytic capacitor of extremely large capacitance, rated in farads rather than microfarads and capable of carrying a charge over relatively long periods.

It needs to be recognised that one of the drawbacks of digital systems is that, generally speaking, they are significantly more power-hungry than analogue units. Currently there are several research projects aimed at finding new ways of generating and storing electric power, both for portable use and for 'greener', less-contaminating vehicles or domestic supplies at a cost comparable to existing supplies. For example, wind and tidal-wave systems and fuel-cells fuelled by hydrogen.

*New Scientist,* 25 March 2006, pp46 – 47, includes an article 'Everlasting Power', by Caroline Williams reporting current work by Jeffrey Cheung at Rockwell Scientific in Los Angeles, USA. By serendipity, Cheung discovered that a bar magnet when coated by a commercial ferrofluid had greatly reduced friction between the magnet and the bench on which it was resting. To quote: "Ferrofluids are simply a suspension of magnetic nanoparticles in an inert liquid of some kind. Pour a little fluid around a magnet, and it quickly migrates to the magnetic poles and stays there, in the same way as iron filings cling to the ends of a bar magnet." The coating of liquid on each end acts as a super-efficient lubricant, around 40 times more slippery than ice.

Cheung realised that there is a wealth of ways this can be exploited: "One of his favourites is to use a ferrofluid-covered magnet as the heart of an electricity generator that needs no input except gentle motion, such as that provided by waves or gentle shaking." Armed with a grant from the Pentagon's Defence Advanced Research Projects Agency, Cheung and his team set about designing a system to provide power for the buoys used for oceanographic monitoring. Existing buoys use battery packs or solar panels to power their monitoring and communications equipment. Clean solar panels work well when the sun is high in the sky, but have to be backed up by batteries. In time, the spattering of guano from sea birds perching on the buoys diminish their efficiency even in bright sunlight. Batteries are bulky, need reliable waterproofing and eventually need to be changed.

In summer 1994, a small, prototype 'self-charging battery' (see **Fig 1**) using Cheung's super-lubricant was tested at the Scripps Institution of Oceanography, California. In a gentle sea, it generated about 0.3W. Chueng believes that with further work to optimise the transfer of wave power to the generator, it will deliver an average of 1W. The eventual aim is to generate not merely watts but megawatts from an energy farm on the ocean. But he also has many hopes for low-power applications including self-powered tyre pressure monitors, computer mice and TV remote controls. The first commercial product is a holster-style mobile phone charger. He says: "The first goal is to have enough power to keep the phone on

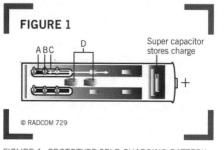

**FIGURE 1**

FIGURE 1: PROTOTYPE SELF-CHARGING BATTERY BASED ON THE USE OF FERROFLUID LUBRICANT. THE SLIGHTEST MOVEMENT CAUSES THE MAGNET HOUSING TO ROCK BACK AND FORTH ACROSS THE COILS, GENERATING CURRENT THAT IS STORED IN A SUPER-CAPACITOR. A: SET OF THREE MAGNETS ON EITHER SIDE OF THE COILS; B: FERROFLUID, STUCK TO POLES OF EACH MAGNET, LUBRICATES MOTION AND FOCUSES THE MAGNETIC FIELD INTO THE COILS; C: MAGNET HOUSING; D: COILS. (SOURCE: *NEW SCIENTIST*)

**TABLE 1**

| Cell type | Nominal Voltage (V) | Energy density (Wh/l) | Power density(W/l) | Self-discharge (% per month) | Cycle life | Comments |
|---|---|---|---|---|---|---|
| Lead-acid | 2.0 | 70 | <400 | 4-8 | 250-500 | Lowest cost |
| Ni-Cad | 1.2 | 60-100 | 220-360 | 10-20 | 300-700 | May exhibit memory |
| Ni-H | 1.2 | 220 | 475 | 30 | 300-600 | Possibly some memory |
| Lithium-ion | 3.6 | 260 | 400-500 | 5-10 | 500-1000 | Intrinsically safe |
| Li-polymer | 3.0 | 150-350 | >350 | <1 | 200-1000 | Contains metallic Li |
| Zinc-air | 1.2 | 204 | 190 | <5 | <200 | Requires air-manager |

**Notes on Table**: Cycle lives are strongly dependent on how the battery is used. Lithium-ion cells with petroleum coke anodes drop fairly linearly from about 4.0V to about 3.0V during discharge. Figures for Lithium-polymer figures are as predicted in 1995 before they reached the market. NiH data apply to 2.9Ah 4/3A cells as manufactured by Duracell and Toshiba. Zinc-air data are for a cell that Zinc-Air Power was developing in 1995 for use in electric vehicles. It must be emphasised that the data give at best only a rough idea of how these cells compare Care must be taken with Lithium cells to avoid the possibility of the cell exploding during recharging, use recommended charger. Cells capable of high discharge currents should be protected against accidental short-circuiting.

Further notes appear in 'TT' September 1995 (*TTS 1995-1999* p52). These and the table are based on material in *Spectrum* (May 1995, p56).

# TECHNICAL TOPICS

standby forever, all you'll have to do is shake it".

*New Scientist* (17 September, 2005, p19) reported the development in the USA of a novel backpack that generates electricity as you walk. It is designed to make life easier for troops, field scientists, explorers and disaster relief workers, etc, by making it unnecessary to carry heavy batteries into the field, yet providing sufficient power to recharge devices such as cellphones etc. The device uses the fact that a walking person moves like an upside-down pendulum. One foot is put down and then the body vaults over it, causing the hip to move up and down by 4 to 7cm. In this device, the motion is exploited by mounting the backpack on a spring loaded plate that is free to slide up and down on rail-like rods on either side. As it moves, a toothed rod fixed to the head plate meshes with a gearwheel on a dynamo, generating up to 7W of electricity. The output increases with walking speed and with the weight of the load in the pack. The device was originally described in *Science,* 9 September, 2005.

'Hydrogen Power Brewed on the Go' by Helen Knight in *New Scientist* describes the development of a new micro-fuel cell that may prove suitable for powering hand-held devices and portable consumer electronics. The difficulty of storing hydrogen or generating it from scratch on a small scale has usually forced developers to focus on fuels less energy-dense than hydrogen, such as methanol. However, a UK firm, QinetiQ, is developing a method of generating hydrogen from pellets of the chemical, ammonia borane, which is almost 20% hydrogen by weight. The pellets are combined with a 'secret' mix of chemicals that react and rapidly heat up when kick-started by gentle warming. QinetiQ has signed a deal with the Japanese firm, Olympus, to develop a working prototype for small fuel cells suitable for consumer devices by 2008. It is claimed that the combined generator/fuel cell will have a target peak power output of 10W and should last between three and five times as long as conventional batteries.

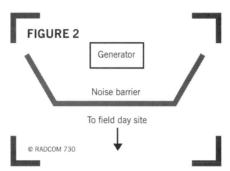

FIGURE 2: A NOISE-SCREEN IS A USEFUL ADJUNCT WHEN USING A TRANSPORTABLE OR EMERGENCY GENERATOR. NOISE SCREENS CAN BE CONSTRUCTED FROM HEAVY CARDBOARD OR CARPET-COVERED PLYWOOD, PREFERABLY HINGED. NEVER ATTEMPT TO BOX-IN A NOISY GENERATOR IN A COVERED OR FOUR-SIDED 'BOX'. (SOURCE: *QST*)

Toshiba in Japan has unveiled two digital music players powered by prototype 100 and 300mW methanol fuel cells. Helen Knight writes: "While 100mW may be fine for MP3 players, which require a fairly steady supply, devices such as cellphones consume very little power while they are idle but can draw several watts while making/receiving a call, requiring a high voltage and high energy density in a short period of time'. QinetiQ claims that the ammonia borane used in the new generator should pose fewer safety risks than methanol, as it is very stable, and, being a solid, will not leak.

*Electronics World* (August 2005, p4) reported that a small British firm, CMRFuel Cell of Cambridge, has invented a new solid-state fuel cell stack structure that reduces the size of the cell by a factor of 10. In the 'Compact Mixed React Cell', the reagents are first mixed and then fed through a porous membrane, making the design thinner, lighter and cheaper. It is claimed by the firm that "We have the ability to create a totally solid-state stack, making it robust and easy to fit in any application that requires a battery source.

The fuel is methanol and CMR is hoping it will be used for such applications as power tools, electric scooters, consumer portable systems and eventually, the automotive market. A working prototype was demonstrated mid-2005, but volume production is not expected until 2008.

There are, of course, established techniques for generating electric power in the field. These include the use of solar cells to charge storage batteries; hand- and pedal- generators, practical for short periods to provide from 10 to about 50W – pedal-operated units with an upper limit of 200W have been developed for use in third-world countries (see 'TT, April 2000). WWII saw the use for clandestine radio of; wood-burning steam generators (see 'TT' February 2003) as well as thermo-electric generators using propane heating, etc. But to generate power in excess of, say, 100W, the prime systems largely boil down to wind-, water- and petrol/diesel generators.

**PETROL-ELECTRIC-GENERATOR LORE.**
For major field operations, such as NFD and DXpeditions, the petrol-electric generator (PEG) remains supreme, although, as pointed out by G3JKY and G0DCG in *RadCom,* January 1998 (with additional notes by G8GEF in 'TT' May 1998), it is now often possible to convert these to use bottled propane gas rather than petrol in order to reduce hazards when filling petrol tanks without interrupting operations. A large range of units is manufactured, ranging from 12VDC units intended for charging vehicle storage batteries to 230V/50Hz or 120V/60Hz transportable units at powers from about 70W up to about 3.5kW.

The choice, use and care of petrol electric generators were discussed in 'TT' October 1999, and earlier in 'TT, November 1987, based on articles by Kirk Kleinschmidt, NT0Z ('How to Choose and Use a Portable Power Generator', *QST,* June 1999) and by Wendell Tulencik, K8OIP, in 'A Few Thoughts on Emergency Power', *QST*, May 1987.

In brief. NT0Z warned that "Generators – like all engine-powered devices – can injure or even kill you if you don't respect them. And unlike your engine-driven garden mower, these powerhouses can electrocute you (or others). Don't be afraid – but do pay attention". Under the heading 'Using Your Generator', he advised: "Regardless of the earthing method you choose, a few safety rules remain the same. Your extension cords *must* interact, have waterproof insulation, three 'prongs' and three wires, and must be sized according to loads and cable runs. For

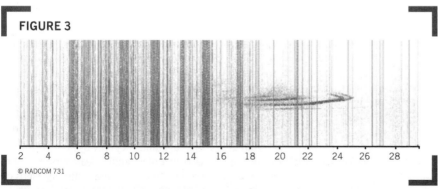

FIGURE 3: G3PLX'S 2 – 30MHz IONOGRAM OF THE ASCENSION ISLAND TO KENDAL, CUMBRIA PATH, AS RECORDED ON HIS OWN EQUIPMENT IN MARCH 2005.

high-wattage loads, use 14 – 16-gauge, three-wire cords… When it comes to power cords, think *big*. Try to position extension cords so they won't be tripped over or run over by vehicles. And don't run electrical cords through standing water or over wet, sloppy terrain."

A problem that can arise when using a PEG at or near maximum rated output is the effect of the significant variation in load presented by an SSB or CW transmitter or transceiver between 'transmit' and 'receive'. A small continuous load such as one or two light bulbs can reduce fluctuations. It is important that the AC frequency should not drop significantly below the rated value due to failure of the governor. A low frequency can cause damage to equipment (overheating of power transformers, etc).

K8OIP stressed the need to provide safe storage of fuel. He provided 10 commandments for PEGs: (1) Use only the fuel recommended by the manufacturer. (2) Pour fuel through a large-mouth funnel with a fine screen to filter out dirt and other contaminants. (3) Keep a waste cloth or paper towelling handy for blotting up spills. Store properly in a covered metal waste bin. (4) Keep a supply of lubricating oil handy. (5) Have some 50ft or 100ft extension cords available. (6) Keep at least one $CO_2$ or dry-powder fire extinguisher ready for instant use when handling fuel. (7) A torch with good batteries is a *must*. Two torches are better than one! (8) Check fluid levels and start the alternator at least once a week. (9) If you have trees close to the house, keep a small chainsaw handy. (10) Use a small trickle charger to keep the starting battery [if any] charged.

Subsequently, C J Chapman added: (11) When you *have* spilled fuel over the generator, allow it to dry thoroughly *before* starting it up! Petrol/oil mixtures take quite a while to dry thoroughly. (12) Check the output voltage [and if possible frequency] *before* connecting your radio equipment. Governors have been known to go wrong.

K8OIP also warned of the dangers of fume and engine heat, pointing out that the exhaust fumes of petrol, diesel, natural gas or propane engines are all lethal and must be properly vented out of any enclosure without relying on natural ventilation. He advised: After use, consider adding a small amount of petrol stabiliser to keep the fuel from oxidising and gumming up the carburettor.

NT0Z admitted that engine noise can be disturbing: set up the PEG in an out-of-the-way area and make a two- or three-sided sound shield from carpet-covered plywood or stiff cardboard (these look like small,

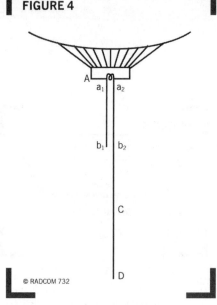

FIGURE 4: ANTENNA FOR BALLOONS AND ZEPPELIN AIRSHIPS AS SHOWN IN THE 1908 TEXTBOOK BY PROF J ZENNECK.

folding room dividers): **Fig 2**. Keep the sound absorber/reflector between you and the generator. Do *not* make a four-sided shield or put the generator into any type of box. Generators need airflow to keep cool.

**HF PROPAGATION – QUESTIONS REMAIN.** My April 'TT' item 'HF Supermodes or Multihop?', suggested, rather rashly, that virtually all very long-distance contacts made by the average amateur station depend primarily on supermode propagation not involving intermediate ground reflection(s). In support of this view I quoted Les Moxon, G6XN, and the work of Hans Albrecht on chordal-hop propagation paths between Australia and Europe. I did point to a possible exception where the 'ground' reflection points were sea or ocean areas.

My views are expertly challenged by Peter Martinez, G3PLX, who writes: "I was interested in your remarks and thought it might be worth looking through some of the oblique ionograms that I have recorded with my Chirpsounder system. This is an extension of the technique that I described in *RadCom*, July and August, 2000. With the present system, I sweep the receiver continuously over the 2 – 30MHz spectrum in synchronism with one of the sounders [listed in the August 2000 issue]. The ionogram shows the path length as a function of frequency. It seemed to me that such ionograms should show very clearly whether some of the longer paths are multihop or chordal.

"**Fig 3** shows one such oblique ionogram which may be relevant to this discussion. It is of the path from Ascension Island to my location at Kendal, Cumbria. It shows that the path is open between 18 and 26MHz on a day in March 2005. The path with the lowest propagation delay is clearly the one with the smallest number of hops (probably two or three). What is particularly clear is the way that paths with higher numbers of hops show on the ionogram. This is a very typical ionogram on this path and is clearly a multi-hop path. If a 'supermode' or 'chordal hop' path were present, there would be no sign of this pattern. Notice that the blackness of the traces represents the signal level, and there is a range of hop-numbers over which the signal strength is fairly constant. There would appear to be no strong case to suggest that progressively larger hop-numbers suffer increasingly severe absorption.

"As the distance gets longer, oblique ionograms like this show the multi-hop traces get closer together in path length, and in the end, for example on ionograms from the antipodes, it's not possible to separate them. But this doesn't mean that these longer paths are chordal, it just means that their number is too great and their path-lengths so similar that the ionogram is incapable of resolving them separately.

"Oblique ionograms like these appear in many textbooks on HF propagation. Some of them do mention chordal-hop paths, but only in the context of paths which symmetrically straddle the equator. You may perhaps argue that the Ascension-Kendal path doesn't fall into this category, but I would counter that you didn't restrict your comments about chordal paths in *RadCom* to trans-equatorial paths!"

Clearly, it is difficult for me to argue that many, if not most, amateur contacts over very long distances depend on 'single-hop' supermode paths such as that provided by chordal hop or by multiple hops between F- and E-layers which can act as a form of waveguide. I note that the Ascension/Cumbria path does straddle the equator although not symmetrically, and the 'ground' reflection points would be over the Atlantic Ocean. I also believe that D-layer absorption would be very low for the range 18 – 26MHz during daylight on this North-South path. There is evidence, as noted in the January 'TT', that supermodes can exist over East-West paths, with no intermediate ground reflections, over distances of up to about 7000km. While G3PLX has certainly dented my confidence in the value of chordal-hop and other supermodes to amateur radio, it remains high for bands lower than about 14MHz, for 'grey-line' propagation, for contacts above the MUF, and for round-the-world echoes etc.

All this suggests that there is still a good case for continuing HF propagation investigations, not only by amateurs, but also by professional researchers. Yet, at the present time, there are disturbing signs of a world-wide cutback of funding for HF propagation research. Dr Brian Austin,

# TECHNICAL TOPICS

G0GSF, who has currently been conducting an interesting amateur research project into low-power 7MHz propagation using small loop antennas with milliwatts of transmitter power for NVIS and medium-distance European contacts, warns of the imminent closure of the ionosondes run by the Rutherford Appleton Laboratory at Chilton and on the Falkland Islands, unless a new source of funding is found by June. This has been precipitated by a decision taken by PPARC (Particle Physics and Astronomy Research).

G0GSF writes: "This has horrified me, especially as the UK has been so crucially involved in ionospheric research ever since Appleton's discovery (along with Breit and Tuve in the USA) in 1924… The UK's role has been absolutely seminal in everything that followed that discovery. For the UK now to close down these two facilities strikes me as verging on scientific vandalism, or worse. As suggested by Sarah James, head of ionospheric monitoring at RAL, I have sent my views to Prof Keith Mason (**keith.mason@pparc.ac.uk**), Peter Warry and Sue Horne, all three at PPARC, Polaris House, North Star Avenue, Swindon SN2 1ET. Copies should be sent to Sarah James (e-mail: **S.F.James@rl.ac.uk**)."

It is worth emphasising that the 75-year data series of regular ionospheric measurements from Slough and Chilton is the key international measure of long-term change in the upper atmosphere.

The latest report is even worse. G0GSF tells me that "All UK research into solar-terrestrial physics is to be significantly cut back – some even curtailed (as at Chilton). A black day for British science."

**MORE ON THE CLASSIC ZEPP.** In the March issue, I noted that the Zepp antenna

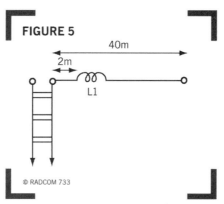

FIGURE 5: THE DL7AB MULTI-BAND MODIFIED ZEPP ANTENNA FOR 3.5, 7, 14, 21 AND 28MHz, WITH L1 BRINGING THE HERTZ ELEMENT INTO RESONANCE ON ALL THESE BANDS. (SOURCE: ANTENNENBUCH)

appeared to have started life as a balloon-suspended antenna invented by an Austrian engineer in 1908.

Niels Rudberg, OZ8NJ, has commented: "The item reminded me of what was written in an old book that I inherited many years ago from an elderly engineering colleague. This book, *Lehrbuch der Drahtlosen Telegraphie* [*Textbook of Wireless Telegraphy*], by Dr J Zenneck (Professor of Physics at the Munich Technical Highschool) is in the German language and was written in 1908.

Pages 193 – 95 are concerned with antennas for balloons and Zeppelin airships (reproduced here as **Fig 4)** ascribed to H Beggerow. It is described as using a Lecher-wire system and was thus intended from the beginning as an end-fed twin-wire resonant feeder system with a quarter-wave section (a1 – b1) and half-wave-radiator (Ab2 – D) and with a2 – D a three-quarter-wave. This is remarkable for 1908, and disproves my suggestion in the March 'TT' that the original version may have intended the unattached wire as a return path/counterpoise. It also shows that the antenna was associated from the start with the Zeppelin airships (by then virtually a generic name for airships in Germany). Thus, in its classic form, it is at least 98 years old!

Bernard Bale, G2ACN, has drawn my attention to the detailed coverage of the Zepp antenna in the German-language *AntennenBuch* by Karl Rothammel, DM2ABK, widely recognised as *the* standard antenna handbook in Germany. My 1961 first edition copy includes sections on 'The Zeppelin-Antenna', 'The Allband Zepp', 'The Double-Zepp' (in effect a centre-fed colinear array with twin resonant feeders), and 'The DL7AB-Allband Antenna'. In the DL7AB version, the radiating element is loaded with a small inductance to bring a 40m Hertz element into resonance on all the five then existing HF bands from 3.5 to 28MHz (**Fig 5**). The inductance, L1, is placed 2m along the wire and comprises five turns, 30mm diameter.

DM2ABK makes the point that, generally, for end-fed Zepps, a feeder length of exactly one-quarter wave should be avoided as this makes it easier to provide well-balanced voltage or current feed to the twin wires at the transmitter end.. He also describes arrangements providing wide-ranging balanced matching to suit either primarily current- or voltage-fed twin feeders: **Fig 6**.

Denis Walker, G3OLM, recalls being much impressed as a schoolboy by Dud Charman's 'Aerial Circus' lectures ('Model Aerials', *Proc RSGB,* No 5, Spring 1949) and later found himself working alongside G6CJ at EMI. He recalls that Dud was able to demonstrate, on his 3GHz table-top setup, the marked reluctance of the Zepp to accept the power offered to it.

He writes "In the late 1960s I came into possession of a small klystron (CV2116) which I was able to get going on 3.25GHz and reproduced G6CJ's model aerial collection. I toured many clubs giving similar table-top demonstrations of what I called 'The Bonsai Aerial Farm'. It always proved very popular.

"In common with Dud, I found the Zepp a most disappointing antenna. Yet there are many users who still sing its praises.

Although, in the March 'TT', you advocate a [very wide] spacing of 6 to 10in, it is clear that a much smaller spacing gives almost equivalent impedance, as shown in G3SEK's 'In Practice' column that immediately follows the March 'TT'.

"The use of too-closely-spaced open twin-line may go some way in explaining the highly-variable claims made for the classic Zepp antenna."

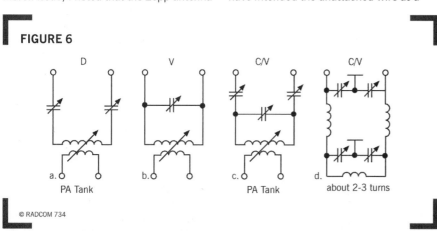

FIGURE 6: BALANCED TUNER CONFIGURATIONS FOR ZEPP ANTENNAS. (A) FOR PRIMARILY CURRENT FEED; (B) VOLTAGE-FED; (C) UNIVERSAL TUNER; (D) COLLINS UNIVERSAL BALANCED π–NETWORK. AN ALTERNATIVE FORM OF A BALANCED π–TYPE TUNER WAS SHOWN IN THE MARCH 'TT'. ALTHOUGH DM2ABK SHOWS THE TUNER COIL IN A, B AND C AS DIRECTLY-COUPLED TO THE PA TANK, FOR PRESENT-DAY EQUIPMENT, USE A COAX CABLE TO THE TRANSMITTER, LINK-COUPLED TO THE TUNER COIL WITH A SMALL NUMBER OF TURNS WOUND ROUND THE CENTRE OF THE COIL.

# TECHNICAL TOPICS

PAT HAWKER
37 DOVERCOURT ROAD,
DULWICH, LONDON SE22 8SS.

G3VA

# Technical topics

*Digits, DDS and the AD9951 chip ♦ More on series-modulated AM ♦ Dynamic power reduction – a correction ♦ Another ATU with balanced output ♦ The Beverage directional antenna*

**DIGITS, DDS & THE AD9951 CHIP.** In the April, 2006 'TT', Colin Horrabin, G3SBI, drew attention to the new Analog Devices AD9951 14-bit DDS chip able to clock at up to 400MHz while requiring significantly less power consumption (and hence cooler running), offering the potential for significantly improved DDS systems. In the May 'TT', I reported briefly that Nyall Davies, G8IBR, had mentioned that he had already found the results with the AD9951 very encouraging. He wrote: "This is a 14-bit DDS chip able to clock at 400MHz and thus will give a considerable reduction in DDS spurs compared to the 10-bit chip used in the 'PIC 'n' Mix' Digital injection system described by Peter Rhodes, G3XJP (*RadCom*, January-May 1999). ...Using a crystal oscillator at 100MHz multiplied by three, I have a phase noise of -139dBc at 2kHz off-carrier with the chip clocking at 300MHz. I now know why I have had to use so much filtering in the past."

Initially, G8IBR clocked the chip at 100MHz and was very impressed with the low level of the spurs: "The addition of 4-bits over the previous DDS contributes an additional 24dB. I use an external multiplier as the internal one does deteriorate the phase noise." **Fig 1** shows his 300MHz oscillator/multiplier clock. G8IBR writes: "I laid out a PCB and used surface-mounted components for the oscillator/multiplier. The FET was a MMBFJ310, the second one a MMBT2222A and I intend to use a BRF93A as the three times multiplier. In the event, I found that I had run out of BRF93 and used a BFY90. a very old transistor, certainly not surface-mount but it worked fine.

"I have built a number of these oscillators using wire-ended J310 and the plastic equivalent of the 2N2222A (MPS2222A). For use in 'TT', I suggest J310. 2N222A, BFR93A or equivalents. Further details including the connections to the AD9951 are on my website, **www.g8ibr.co.uk** and, in due course, if requested, I can make the PIC circuit and software available.

"One of the problems that needs to be watched is that the frequency generated is given by: Frequency = (Tuning word x clock)/$2^{32}$. I do my calculations using 24-bit arithmetic. This means that as the clock frequency goes up, my frequency accuracy can deteriorate to ±25Hz, although this remains acceptable for normal usage."

G8IBR is now working on moving the digital processing of his PIC 'n' Mix receiver to nearer the antenna with the hope of eliminating the IF strip. He has set out his thoughts in a detailed article on digital receivers, starting 'Towards the antenna', and including sections on 'Overall design'; 'Noise figure'; 'Third-order Intercept (IP3)'; 'About the Digital Down-Converter'; 'The Hardware'; 'The Achilles Heel'; 'Finally'; and 'Appendix'.

To quote briefly from the introduction: "Over the past 10 to15 years we have seen digital processing move into commercial amateur radio equipment to the extent that almost any transceiver will have its compulsory digital signal processing. We have seen it move into home-built equipment particularly with Peter Rhodes's PIC-a-STAR. This has provided the ability to select filtering and other DSP functions using audio or low IF processing. We are still dependent on a post-mixer crystal filter to remove large adjacent signals from the IF amplifiers and give receivers their large signal handling ability. When receiving a weak signal we have to provide 100dB or more of gain somewhere in the system and a large signal plus 100dB is too enormous to handle, so we filter the large signal out. These requirements dictate the configuration as requiring a mixer local oscillator and crystal filter. Any audio frequency processing DSP is only trimming the filter bandwidth below that already provided by the crystal filter and any very large, very close-in signals that we desire to get rid of will have their effect on the IF amplifiers.

"We are now able, or close to being able, to eliminate these for an HF amateur-band receiver. The exploration of this is the purpose of this article. It is probable that much commercial equipment will go this way and make many useful devices available, but the amateur requirement is particularly stringent. The target then is to make an RF sampling receiver covering the amateur bands up to 30MHz. ...The overall line-up becomes as in **Fig 2**."

After detailed discussion of the various stages, G8IBR comments: "The Achilles heel of any digital system is in its overload characteristics. Once an A/D converter exceeds its maximum, the data becomes corrupt and the result, performance-wise, is catastrophic. An analogue system does degrade more gracefully. The A/D converter has to handle all the signals within the band-pass filter. On the 7MHz band that means all the signals in the adjacent broadcast band. Setting the requirement at S9 +60dB is marginal as there are quite a number of stations that approach this figure..."

G8IBR concludes that the specification outlined in his article should provide a very commendable performance,

FIGURE 1: CIRCUIT DIAGRAM OF THE OSCILLATOR / MULTIPLIER USED BY G8IBR TO CLOCK AN AD9951 DDS CHIP AT 300MHz.

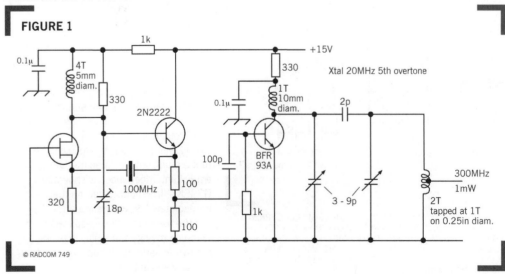

**FIGURE 1**

# TECHNICAL TOPICS

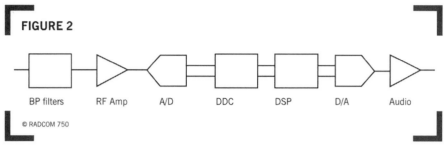

FIGURE 2: OVERALL BLOCK LINE-UP OF A RECEIVER IN WHICH THE INCOMING RF SIGNAL IS DIGITISED AT THE SIGNAL FREQUENCY RATHER THAN AT AF OR A LOW IF, AS DISCUSSED BY G8IBR.

adding "I would like to build [such a project] but my doubts are not on the technicalities but the time and effort that it would involve. Could I ever finish it and remain married? I am still debating with myself whether or not to plunge in, but feel that I should write up the ideas at this stage so that others can benefit from them."

**MORE ON SERIES-MODULATED AM.** 'TT' May, 2006 (pp102 – 103) included an item 'Ancient Series Modulation' in which I suggested that there is still a role in amateur operation for amplitude modulation, noting that there are still several active AM nets on 3.5 and 7MHz. Indeed the 'Box 25' twice-weekly net on 3.5MHz in which I participate includes a wartime colleague, Gerry Openshaw, G2BTO, who makes very effective use of a 25W AM Panda Cub transmitter even though some of us have SSB-only transceivers. It should not be forgotten that AM can be received as single sideband although better performance requires the reception of both sidebands, and preferably synchronous detection (possible on such receivers as the excellent AOR AR7030).

Dave Gordon-Smith, G3UUR, comments: "I was interested to see the section on series-modulated AM. This method of generating AM goes back to about 1924 and was popular with early AM operators because it didn't require expensive chokes or modulation transformers. It's also popular amongst the AM community in the States these days for the very same reason. A number of designs for low-power, series-modulated transmitters have appeared on the internet, and also in magazines devoted to vintage radio over the past ten years.

"The transmitters used to expand the BBC's national network in the 1930s used series-modulation. It was the most economical way of getting, with good linearity, the wide AF response they required. None of the modern [amateur] designs use any special techniques to overcome the modulation percentage limitation that is inherent with the use of identical valves in the PA and modulator as this requires extra hardware. In the 1930s, I believe they used to feed some additional current from the HT supply through to the modulator valve via a series resistor-choke circuit that bypassed the PA to achieve 100% modulation [see below – *G3VA*]. This increased the voltage across the modulator valve at the expense of the voltage across the PA, thereby allowing the voltage on the PA to reach double its standing value on peaks of modulation.

"You may be unaware of the fact that AM has never died on 160m. It has survived remarkably well through the years because it has continued to be used for local nets, particularly some club nets that did not transfer to 144MHz FM in the 1970s. A couple of years ago I wrote an article about AM activity in the UK for the American vintage-radio magazine *Electric Radio*. The AM activity on 80m these days is not new; it's just that more stations are realising how good AM sounds, and how much fun it can be. In fact, AM activity on 80m never totally died away and there was a revival that started in 1984 with G3LEO, G4XWD and a few others using old military gear that just grew and grew over the years. It has now become greater than the critical size needed to keep it going. AM has also been popular on 28MHz during the peak years of the last two solar cycles. At other times, 14,286kHz is used for transatlantic AM contacts. It is useful to have a carrier to squash the noise, particularly on 80m and 160m – and it is so much nicer to listen to AM than it is to SSB. Personally, I find it a strain listening to most SSB signals. …I'll continue to use my restored vintage and valve equipment on the amateur bands. …I learn more from playing about with these older rigs than I ever would shelling out several thousand pounds for a 'black box'".

By chance, soon after writing the May item, I came across a detailed article 'Series Modulation', by E B Vass, G8AD, in *The T & R Bulletin* (original title of *RadCom*), July 1938, pp2 – 5. This included a review of amplitude modulation techniques, including Choke (or Plate Modulation); Grid Modulation; Suppressor Grid Modulation; and then concentrated on Series Modulation. It described a relatively simple crystal-controlled transmitter CO-FD- and Series-Modulated PA comprising a LS5 PA triode valve in series with a PP3/250 triode as the modulator valve. Both these were directly-heated triodes and it was stressed that the PA filament needed to be fed from a separate transformer winding of low-capacitance to the primary as otherwise a high

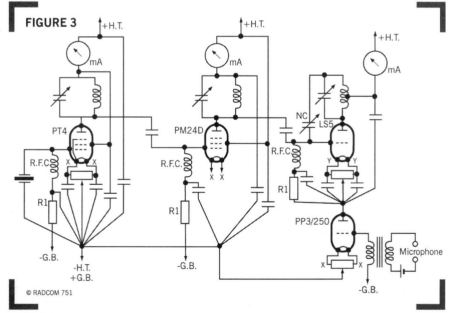

FIGURE 3: CIRCUIT DIAGRAM OF THE SERIES-MODULATOR TRANSMITTER DESCRIBED BY G8AD IN 1938. R1 IS 10,000Ω OR AS NECESSARY TO PROVIDE AN RF-DERIVED NEGATIVE BIAS OF ABOUT 42V (OR AS NEEDED FOR LINEAR OPERATION OF THE PA VALVE). THE ABSENCE OF ANY AF AMPLIFIER STAGE WAS POSSIBLE BECAUSE OF THE HIGH OUTPUT OF A CARBON-BUTTON MICROPHONE, BUT WOULD HAVE BEEN NEEDED FOR BETTER-QUALITY MICROPHONES. ALL FIXED CAPACITORS 2nF.

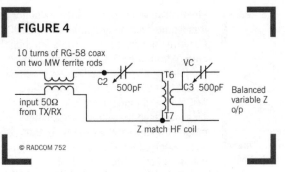

FIGURE 4: ATU PROVIDING BALANCED OUTPUT, AS USED BY 9H1FQ WITH BALANCING 1:1 TRANSFORMER PLACED BETWEEN THE TRANSMITTER AND Z-MATCH.

capacitance would limit the high-frequency AF response. The use of a carbon microphone with its high output meant that no audio amplifier was necessary but, of course, if a better quality microphone were used, a single-stage AF amplifier could have been used. **Fig 3** shows the three-stage transmitter implementing series modulation as described in 1938 by G8AD as a straightforward implementation of series modulation, using a PA bias of approximately -42V derived from the drive by R1 (10,000Ω). It was stressed that both the PA and modulator valves should have linear characteristics and be operated at a point mid-way along the linear portion of their characteristics to achieve optimum quality and a modulation depth approaching 100%. G8AD also showed how series modulation could be applied to a push-pull PA by using two modulator valves in parallel:

Neither of G8AD's designs used identical valve types for both power-amplifier and series-modulator, and no additional circuitry (see G3UUR's letter above) was used or apparently needed to achieve full depth of modulation. Indeed, he wrote "It must be emphasised that a low-powered transmitter, fully modulated, is far superior to a high-powered transmitter in which only part of the carrier is modulated." However I suspect that the modern solid-state series-modulator using identical FETs, as shown in the May 'TT', might give a greater depth of modulation with a greater share of the supply voltage across the modulator FET, along the lines noted by G3UUR.

Where a pentode valve, with its suppressor grid brought out to its own pin, is used, suppressor grid modulation forms an alternative low-cost approach, as in the wartime TT1154 with the STC PT15 valve. But series modulation should provide a higher audio quality with a modern microphone and audio amplifier.

**DYNAMIC POWER REDUCTION – A CORRECTION.** An unfortunate chain of events, primarily brought about by my mislaying a letter written in 2003 by Brian Horsfall, G3GKG, has resulted in an apology to him from Dave Gordon-Smith, G3UUR. As I noted in the May, 2006 'TT', I included a description of a dynamic power reduction circuit contributed by G3GKG. Then in the August 2003 'TT' I reported that Stan Brown, G4LU had pointed out that "the current transformer calculations [in the item] are theoretically incorrect" together with G3GKG's response " the statement on which my calculations are based are true enough for all practical purposes".

In 'TT' May 2006, in presenting details of an alternative power reducer, G3UUR and G4LU criticised G3GKG's item on the grounds that he had not included in the calculation the coefficient of coupling of his ferrite transformer". By this time I had mislaid G4LU's letter, and was unable to check whether this was really the case.

As a result, G4LU complained: "What I commented on was G3GKG's statement that the secondary current is determined solely by the turns ratio of the transformer and will deliver the reduced secondary current 'through any resistive load connected across it.' This is patent nonsense and defies the principle of conservation of energy. …The secondary current will depend on the current that the induced voltage drives through the impedance of the secondary winding in series with load resistance. That is fundamental Ohm's Law and there is nothing esoteric about it. The secondary current will decrease as the resistance is increased and, furthermore, it will become frequency-dependent. That applies whether the coupling factor is unity or less. The principles of a current transformer at RF are precisely the same as those at power frequencies and the current ratio and turns ratio are related only when the secondary is short-circuited. My experiments have shown that the smaller one makes the load resistances compared with the secondary inductance of the transformer, the nearer the secondary current becomes related to the turns ratio and the more constant is the output current/voltage with changes in frequency with load resistance. That is fundamental Ohm's Law and there is nothing esoteric about it."

G3UUR writes: "Having read G4LU's recent letter: I completely agree with him. I had assumed that his criticisms were due to G3GKG's omission of the coefficient of coupling from his calculations. I don't think you went into detail about G4LU's criticisms and I jumped to conclusions. You have to have a value of secondary inductance high enough compared to the load resistance at the lowest frequency in use to get it to act as a frequency-independent transformer above that frequency.… You may have noticed that I always specify that the reactance of the winding should be a minimum of 10 times the load resistance at the lowest frequency of use in my designs. Adhering to the general rule is particularly important in the two-transformer VSWR current sensor because the transformers will appear to give good balance and work well even when their inductance is far too low, but the sensor will be presenting a reactive load to the transformer, rather than a mainly resistive one, that isn't indicated by the sensor itself. This situation can give can give a solid-state power amplifier a very

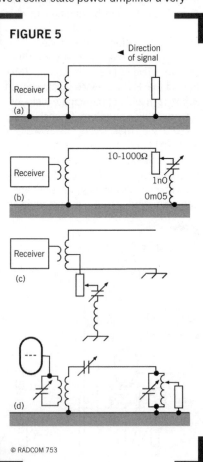

FIGURE 5: SOME OF THE MANY POSSIBLE VARIATIONS OF THE 'BEVERAGE' OR 'WAVE' RECEIVING ANTENNA. (A) AS USED BY SOME AMATEURS FOR DX RECEPTION ON 1.8 OR 3.5MHz. (B) AND (C) ARRANGEMENTS SUGGESTED BY HAROLD BEVERAGE IN 1923 FOR USE ON WAVELENGTHS "BELOW 450 METRES", WITH (C) A VARIATION PERMITTING THE TERMINATION TO BE AT THE RECEIVER END. (D) ARRANGEMENT ILLUSTRATED IN THE 1931 EDITION OF THE CLASSIC *ADMIRALTY HANDBOOK OF WIRELESS TELEGRAPHY*.

## TECHNICAL TOPICS

### FIGURE 6

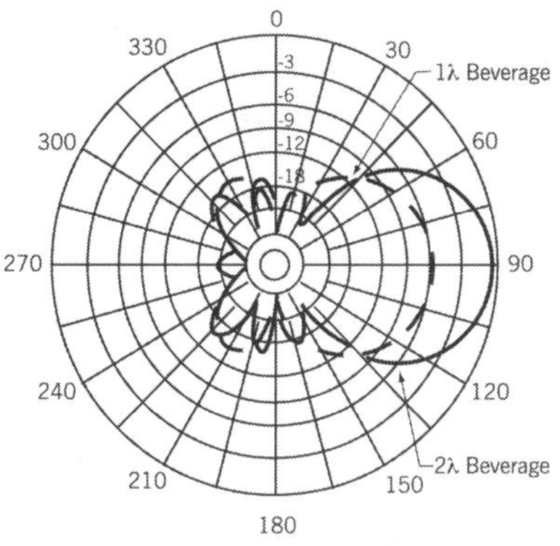

FIGURE 6: DIRECTIVITY PATTERNS OF ONE- AND TWO-WAVELENGTH BEVERAGE ANTENNAS SHOW A BROAD FORWARD LOBE AND GOOD REJECTION OF SIGNALS FROM THE SIDE AND BACK. AZIMUTH PLOT AT ELEVATION ANGLE OF 10°. FREQUENCY 1.83MHz. MAXIMUM GAIN 8.52dBi. (SOURCE: ARRL ANTENNA BOOK)

hard time, even when the VSWR indicator shows a perfect match! ...Apologies to Stan, G4LU."

### ANOTHER ATU WITH BALANCED OUTPUT.
In connection with recent 'TT' notes on the classic 'Zepp' antenna, several forms of roving balanced output from an ATU have been noted, Paul Debono, 9H1FQ, sends along the further configuration shown in **Fig 4**.

He writes: By putting the balanced 1:1 transformer at the input [to the Z-match tuner], it will not only reduce coax radiation, but the losses are much lower than when it is connected at the output. The ATU uses a standard Z-match coil [and variable capacitors] and is capable of coping with complex reactance and impedances and provides further balancing and isolation."

### THE BEVERAGE DIRECTIONAL RECEIVING ANTENNA.
To mark the death in 1993 at the age of 99 years of the professional (RCA) antenna engineer, Harold H Beverage, one-time W2BMI, a 'TT' item – 'Beverage and His Wave Antenna' – was included in 'TT' June 1993 (see also *Antenna Topics* p305). This traced something of its history and early use, by Paul Godley, (W)2XE in Scotland during the pioneering 'Trans-Atlantic Tests' of 1922. I included a number of diagrams showing various configurations of his travelling-wave antenna, some taken from his 50-plus-page paper 'The Wave Antenna – A New Type of Highly-Directive Antenna', spread over five issues of the *Journal of the AIEE* (March – July, 1923): **Fig 5**.

Recently, *QST* (April, 2006, pp33 – 36) has carried an excellent article 'A Cool Beverage Four-Pack', by H Ward Silver, N0AX, sub-headed 'Beverages are highly touted as receiving antennas – learn how they work, what they offer the DXer, and how to double their benefit by creating a switchable Beverage array'. It also notes that the American firm DX Engineering "has packaged all of the transformers and switching circuitry and marketed this as the 'RBS-IP Reversible Beverage System'".

N0AX points out that the Beverage provides good signal-to-noise performance, cutting down interference and static rather than strong signals. The Beverage needs to be close to the ground, instead of high in the air, and is best over ground with a medium-to-poor conductivity: "The trade off is length – Beverages need to be at least three-quarters wavelength long to provide useful performance." Optimum performance needs a length of some five wavelengths but even a two-wavelength Beverage shows a maximum directivity gain of some 8.5dBi and, more importantly, very good rejection in the reverse direction on the low-angle vertically-polarised component of incoming DX signals: **Fig 6**. N0AX illustrates antennas of 250 and 300ft long intended for use on 3.5 and 1.8MHz.

**Fig 7** shows the arrangement of a reversible Beverage

### FIGURE 7

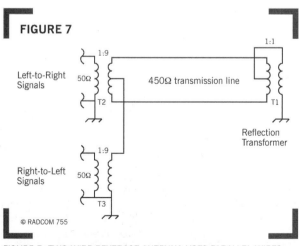

FIGURE 7: TWO-WIRE BEVERAGE ANTENNA USES PARALLEL WIRES (EG LADDER-LINE) AS BOTH AN ANTENNA AND AS TRANSMISSION LINE ALLOWING, WITH SUITABLE SWITCHING, RECEPTION IN EITHER DIRECTION. (SOURCE: *QST*)

### FIGURE 8

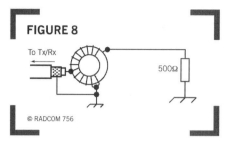

FIGURE 8: AN ALL-BAND 'TERMINATED LONG-WIRE' ANTENNA (HORIZONTALLY-POLARISED) AS USED BY G3SRO IN THE 1970s. ABOUT 400ft LONG, 40ft HIGH, TERMINATED BY NON-INDUCTIVE HIGH-WATTAGE RESISTOR EARTHED TO AN 8ft ALUMINIUM STAKE VIA A VERTICAL DOWNLEAD. FED BY STEP-UP BIFILAR-WOUND 1.5in TOROID TRANSFORMER WITH 3 TURNS INPUT AND 7 TURNS OUTPUT. SWR SHOULD BE LESS THAN ABOUT 1.8:1 ON ANY BAND BETWEEN 3.5 AND 28MHz.

antenna using 450- [or 300-] Ω ladder line and N0AX shows how two such arrays at right angles can be switched to receive signals from all four quadrants. For anyone interested in 1.8, 3.5 or 7MHz DX reception with space to erect a roughly two-wavelength span, N0AX's article should prove a useful guide. For example he writes: "Beverages are not fussy antennas. You can change their heights to clear obstacles, lay them on trees or other non-conductive supports, run them over uneven ground and still get good results. As you gain more experience and want to get the last drop of performance out of them, you'll want to pay more attention to balance symmetry and optimising terminal impedances. When you are getting started, it's more important to get the antennas up so you can begin to learn about them." Remember, the height of the antenna need be only about 8-10ft and results will not be improved by raising it. Of course, a high long-wire terminated antenna can form a very useful horizontally-polarised transmitting/receiving antenna, but this is not a Beverage antenna. 'TT' has several times drawn attention to an all-band terminated long-wire antenna first described by G3SRO in 1972: **Fig 8**.

### HERE & THERE.
In the May 'TT' (pp103 – 105) item on reduced radials for monopole or ground-plane antennas. Nick Brooks's callsign was wrongly given in the text as G4BMI but correctly as G4BMH in the caption to Fig 5. His average ground parameters should have been 0.005, 13 (0.003, 20 quoted) and the space saved by the double-folded counterpoise has a space saving of 70%, not 80%, ie down to 30% of the original size. Some pertinent comments on ground induced losses and a further suggested system of folded counterpoise/radials have come from G3UUR, but will have to wait.

# Technical topics

TECHNICAL TOPICS  
PAT HAWKER, MBE  
37 DOVERCOURT ROAD,  
DULWICH, LONDON SE22 8SS.  
G3VA

*Rapid-acting circuit-breaker* ♦ *Verticals with compact radials* ♦ *More on Windom-type antennas* ♦ *More on alternative power sources* ♦ *Understanding polyphase networks*

**A RAPID-ACTING CIRCUIT BREAKER.** There are times when it appears that the era of amateur radio experimentation in fields *other* than 'digital', based on commercially-available chips, is retreating into past (happy) days. But then comes evidence that experimentation in circuitry and antennas still survives, if less evident than in some past decades.

Clearly, John Lien, LA6PB, still enjoys experimentation and in finding solutions to long-standing problems: He writes: "As experimenting radio amateurs, we often cause a short-circuit and are thankful that the system is protected by a fuse. But then the question arises, Do we have a replacement fuse at hand and is it easy to fit?'. Often the answer to one or other or both of these questions is 'No'.

"Most transceivers have a 13.5V output jack, useful as a power source for our many external devices. On my transceiver, this outlet is protected by an internally-fitted 500mA fuse. After having dismantled my transceiver for the nth time to replace the fuse, I decided that a quick-acting circuit breaker might provide a solution to my fuse problem.

"After some experimentation, I came up with the circuit shown in **Fig 1** using only junk-box components. At its heart is a fast reed relay.

"The most important electrical factor for any relay is the number of ampere-turns, that is the current through the relay coil multiplied by the number of turns. In my case, I found a very small reed relay manufactured by Hamlin, 5V, R = 500Ω, contacts NO, 0.5A, 200V with closing and opening times less than 1ms. It dawned on me to provide this relay with an extra coil to sense the current to be protected.

"To decide how many ampere-turns would be necessary to close the relay, I first wound 50 turns around the original relay coil and found that the contact closed at 660mA, In other words the relay needs 33 ampere-turns to close. I decided that the relay should be made to close at 400mA, so as to protect the 500mA fuse. Hence I wound on the relay a coil of 82 turns of 0.55m (AWG nr 24) copper wire. Tested, it duly closed the relay contact at 400mA. The resistance of the coil is negligible. It is important that the magnetic field developed by the two relay coils should be in phase, in order that the resulting field is not weakened.

"From Fig 1, it can be seen that when voltage is applied, the green LED lights. When the current through the extra relay coil, L, exceeds 400mA, both relays K1 and K2 will close and be latched. Relay K2 is a small 12V relay, R = 500Ω, contacts NC/NO, about 2A. When the relays close, the current is shut off, the red LED lights and the green LED turns off. On closing the push-button switch, the circuit breaker is once again ready for operational use.

"It is not necessary to connect diodes across any of the relay coils. The non-critical value of the capacitors is 1000µF, ensuring that the relays always get the (almost) full voltage.

"While testing, I found that even a 100mA fuse could withstand a direct short-circuit through this device. I feel that the use of an extra current-sensing relay coil could also prove useful in other applications where there is a need to limit the current."

LA6PB's design has been previously published in the Norwegian journal *Amatørradio*, March 2006.

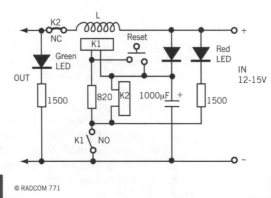

FIGURE 1: LA6PB'S FAST CIRCUIT BREAKER FOR SUCH APPLICATIONS AS PROTECTING A 500mA FUSE MOUNTED INTERNALLY IN THE 12V SUPPLY POWERING EXTERNAL UNITS.

**VERTICALS WITH COMPACT RADIALS.** Dave Gordon-Smith, G3UUR, writes: "I was interested to see G4BMI's piece on the use of a single [space-saving] folded radial for use with quarter-wave verticals in the May 'TT'. I'm surprised that he finds the performance is so good at low heights. Is this really the case, or is it that *EZNEC* just can't model the real situation adequately? G4BMH's design is quite clever in the way the resultant current in the folded part of the single radial almost balances that in the single-wire part to minimise the ground loss caused by magnetic field induction at low heights. However, there is a fundamental problem with the different velocity factors in the folded and straight parts of the radial. I doubt whether the ratio of the two parts of the radial could be adjusted to obtain minimum ground loss from magnetic induction at the same time as obtaining the best omnidirectional characteristics. Also, the antenna is quite asymmetrical, and the electric field induction loss could be quite considerable at low heights. I believe that *EZNEC* is inferior to *NEC-4D* (the professional version) and it seems even *NEC-4D* has trouble accurately modelling unusual antennas close to real ground.

"I suspect that, in practice, there will quite considerable ground loss at low heights with the single-wire folded radial, and only practical experimentation would establish how much. Again, it comes down to careful measurements of the feed resistance or bandwidth versus height above ground to ascertain the real effect of ground loss. But, even if the ground-plane (GP) has to be elevated to reduce this and get good DX performance, if only one-band operation is required, G4BMH's approach is a lot simpler than putting in an extensive earth system. However, since the single-radial GP system still has to be supported on either side of the radiator base, and there may need to be a compromise between the azimuth pattern and ground loss, I would still go for a two-radial GP using double-folded radials (three parallel wires connected in series electrically (see **Fig 2**) and balance up the radials by trimming their lengths while monitoring the RF current in each using a clip-on current transformer.

"Fibreglass fishing poles are readily available these days in lengths up to 33ft and would be a very convenient way of supporting a 14-gauge wire ground-plane antenna on 7MHz. Two 5m (16ft) fishing poles attached with conduit clamps to a wooden cross-piece at the base of the radiator could be used to support the folded radials if no convenient trees or posts were available. The spacing of the folded radial wires could be as little as one inch, or even slightly less. Since each radial is a single length of wire folded back on itself twice, the upper section could be taped to the fishing pole along its whole length. The middle section could be looped back from the far end of the pole, where it is very thin, and taped most of the way back along the thicker part of the pole. The bottom section could be slung below the other two using nylon cord at both ends to hold it in place. In order to try to get the feed impedance nearer to 50Ω, the length of the radiator could be increased beyond a quarter-wavelength and the radials trimmed to restore system resonance."

MORE ON WINDOM-TYPE ANTENNAS. I hesitate to refer again (see 'TT' April 2006, pp72 – 74) to the 'Off-Centre-Fed Antenna (OFCA)' as this form of so-called 'Windom' antenna has been well-covered by Peter Dodd, G3LDO, in his June 'Antennas' column. However, to correct a typographic error and to add further information I am indebted to Martin Hengemühle, DL5QE, for noting that my reference to a Windom-type antenna popular in Europe as the 'TD4' should have been 'FD4'. He points out that the FD4 has been produced and marketed for some decades by the German firm Fritzel with FD4 standing for 'Fritzel Dipole 4-bands' (3.5, 7, 14 and 28MHz) with a smaller version, the FD3, for the 7, 14 and 28MHz bands. It has a 50Ω coax feeder off-centre-fed to the radiating element via a 1:6 toroidal transformer. DL5QF adds: "It is quite popular in Germany despite its tendency to produce high RF in the shack with some feeder lengths. It is used primarily on 3.5MHz and rather less so on 7MHz. It tends to be less susceptible to rising SWR at band edges, especially on 3.5MHz than conventional half-wave dipoles, although whether this is due to the influence of the 1:6 balun or to multiple-resonances etc, has yet to be discovered by experiment."

DL5QF has carried out a series of experiments using scaled down 50MHz and 144MHz versions of FD4-type antennas in an attempt to achieve "better behaviour" with reduced outer-braid radiation from the feeder. For example, he fitted ferrite 1:1 choke baluns on the feeder, but was unable to obtain significant reduction of the radiation. In brief, he believes that: "Any FD4-like version of the Windom antenna (and possibly other versions) is not an antenna that can be easily 'tamed'. While, in remote areas, feeder radiation may well add some wanted low-angle radiation, this may not be a good thing in rural or urban areas [ie may cause RFI problems, although these may be less of a problem even in urban areas than suggested by DL5QE – G3VA]. However, several friends have found no evidence of feeder radiation on 3.5MHz when using short feeders (less than 0.1-λ to 0.2λ)…"

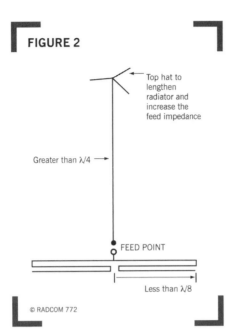

FIGURE 2: QUARTER-WAVE VERTICAL ANTENNA WITH TWO LINEAR-LOADED RADIALS FOR USE WHERE ONLY A LIMITED SPAN (LESS THAN A QUARTER-WAVE) IS AVAILABLE FOR THE RADIALS.

In the April 'TT', it was noted that all versions of off-centre-fed antennas, including both the original single-wire transmission line and the twin-wire 300Ω version, are inherently prone to at least some radiation from the feeder. To my mind, it would seem that the single-wire, single-band version offers the best possibility of reducing this to a minimum if adjusted in accordance with the 1929 advice given by Windom based on the work at Ohio State University (see the April 'TT'). Where one or more baluns are used in conjunction with twin-wire or coaxial feeder, there will also be power losses brought about by reactance. This would, for example, account for the finding that the FD4 shows a broader bandwidth than the conventional half-wave dipole.

The April 'TT' noted that the original single-band antenna required the tapping point to be adjusted to suit the diameter of the single wire feed-line. The later 'one-third' tapping point for dual- or multi-band operation still took the wire-diameter into account. An important contribution was made by C F Turner, ST2CM, in The T & R Bulletin, July, 1938, pp6/7, with an added note by 'Dud' Charman, G6CJ. This confirmed the use of the 'one-third' tapping point as providing operation at fundamental and harmonic resonances, and was based on a practical test of a 33ft top using 12SWG wire supplied by a feeder of 18SWG and operated on 14 and 28MHz: "Using a thermo-milliameter shunted across about 6in of feeder, a comparison of the feeder current was possible at various points throughout its entire length. No variation could be detected on either frequency, indicating that a fair approximation to a correctly-matched system had been obtained. Results, when compared with the 33ft centre-fed Hertz, was very gratifying. G6CJ pointed out inter alia that "The chief difficulty with the multiband aerial has still to be met, namely, that the length of top varies with the order of the harmonic.

"There are two ways of dealing with this. The first method is to employ a longer top, so that it is only used on the harmonics as the length tends to a more finite value. Our own experiments using a 136 – 138ft top confirm this. The second method is to make the tuning of the top very flat, so that the length is of less importance. This can be effected by the converse of ST2CM's method, increasing the size of the top, rather than by reducing the diameter of the feeder, possibly by employing two wires in parallel. It should be quite possible in this way to make a Windom which is practically aperiodic."

MORE ON ALTERNATIVE POWER SOURCES. The June 'TT', in discussing non-mains power sources, noted that for extended field operation with transceivers rated at the order of 100W or more, the Petrol Electric Generator (PEG) remains supreme despite such safety problems as attempting to refill the fuel tank without interrupting operations and the significant nuisance caused to operators and any local residents by the noise of the usual two-stroke engine. One reader even points out that the use of carpet-covered plywood to form a sound barrier (as suggested by NT0Z and shown in the June Fig 2), can present a hazard if petrol/oil is spilled, since it will take some time to dry out.

It was thus with considerable interest that I read in QST (June 2006, pp44 – 45) an article by Kim Owen, KO7U, 'No Fuelling: Field Day with Hydrogen', sub-headed: 'A hydrogen fuel cell that powered a Utah club's CW station last FD was so quiet it shared quarters with the ops. The half gallon of water it generated as waste was a bonus."

KO7U relates how, in June 2005,

# TECHNICAL TOPICS

KK7APR operated in the Wasatch-Cache National Forest at a height of 8978ft, powered by a hydrogen fuel cell of the type more descriptively called a Photon Exchange Membrane (PEM) fuel cell and manufactured by Ballard Power Systems, Inc of British Columbia, Canada (www.ballard.com). This cell is rated as delivering 1200W or 46A at 26VDC for a minimum of 1500 hours, which is just over two months. Unloaded, the output voltage is about 45VDC. The club station used a DC/DC converter for regulation that met the specifications of a typical HF radio running 100W with an input voltage range from 20 – 50VDC and an output of 13VDC at 20A continuous. Unfortunately, the converter was designed for use with a single transceiver although the fuel cell could have provided more than enough power for two HF stations. The fuel was contained in a 'K-size' bottle (about 4ft high) of hydrogen gas at a pressure of 2000psi. KO7U wrote: "To meet the lower input pressure requirements of the fuel cell, we used a regulator and set it at about 20psi. The cell will accept gas pressure at 10 to 25 lb/in$^2$. We consumed about 1100psi from one bottle over the 24-hour period." We expected the output power performance to be degraded because of the thinner air (oxygen) at the high altitude. On site, we could not see noticeable differences from the performance we had seen at an earlier test at an altitude of 4500 feet. We felt that the heavy duty cycle and switching of CW would be the worst environment that a fuel cell would see in amateur radio"

The type of fuel cell used measures less than 24 x 12in, is just over a foot high and weighs about 29 pounds, much less than a full size conventional battery. It proved so much quieter than a generator that it was kept in the same trailer as the CW equipment and operators. It requires a computer and 18 – 28V DC to start up and shut down. Software that comes with the cell provides ample information on the various parameters such as output current and voltage, fuel consumption, temperature, oxygen and clear text error codes. There are built-in safety features to protect both the system and people. For example, a hydrogen detector will shut down the system if dangerous levels are found. The cell will operate from 37 – 86°F (3 – 30°C).

It was in 1999 that 'TT' reported promising development work on proton exchange membrane fuel cells at Ballard Power Systems (*RadCom* March, 1999 see also *Technical Topics Scrapbook*, 1995 – 1999, p204) and by American Power Corporation ('TT' August, 1999 see also *TTS*, 1995 – 1999, pp 290 – 291). Both projects were targeted at electric vehicle applications, but APC was also developing packages for domestic power units about the size of a refrigerator, see **Fig 3** reproduced from the 1999 item.

In the March. 1999 issue, I quoted Tom Gilchrist, writing in the November 1998, *Spectrum* as explaining: "The PEM type of

**FIGURE 3**

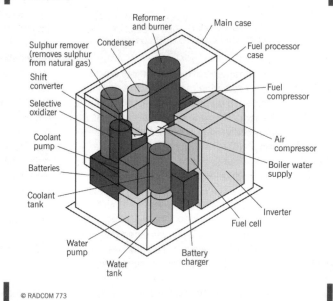

FIGURE 3: THE 1999 PROTOTYPE OF AMERICAN POWER CORPORATION'S PACKAGED DOMESTIC POWER SYSTEM PROVIDING SOME 3kW AT 120V AC (10kW PEAK LOAD FROM THE EIGHT BUILT-IN LEAD-ACID BATTERIES). IN THE AUGUST 1999 'TT', IT WAS STATED THAT APC HOPED THAT, WHEN IN FULL PRODUCTION, IT EXPECTED THE PRICE WOULD BE OF THE ORDER OF $5000, INSTALLED. SIZE ABOUT THAT OF A LARGE REFRIGERATOR (ABOUT ONE METRE CUBED). AT THE TIME, I THEN SUGGESTED THAT SUCH SYSTEMS COULD SUPPORT FULL-POWER AMATEUR OPERATION FROM REMOTE SITES OR HOMES NOT CONNECTED TO THE NATIONAL GRID.

fuel cell, so-called because a hydrogen ion is just a single proton, offers relatively benign characteristics as a power source for electric vehicles. Operating at about 80°C and employing a thin plastic sheet as its electrolyte it is easy and safe to handle in manufacturing and in later use. Unlike some other electrolytes, its solid plastic membrane can tolerate a modest pressure differential across the cell, making for easy pressurisation, which increases power density, simplifies the rest of the system and reduces cost... other attractions include low emissions, high conversion efficiency, and fast start-up and fast response to transients."

Clearly, the PEM fuel cell represents at last a serious and practical, if still emerging, contender for Field Day and emergency communications, although I remain uncertain as to how the overall costs compare with petrol electric generators and availability in the UK.

A truly 'green' source of limited power is the solar array of photovoltaic cells that can be an effective means of keeping batteries charged. Large solar arrays are also being used to supply domestic power in California in hybrid systems where there is also a connection to the mains supply. At night, the domestic electricity is supplied from the mains; in daytime, the solar array takes over with any surplus fed back into the mains, reversing the meter reading. The solar array is thus a 'green' source of power that can also reduce the cost of the mains supply, although it apparently takes some six years or so to justify the capital cost of the solar array.

The bugbear remains the cost of photovoltaic cells and the relatively low conversion efficiency. Despite much research effort, including a major Japanese government-sponsored project aimed at reducing cost to around $1 per watt, solar energy costs remains relatively high, with conversion efficiency of most cells now in use still of the order of 10%, often less.

A new approach utilises emerging nanotechnology. A three-page feature article in *New Scientist* (27 May, 2006, pp45 – 47), 'Two For the Price of One', is sub-headed 'Make solar cells as small as a molecule and you get more than you bargained for. Could this be the route to limitless power, asks Herb Brady'. To quote selectively from the introductory remarks: "In all solar cells now in use – in everything from satellites to pocket calculators – each incoming photon contributes at most one energised electron to the electric current it generates. Victor Kilmov, a physicist at Los Alamos National Laboratory in New Mexico, has broken through this barrier. He has shown that by shrinking the elements of a solar cell down to a few nanometres, or millionths of a millimetre, each captured photon can be made to generate not one, but two or even more charge carriers. Producing this multiplicity of electrons – an achievement that has been replicated by a group at the National Renewable Energy Laboratory (NREL) in Golden, Colorado – is a remarkable piece of physics. If the effect can be harnessed, it could change the whole energy debate by making solar power much more efficient and economical. While there are many ongoing efforts to improve solar efficiency – by concentrating sunlight, for example, or by making it easier for electrons to move around within a cell – the new approach is unique in that it gets to the very root of the process and also complements other methods. For decades, photovoltaics have been stranded on the effete fringe of energy technologies – ideal for niche applications such as satellite, but not economically competitive here on Earth."

Herb Brady's article concludes; "If each photon can generate multiple charge carriers, the overall power efficiency of solar cells could be dramatically increased. The world record for a ground-based cell is 24.7%, achieved at the University of New South Wales. Klimov predicts that the multiple-carrier generation could one day yield a cell with double that efficiency, approaching 50 per cent… With more work, the chips cranking out extra electrons in New Mexico and Colorado could one day bring a bright solar future for us all."

UNDERSTANDING POLYPHASE NETWORKS. W J (Pim) Niessen, PA2PIM, writes: "Over the past 30 years, polyphase networks have been mentioned several times in 'TT' and many aspects of these fascinating networks have been highlighted. However, little has been mentioned in the amateur literature about the insertion loss; an exception is an article by Yoshida Tetsuc, JA1KO, in 'Polyphase Network Calculation using a Vector Analysis Method' (QEX 1995, pp9 – 15).

"What I have not seen mentioned anywhere yet is that the insertion loss of a polyphase network depends on the signal direction through the network. It is a pity that, say, 9dB of loss is introduced in the signal path of a DC (direct conversion) receiver which can be avoided. All it takes is inserting the network the right way round. Indeed, designs have been published where these improvements are possible without sacrificing other performance parameters.

"These aspects, together with other useful tips are described in my paper 'Understanding and Designing Sequence Asymmetric Polyphase Networks', sub-headed 'The functioning of Polyphase networks described, a spreadsheet macro to design them, and useful tips to get an optimum result' (Version 3.5, 2006-03-01, pa2pim@amsat.org)".

This is an extensive 23-page, English text paper, including Appendix, 15 diagrams and list of 25 references including six to 'TT' items from 1973 to 2001 To extract a few brief tips from his Section 8 'Ten Tips for Designing Polyphase Networks: "8.3 Place the most accurate components at the output… 8.4 Insertion loss depends on the signal direction through the network. For a network with the same resistors in each section, the insertion loss will be minimum where the smallest capacitors are put at the output of the network, ie the smallest RC product must be put at the output. For a network with the same capacitors in each section, the insertion loss will be a minimum when the largest resistors are put at the output of the network, ie the largest RC product must be put at the output. The RC products for each section must either increase or decrease from input to output for minimising the insertion loss. A measured example is shown in **Fig 4**".

PA2PIM's paper contains a number of other ways of achieving maximum performance using polyphase networks, as well as much fundamental design procedure, etc.

HERE & THERE. Lawrence E Stoskorp, N0UU, comments: "'TT', April, 2006, p72, has another note on hybrid cascode power amplifiers. The spirit of building continues. However – maybe I'm preaching to the choir – the main points in using this circuit were ignored in the brief item [from G3DXZ]. With the input properly designed in Class-AB mode, the input impedance can set the linearity of the whole circuit at a fixed gain. This gives quite a few variables to set in a very low-noise, unilateral amplifier. The FET portion is well discussed by Hayward in his *Introduction to Radio Frequency Design*, pp215 – 218. It deserves mentioning again."

John D Wightman, ZL1AH (formerly G3AH) writes: "I am flattered that my ramblings ['TT', April 2006, p75] found space in the column. However there is a small problem about which you may get some feedback. Someone will point out that the Marconi SW8B transmitters would not operate on 32MHz. When I wrote to you, I must have been having a 'senior moment', because I meant to say that I worked VOAC on about 8MHz [possibly about 9.4MHz ie the 32m given in his original letter – *G3VA*]."

Roy Mander, GW4DYY, noted the small low-power wave generator described and illustrated (Fig 1) in the June 'TT' (p71) utilising the new super Ferrofluid lubricant. He reports that a basically similar generator (without the special lubricant) is used in a self-charging torch that he picked up recently at a car boot sale for £1. He writes: "It works very well indeed and appears to be basically similar to Chung's self-charging of a super-capacitor 'battery' by passing a magnet through a solenoid, The torch has never gone flat on me. If it appears to be a little 'low', a few quick shakes restores it to its original brightness as the magnet slides through the coil. The exterior switch is magnetic rubber that closes an internal reed switch. The whole is completely watertight and has a bull's eye lens to focus the beam from the high intensity LED fed through a full-wave rectifier."

**FIGURE 4**

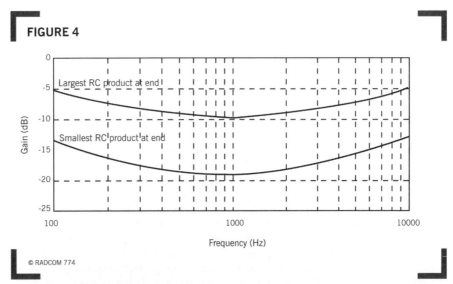

FIGURE 4: THE INSERTION LOSS OF A POLYPHASE NETWORK WITH EQUAL CAPACITORS DEPENDS ON THE SIGNAL DIRECTION THROUGH THE NETWORK (SEE TEXT). IN THIS EXAMPLE, THE DIFFERENCE AMOUNTS TO SOME 9dB (PA2PIM).

# Technical topics

PAT HAWKER, MBE
37 DOVERCOURT ROAD,
DULWICH, LONDON SE22 8SS.

G3VA

*Alford circuit for balanced feeders* ♦ *Hazards and safety hints* ♦
*SMD jig revisited* ♦ *Beverage antenna lore* ♦ *Amplitude
modulation efficiencies* ♦ *Revisiting the conjugate match* ♦
*Past glories and future hopes*

**ALFORD CIRCUIT FOR BALANCED FEEDERS.** The increasing use of balanced feeders for multiband doublets and other centre-fed antennas as well as for the classic Zepp antennas has been reflected in several recent items in 'TT' (March, June and July 2006). Most of these arrangements are well-known, but now Stan Brown, G4LU provides information on the little-known Alford network. He writes:

"One circuit that has not yet surfaced in the transformations for balanced feeder working is the Alford Circuit. I first encountered this arrangement about fifty years ago in a commercial transmitter. I then used it at a former QTH, where I was able to employ a dipole antenna with balanced feeders. The circuit is shown in **Figure 1.** If each capacitor is proportioned so that the magnitude of their reactances is twice that of the inductance, L1, then the output voltages will be balanced. The input of the basic circuit will present a load impedance which is partly reactive and a second inductance, L2, is required to cancel this out. L2 can also serve to make the overall circuit into an 'L' arrangement to bring the input down to a low impedance suitable for a co-axial cable link to the transmitter/receiver. This arrangement assumes that the balanced antenna is presenting an essentially resistive load to the circuit; if this is not the case the second reactance (L2) may need to be capacitive instead of inductive. In my installation this was not necessary; all the amateur bands were covered with two variable inductors and switched capacitors.

"There is one snag – and what circuit does not have at least one? There is an anomalous condition where the circuit can be adjusted for equal voltages on the feeders but they are not completely opposite in phase. The solution to find the correct adjustment is quite simple. Each feeder is provided with a diode voltmeter connected via low-value capacitors (say 5 – 10pF). If the voltmeter connections are joined together both meters will read zero if the circuit is correctly adjusted; if not, they will show a residual reading. Readjustment will provide the correct output.

"In amateur use, it may be more convenient to replace inductors with capacitors and vice-versa but the same principles will apply."

**THE HOBBY, THE HAZARDS, THE SAFETY HINTS.** As ever, the times they are a-changing. Few would dispute the fact that Amateur Radio is today far less hazardous than in the days when high-voltage electric shocks from the mains supply or from high-voltage PSUs were frequent; when it was said you could spot an operator by his RF or soldering burns. Today, there must be many amateurs who have never experienced such incidents, protected by the battery of regulations governing electrical installations, hazardous chemicals etc. And if at times, it seems, as in the case of lead-free solder, that bureaucracy may have gone a little too far, it must be to our advantage that the hobby is now relatively free of risks. But it needs vigilance to maintain this state – there are still a few hazards lurking ready to strike the unwary.

Geoff Turner, M3FFT, adds to the "12 commandments" for safe operation of Petrol-electric generators ('TT', June 2006). He writes:

"*Sound Proofing:* Whilst carpet will undoubtedly absorb a little noise, it will more readily absorb oil, petrol and other inflammable substances and will become a very real fire hazard in its own right. On a field operation the results could be devastating. Much better to use acoustically engineered room insulation foam which is self-extinguishing and will not absorb inflammable liquids but, sadly, it is not cheap.

"*Fire Extinguishers:* If you have to use a $CO_2$ fire extinguisher for the first time and it is for real, what often happens is that the noise from the extinguisher is so sudden and so loud that the extinguisher is dropped, causing injury to the nearest foot and having little effect on the fire. If it is used correctly, the greatest danger is re-ignition of a liquid fire, as it does little to cool the original heat source. Much better to use a multi-purpose dry powder extinguisher, but note that the powder could damage your rig so keep it well away from and upwind of any potential fire source. For grass fires a couple of buckets of water are cheap and quite effective, particularly as they do not require you read the instructions before use, although once again your rig could get damaged.

"*Fire Safety Advice:* This is available from **www.firekills.gov.uk**"

Personally, following some problems with aged mains-transformers, I keep a smoke detector, fire blanket and a fire extinguisher readily available in the shack. A fire blanket has the advantage that it can put out small fires without causing additional damage to the equipment.

A safety caution when dealing with high voltages that remains as true as ever is summed up in the old adage: "*Its volts that jolts, but mils [mAs] that kills*". The danger of electrocution or serious injury stems from even quite low current flowing through or near the heart and causing fibrillation. The current flowing will depend on skin resistance and the area of contact; with damp or wet skin contact the body current may result in fibrillation and could prove fatal with a voltage as low as 50V, if the current path is through the heart. On the other hand, with dry skin and limited contact area, a 1000V shock, may give the unwary recipient a hefty kick but no real damage other than temporary trauma. A traditional precaution when working with high voltages is always to keep one hand in your pocket to eliminate the risk of a current flowing through the heart. In workshops etc, it is prudent to stand on a rubber mat or at least wear rubber-soled shoes.

Workshop chemicals can be a safety hazard. Solder flux can induce a form of asthma. Good ventilation is the answer.

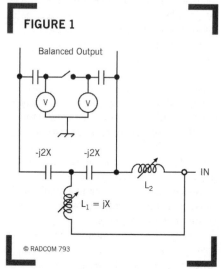

FIGURE 1: THE ALFORD CIRCUIT FOR PROVIDING BALANCED OUTPUT FROM AN UNBALANCED (COAXIAL) INPUT AS DESCRIBED BY G4LU

## TECHNICAL TOPICS

But undoubtedly, the most severe hazard faced by radio enthusiasts in the present era of factory built equipment comes from working on the erection, adjustment, maintenance or dismantling of antennas and the associated ladders, poles, masts, towers, rooftops, etc. Precautions when using ladders are widely published. Professionals are expected always to use climbing harnesses when working at height.

**SMD JIG REVISITED.** When it was starting to become clear that constructors would need to come to terms with surface mounted devices (SMD), 'TT' November 1993 (see also *TTS 1990 to 1994,* pp 237-239) included some basic advice mostly based on an informative article in *Electronics Australia.* I also gave details from G3ZHE of a home-made jig to hold the miniature devices to a PCB for soldering. Colin Horrabin, G3SBI has recently pointed out that this is continuing to prove useful to home constructors, and believes it should be brought again to the attention of readers who may have more recently begun to cope with SMD.

To quote the 1993 notes provided by Albert Heyes, G3ZHE: "Your [earlier] comments on surface mounted construction raised a smile. I'm 58 years old and find that as I get older the electronic bits get smaller! But I have built a number of SMD items over the past year for QRP operation. They work first class and now I find pushing component wires through holes seems all wrong.

"To make SMD construction easier, I have made a jig (**Figure 2**) for holding the components to the PCB. It resembles a small gibbet made of double-sided PCB and mounted on a small wooden base. I aligned two 2BA nuts and soldered them to the arm, then used an old knitting needle suitably weighted at the top. The circuit board is then placed with the component under the needle point. One end of the component can now be soldered using a small iron. The surface tension of melting solder will move small components if they are not pinned down."

### G3NOQ ADDS TO BEVERAGE ANTENNA LORE.
The July 'TT' item on the classic Beverage directional receiving antenna reminded Alan Boswell, G3NOQ of work he did on this type of antenna some 15 years ago, using NEC modelling. He writes:

"These antennas work well as travelling-wave structures a bit like the rhombics, with a forward beam that might be 25-30-degrees wide, depending on the length in wavelengths. As mentioned in July, the

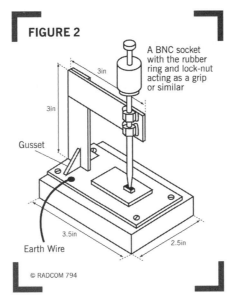

FIGURE 2: G3ZHE'S HOME-MADE JIG MAKES IT EASIER TO HOLD SMD DEVICES TO THE PCB WHILE SOLDERING. AS FIRST DESCRIBED IN 1993.

gain is lowish, around –25 to –10dBi, but this is acceptable for receiving at frequencies where the external noise is much higher than those levels (10MHz and below). Like rhombics, they are good wideband antennas providing an octave or more of performance, and they have the advantage of needing to be only a few feet off the ground in the HF region.

"I found that Beverages work well with poor ground of 2-10 mS/m conductivity, but this is not at all critical, as NOAX reported. In fact, the only type of ground on which a Beverage is useless is a perfectly conducting ground plane!

"For anyone who has plenty of space, two or more Beverages can be arrayed side-by-side and spaced apart by half a wavelength or more at the highest operating frequency. The beam can be steered some way either side of the geometrical axis by phasing, but of course it cannot be steered beyond the limits of the single-element azimuth-pattern. Like a rhombic, the Beverage is of most use for receiving from a particular range of directions."

The main difference between the low Beverage and the much higher rhombic antennas is that the rhombic has not only directivity but also high gain and is thus suitable for transmission as well as reception, and is a highly effective antenna at HF, VHF and even UHF where the space required even for 5λ sides is very modest.

### AMPLITUDE MODULATION EFFICIENCIES.
Recent 'TT' items on series modulation may have left some readers wondering why, with its significant economic advantages over high-level anode (plate) modulation, it was not more widely used by amateurs in the post-war period, yet still deserves consideration. It is partly a matter of efficiency in terms of RF output relative to DC input. For many years, UK amateur transmitter powers were limited by the DC input to the power amplifier, rather than output power. High-level anode modulation permits use of the high efficiency of Class C amplifiers with an RF output (including carrier and the two sidebands) of around 70% of DC input. This is well above the conversion efficiency of a Class A or Class B linear amplifier. High-level anode modulation thus gives considerably greater talk power than is possible with a linear amplifier. It was noted in the TT items on series modulation that the bias needs to be set for what in effect is linear amplification. Thus, for any given type of PA valve run at the same DC input, greater 'talk power' can be generated in the two sidebands of AM with anode modulation than with series modulation. But one would add that since amateur transmitter power regulations no longer depend on DC input power, this is perhaps now of rather less practical significance than was previously the case. What may be equally important is that series-modulation can provide near 100% modulation with low distortion.

Brian Johnson, G3LOX, former BBC engineer and then producer/writer, has reminded me that the wartime BBC "Group H" low-power, low-cost, medium-wave transmitters used series modulation. In practice, they suffered from the fact that the heater/filament of the RF valve being at half the EHT resulted in spark-over across the ceramic valve base pins due to metallic dust etc. This required that each night the bases required very intensive cleaning in readiness for the following day. Quite a chore for the station engineer!

Tony Preedy, G3LNP, is concerned that the recent items about series modulation may mislead readers into thinking that this is a modulation system that is comparable to high-level plate modulation in terms of efficiency (see above). He writes: "In fact, series modulation comes in a category known as Efficiency Modulation Systems which includes control-grid and suppressor-grid modulation. This term arises because they rely on simultaneously changing the plate current and the efficiency of the RF amplifier in order to achieve linear modulation characteristic.

"Modulation efficiency is that of the Class C RF amplifier at modulation index of 1, or 100% positive modulation, and this falls pro-rata as the modulation index is reduced. Thus at modulation index of 0 or un-modulated, the plate current is half what it was at 100% modulation and so is the plate efficiency because the plate load resistance is now too low to efficiently load the amplifier. (The Doherty system [used by a few amateurs in the 1950s, see below] overcame this problem by modulating the load impedance). Hence the output power with series modulation is 25% of that at PEP. The average plate efficiency of these old series systems did not exceed 30% and that required vast water cooling systems and oversized tubes (valves). Compare this with plate modulation in which the plate efficiency of the RF amplifier is maintained at typically 80% or so by simultaneously varying both the plate current and plate voltage to achieve linear modulation.

"The modern way to plate modulate a tube amplifier without the problems associated with iron cored components is to pulse width modulate (PWM) the plate supply voltage. Perhaps this is where the AM enthusiasts should be experimenting if they want to get anywhere close to SSB audibility performance."

Checking my volumes of papers presented at International Broadcasting Conventions in the 1970s and 1980s, I find that, in 1978, Marconi engineers described "A completely self-contained 50 kW MF Transmitter" with the final RF amplifier using the Doherty system using two 4CX35000C valves. **Figure 3** shows the basic Doherty amplifier configuration. It was claimed that this B6034 transmitter represented a significant advance in transmitter reliability and simplicity.

PWM (also known as pulse duration modulation or PDM) was one of the series of pulse modulation systems developed by the great Alec Reeves that culminated in his Pulse Code Modulation that has become the basis of all modern digital systems. In the late 1970s and early 1980s, there began to appear high-power (250 and 500kW) MF and HF broadcast transmitters using PWM/PDM At the 1978 IBC, AEG-Telefunken presented a paper on "PDM transmitters" emphasising the very high power conversion efficiency that was offered by its PANTEL system of PDM-anode-modulation. It was explained that "in a PDM [PWM] system the duration of pulses (duty cycle) is varied as a function of AF voltage. The relation between the instantaneous value of the AF voltage and the pulse duration is linear. The audio frequency can be obtained from the PDM oscillation by filtering out through a low-pass filter with a cut-off frequency corresponding to the highest audio frequency."

To quote from a paper presented by Marconi engineers at the 1984 IBC ("Trends in high power broadcast transmitters"): "The search for systems resulting in improved overall transmitter efficiency has gone on since the drastic energy price rises of the 'seventies. Prior to that period two main forms of high level [amplitude] modulation were employed. The Class B modulator favoured by most manufacturers. ... Similar overall efficiencies at MF were obtained using the Doherty modulation principle. This technique in the HF bands has not been widely used due to the difficulty of realising the λ/4 network and maintaining quadrature phase at the high end of the HF spectrum.

"During the late sixties and early seventies, manufacturers in the USA, Germany and the UK began to develop PWM techniques. These developments eventually manifested themselves in series modulated PWM systems by Gates-Harris, and the Telefunken Pantel system ... A shunt-fed PWM system known as Pulsam was developed by Marconi followed by a DC-coupled system known as Advanced Pulsam. Brown Boveri developed a pulse step modulation system [described at the same 1984 IBC]. The resulting performance of these various systems improved the modulator efficiency to about 90% and hence the HF transmitter overall efficiency to a figure between 60 and 70% [compared to about 50% for a Class B modulator system]."

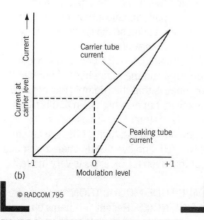

FIGURE 3: (A) BASIC CIRCUIT FOR DOHERTY AMPLITUDE MODULATION. (B) THE CARRIER AND PEAKING TUBE (VALVE) RELATIONSHIP OF DOHERTY MODULATION SYSTEM. (SOURCE: *IEE CONFERENCE PUBLICATION NO. 166, 1978*)

It should be appreciated that 'overall efficiency' includes the total energy used by the complete transmitter including all stages, cooling system etc.

I have never come across a practical design for an amateur HF AM transmitter based on PWM techniques but clearly, as suggested by G3LNP, there seems plenty of scope for pursuing work on these systems for amateur AM operation. I wonder if any reader could start the ball rolling with a simple practical design for QRP?

**THE CONJUGATE MATCH REVISITED.** For more than a quarter of a century, controversy at a high level has raged on whether optimum power output from a transmitter is delivered only when a conjugate match exists. The debate, involving highly respected professional engineers and scientists, including Walter Maxwell, W2DU, Dr John Belrose, VE2CV and Collins engineers Warren Bruene, Tom Rauch and my former IBA colleague Tony Harwood, G4HHZ *et al* has filled many columns in *QST, Communications Quarterly, RadCom, QEX, Electronics & Wireless World* etc. Some years ago the stage was reached where many editors decided that a line must be drawn on further contributions, with readers left to decide for themselves where they stand. I hesitate to re-engage 'TT' in this important, if esoteric, controversy but the topic still exercises Dave Gordon-Smith, G3UUR, who believes that there are fallacies in the case presented so far on both sides of this long-lasting debate. He writes:

"The reverse VSWR experiment (Bruene's technique) conducted by Tom Rauch to measure the output impedance of a tuned HF power amplifier, and described by Belrose, Maxwell and Rauch in their article published in *Communications Quarterly* (Fall 1997, pp25-40), convinced many amateurs that a conjugate match is a physical requirement for maximum power transfer in tuned HF amplifiers. However, the resistive component of the output impedance has to be non-dissipating in order to satisfy the high efficiency of power transfer, and this is not consistent with being able to detect its presence with a technique that relies on steady-state signals, such as VSWR measurement.

"This inconsistency has led me to question the validity of Bruene's method, and also to devise a more direct and obvious way of measuring the output impedance of a tuned power amplifier. This method uses a switching MOSFET to dynamically change the value of the load resistor for one half of a cycle in every ten cycles, and maintains the energy level of

# TECHNICAL TOPICS

T9 TABLE MICROPHONE AND THE NEW PT8000 HF/VHF TRANSCEIVER BY HILBERLING GMBH.

the pi-network as close to its normal state as possible during the test. The results for a wide range of impedance levels are presented in graphical form in **Figure 4**. They show quite conclusively that the output impedance is significantly less than the load impedance for all loads except those below about 100Ω. Therefore it follows that a conjugate match is *not* a necessary requirement for maximum power transfer in tuned power amplifiers, as long claimed by Maxwell and Belrose.

"An energy analysis of the pi-network circuit can be used to predict values of output resistance very close to those measured, so it would appear that the output impedance of a tuned power amplifier is determined by the Q and transformation ratio of the pi-network and *not* the valve (tube) characteristics, as claimed by those who oppose the idea of a conjugate match in tuned power amplifiers. Their claim regarding the valve (tube) is misguided too. This is shown in a second article that I have put on the web. This indicates that the resistive component of the output impedance of a broadband power amplifier is related to the conduction state of the power device at the peak of the RF waveform, and not some resistance derived from the slope of its characteristic curves. This suggests that the anode (plate) resistance and its time-averaged variants are not as relevant to the output resistance in the power case as they are in voltage amplifier applications. Further details of the experimental methods used, and other information relating to this topic can be found at http://mysite.orange.co.uk/g3uur."

**PAST GLORIES – FUTURE HOPES.** One of the saddest features of the recent decline in the manufacturing industries in the UK has been the virtual disappearance of a home-grown HF radio communications industry. Such famous brand names as Marconi, Plessey, Eddystone, Racal, Redifon, GEC, BCC, STC, KW Electronics etc have virtually vanished. The always interesting "Lighthouse – The Magazine of the Eddystone User Group" finally closed with Issue 96, April 2006. EUG (which continues as a Group) was founded in 1990 by Ted Moore, G7AIR with (subsequently) "Lighthouse" compiled and edited by Graeme Wormald, G3GGL. Over the years the magazine has provided an in-depth insight into the products and activities of this Birmingham firm with its long history of providing excellent quality components for the radio amateur. Such 1930s kits as the "All-World Two" receivers gave way to the post-war production of communications receivers for the amateur, SWL and professional markets.

Modern amateur radio could hardly exist without the Japanese firms and Far East production facilities. Nevertheless, it is much to be welcomed that a European firm has again entered the Amateur Radio market with what appears to be a high grade HF/VHF transceiver. This created something of a sensation when launched at the recent DARC Convention and Exhibition at Friedrichshafen. Gian Moda, I7SWX has provided (via G3SBI) brochure details of the new PT-8000 family of transceivers from Hilberling GmbH which has been developed and will be manufactured in Germany. It is claimed that this design represents the leading edge of RF-technology and incorporates technologies never seen in ham radio until now.

For example, the PT8000B can provide an output of 600 watts PEP (IM3 typically –36dB/PEP) from two SO3923 MOSFET devices in push-pull with a diplexer/LPF at its output and 100V drain voltage. The receiving system covers 9kHz to 52MHz and 142-172MHz. An auto-tune preselector for 1.8 to 30MHz exploits an unusually large toroid to achieve large signal tolerance. Two autonomous receivers use newly developed GaAsFET hybrid preamplifiers. The RF-PCB cards are completely shielded modules with 50Ω coax connectors, manufactured to industrial (professional) standards. IP3 performance in amateur bands between 1.8 and 52MHz is given as typically 39dBm at 20kHz spacing (narrower spacings are not listed so oscillator phase noise cannot be assessed). Crystal ladder filters are extensively used, including four 8-pole roofing filters and no fewer than fourteen 16-pole ladder filters (their shape factor at 2.7kHz is 1.3). The audio DSP improves performance of the crystal filters at their flanks, provides narrow bandwidths down to 50Hz, and includes noise reduction facilities.

The above figures are based purely on the information provided in the manufacturer's brochure and it will be interesting to see how the rig stands up to critical review, though there can be no doubt that this is a most impressively piece of what appears to be virtually 'professional-standard' gear.

**HERE & THERE.** Since July 1, the German DK0WCY beacon station (10,144kHz continuous, 3579kHz mornings and early evenings, 0630 – 0800 and 1500 – 1800 local UK time) has given a much extended ionospheric propagation service using automatic updating of information from Ruegen in addition to the long established data from Kiel and Boulder. It now updates every 10 minutes in Morse or (on 10MHz only) with RTTY at 10 and 50 minutes past the hour. The general call now includes the current MUF at Ruegen and the bulletins give the current FoF2 (critical frequency). At times the MUF etc may be given as NA (not available) but overall DK0WCY provides an excellent and informative service.

Sadly, also on July 1, Brian Bower, G3COJ, one of the leading British amateurs in the DXCC honours roll, occasional contributor to 'TT', a career in BBC Designs Department and a mainstay of the BBC Ariel group nets, became a 'silent key' – *RIP.*

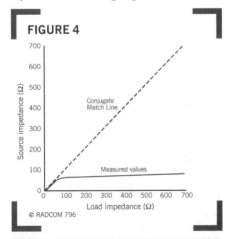

FIGURE 4: GRAPH OF SOURCE IMPEDANCE VERSUS LOAD IMPEDANCE FOR A TRANSMITTER WITH A PI-NETWORK OUTPUT CIRCUIT MATCHING INTO A PURELY RESISTIVE LOAD. (SOURCE G3UUR)

# Technical topics

**TECHNICAL TOPICS**
PAT HAWKER, MBE
37 DOVERCOURT ROAD,
DULWICH, LONDON SE22 8SS

G3VA

*The law of unexpected consequences* ♦ *Third hands and jigs* ♦
*New low-spurious DDS chip* ♦ *Sporadic-E and HF NVIS* ♦
*Ultra-broadband ships' aerials* ♦ *Secret listeners of WW2*

**THE LAW OF UNEXPECTED CONSEQUENCES.** One of the dangers of introducing changes into established systems is that, even when well-intentioned, they can result in unforeseen consequences. When the results are beneficial we call this serendipity – the happy accident. But all-too-often the results may prove counter-productive or even disastrous. Such, I personally believe, may be the result of OFCOM's proposal to dispense with the need for UK amateurs to keep running logs (either in books or as electronic data).

A well-kept log is a key factor in technical experimentation; for the study and evaluation of propagation, antennas, RFI investigations and, if less technically, for the exchange of QSL cards, operating awards, contests etc. And as an aide mêmoire to future recollection of enjoyable radio contacts, It is surely not old-fashioned to feel that 'The Log' has always been, and still remains, at the heart of Amateur Radio.

Personally, I would feel lost without access to my near 20 log books since 1938 or even one kept as a keen listener during 1937-38 while 2BUH. The books record all the thousands of local and DX *contacts* I have ever made often together with some indication of the rig, the approximate power, type of antenna etc.

I certainly intend to continue keeping a log and can only hope that other UK amateurs do so. I appreciate the good intentions of the OFCOM officials. They clearly feel that they are fulfilling their remit of ensuring optimum use of the radio spectrum with minimum bureaucracy. But surely not by abandoning the "self-training" aspects of our hobby in favour of it becoming a leisure-time fun activity!

Admittedly, change is inevitable if the hobby is to survive and flourish in these days of effective, low-cost personal communication via internet. Amateur experimentation is everywhere under threat. Ian Pawson, standing down after ten years of editing the excellent *CQ-TV, (Journal of the British Amateur Television Club)* writes, *inter alia*: "During these ten years, the membership has dropped from 2400 to its present 800. It would seem that there is a decline in building electronic equipment in general and amateur TV gear in particular. I just hope that this decline is not terminal. Why would anyone want to go to the trouble of obtaining an amateur licence when they could buy a webcam, attach it to a PC, and then exchange pictures with anyone, anywhere in the world – this has not helped interest in our hobby."

**'THIRD-HAND' CLAMPS & VICES.** The "Hints & Kinks" column of *QST* (July 2006, p 56) under its new editor, Larry Wolfgang, WR1B, includes a familiar item from Jack Rosen, KA8LFX, but may be new to some readers. It is the dodge of winding a stretched rubber band tightly round the handles of a pair of thin-nose pliers to form a vice for holding small items during assembly or for holding wires etc while you solder them: **Figure 1(a).**

In 'TT' August 2002, I included two methods of using wooden clothes pegs to form 'third-hand' clamps. My own tried and tested method is that shown in **Figure 1(b).** As noted then I devised this for use with my "Helping Hands with Magnifier" (available from a number of UK distributors including Waters & Stanton, H-601). As I wrote in 1992: "Normally these use crocodile (alligator) clips to hold wires or small parts in place [while soldering] also forming a useful heat sink, but there are occasions when a firmer, insulated grip is needed. I find the existing (adjustable) clips will firmly grip the peg which, in turn, can provide a tight grip on a small PCB etc by tightening the screw."

**FIGURE 1**

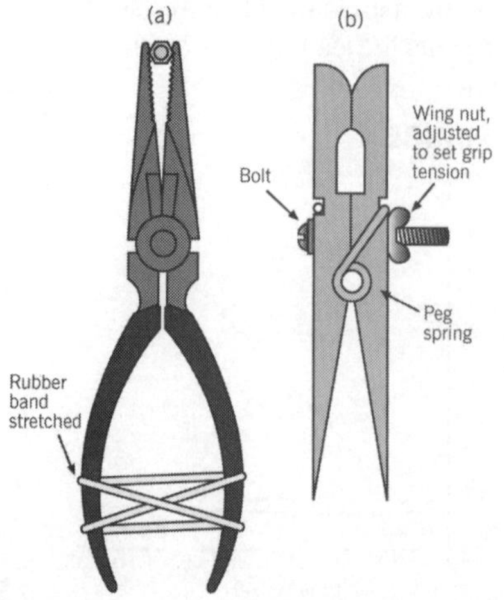

FIGURE 1: IMPROVISED "THIRD-HAND" CLAMPS AND VICES CAN PROVE USEFUL SOLDERING AND CONSTRUCTION AIDS.

**THE AD9951 CHIP AND DDS SPURII.** In the April, May and July, 2006 'TT', attention was drawn by G3SBI and G8IBR to the very attractive features of the recently introduced Analog Devices 14-bit DDS chip type AD9951 in terms of its phase noise performance, low level of spurii and its low power consumption. Further evidence of the improved performance compared with the earlier AD9851 devices comes in a package of information from G3SBI based on work by Martein Bakker, PA0AKE, drawing on the collaborative work of Giuliano, I0CG, and Giancarlo, I0GLU, displayed on http://www.radioamatore.it/iOcg/add9951.html.

PA0AKE writes to G3SBI: "While waiting for the 9MHz filter parts, I have made myself useful in attempting to measure the phase noise of my AD9951 DDS. So I built a little 600Hz-wide 4-pole crystal filter in the 3.5MHz band using cheap crystals to do the reciprocal mixing trick described for the CDG2000. This filter performed better in reality than in simulation: only 3dB loss! So that gives good hope for the 9MHz roofing filters. I have the following phase noise results for the AD9951 VFO: 2 kHz off – 140dBc/Hz; 3kHz off – 145dBc/Hz; 5kHz off – 150dBc/Hz; 10kHz off – 155dBc/Hz. This is about 10dBc/Hz better than the HP8604B! It seems to be in line with the maker's AD9951 datasheet.

"I made my measurements (above) without the Spectrum Lab PC program. I gave just enough attenuation until I could not see or hear much difference in the signal/noise ratio of the small signal with or without the large signal being applied. This is not very precise, but I feel is likely to give more pessimistic results than would a more precise procedure. As soon as I have a 'shack-PC' I will repeat my measurements with the Spectrum Lab software to obtain more accurate results.

"The designers of the DDS boards that I am using – I0CG and I0GLU – have published a single figure – 120dBc/Hz @ 1kHz – that seems to correlate with my measurements.

"I am rather impressed with these results. Now the only problem remaining

# TECHNICAL TOPICS

with the DDS solution is to remove the spurii. I have a few bad ones in some bands. I think I will just make a sharp three-resonator band filter for each band and will test that solution first. To go to the VCO-PLL approach would only destroy the good phase noise figures. An alternative approach might be a single PLL-based tracking filter, but that is probably too complicated and might not outperform nine separate band-pass filters."

G3SBI comments: "Everyone seems to be using the 500MHz external oscillator option rather than the internal (maximum 400 MHz) oscillator of the AD9951 DDS chip [see for example G6IPR's 300MHz oscillator in the July 'TT' – G3VA]. What is really interesting about the Italian work shown on the web are the two spectrum analyser plots at 42.8Mhz output. **Figure 2(a)** shows the use of the earlier AD9851 chip; **Figure 2(b)** shows the marked improvement with the AD9951.

"However, PA0AKE does point out that spurii in high dynamic range applications remain a problem. But you would be very unlucky to have a spur sitting on a high amplitude signal so that in practice this may not prove to be a serious problem. Martien, like me, is a purist and will not be happy until he has found a good technical solution.

"Perhaps, in the not distant future, we shall see the appearance of 500MHz 16-bit DDS chips using the new low-voltage transistor technology. When this happens it may well spell the end of the hybrid DDS/PLL approach to low-phase noise local oscillators."

The Italian website of I0CG and I0GLU outlines a direct digital synthesis (DDS) unit based on the AD9951/54 family capable of generating a frequency-agile analogue output sine wave up to 200MHz with a fine tuning resolution of 32 bits. It comprises a 2-layer PCB for assembly inside a metal box (90 by 35mm) that includes room for the 3.3V digital Vdd regulator; the 1.8V analogue Vdd regulator; quartz or on-board/external oscillator; low-pass filtering and linear amplifier. The unit can be interfaced to a micro-based controller through a suitable 6-pin connector. It seems likely that this may be marketed as a kit.

Colin Horrabin, G3SBI, has passed along an e-mail received from Bill Carver, W7AAZ, who knows of the Italian work: "I am using *exactly* the same [Italian] DDS that Martien is using. I take output from 78 to 100.08MHz, divided by various even integers to produce square wave LO for the H-mode mixer. I have a seven-pole band pass filter for this 78-100.08MHz (corner

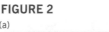

**FIGURE 2**

(a)

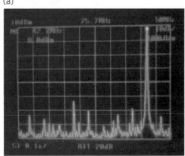

(b)

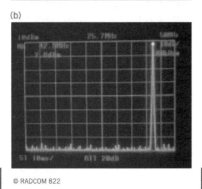

© RADCOM 822

FIGURE 2: (a) SPECTROGRAM OF THE OLD AD9851 DDS WITH OUTPUT AT 42.8MHz. (b) NOTE THE MARKED REDUCTION OF SPURII IN THIS SPECTROGRAM WITH THE NEW AD9951 WITH 50MHz FREQUENCY SPAN AND OUTPUT AT 42.8MHz.(SOURCE: http://www.radioamatore.it/i0cg/add9951.html)

frequencies –2dB); this filter completely removes all 'self-generated' spurii (i.e. spurs that can be heard without *any* signal applied to the receiver).

"However, I am replacing the 100MHz x 5 clock with my own 134MHz x 4 scheme, I found and others confirmed, 600MHz content in the clock signal at –57dBc. My scheme has balanced diode doubler, 7-pole LC filter at 268MHz, MMIC, second 2-pole 536MHz helical filter. I have not yet completed this and do not know whether it will improve the spur level."

G3SBI adds: "W7AAZ's comments about wide-band spur levels not raising the noise floor of the receiver with no signal present are interesting, This was always one of the problems with earlier DDS chips. Some of these recent developments in DDS technology do not improve the situation with up-conversion receivers but they certainly do with down-conversion rigs."

**SPORADIC-E AND HF NVIS.** On July 2, 1938 there marked an important development in the UK amateur use of the 56MHz band. Throughout the 1930s progress had been made, if slowly, in extending the range of contacts on this VHF band. This was mostly by operating simple super-regenerative equipment from the top of high hills to achieve maximum line-of-sight distances, but there was a continuing hope that one day the band would open for DX contacts via the ionosphere. In 1937, near the peak of a sunspot cycle, there were reports that 56MHz signals from G5BY had been heard in America, and the London 405-line television signals on about 40MHz had been seen by RCA engineers.

But then on July 2, 1938, there occurred a confirmed two-way contact between G5MQ and an Italian station I1RRA with signals at good strength. As *The T&R Bulletin* put it "56MHz yields to persuasion". That summer numerous harmonics of commercial stations were heard in or around the 56MHz amateur band. But it was soon recognised that the Italian contact must have been made – surprisingly – by reflection from the lower ionospheric E-layer, usually considered far too weakly ionised to reflect 56MHz signals.

It was not until the early post-war period that UK amateurs began to investigate more deeply these occasional (sporadic) E-layer reflections. They also became of interest to TV engineers because of interference to Band 1 programmes by co-channel stations in continental Europe.

For many years an enthusiastic Ron Ham and also a team at the University College of Wales monitored the appearance of sporadic-E propagation which affected VHF signals on 28, 50, 70 and occasionally 144MHz. Although such events proved unpredictable, it was shown that in the UK they occurred mainly between May and August, most often in June and July. There is also a weaker winter peak in December and January. February to late April is the least likely time for sporadic-E events. For years, mystery surrounded – and to some extent still surrounds – the nature of sporadic-E. But much is now clear.

I reported in 'TT' February 1985 that Dr E B Dorling, of the Mullard Space Science Laboratory of London University had described in a letter to *Wireless World* (April 1978) how knowledge of this curious phenomenon had been much increased by a combination of ground-based and rocket observation. He wrote:

"Sporadic-E was first seen to occur in the way it does, that is as very thin intense layers of ionisation, by a British Skylark rocket flown from Woomera [Australia] in 1958. By 1966 an association between these layers and sharp reversals in wind direction at high altitude had become recognised. Wind measurements in the very rarefied atmosphere up to 150km or so revealed that a surprising pattern of wind reversals with height can occur. What is more, the measurements showed that the pattern often descends slowly over a period of hours, with, for example, a sharp wind shear first appearing above 150km height, then moving downwards to below 100km before fading. The cause of this rather unexpected wind structure appears to be propagation of atmospheric waves horizontally over great distances.

"The sharp wind shears are at the root of the sporadic-E layers, though in a rather complicated way. The winds, tenuous though they are at such heights, act to move the ions and electrons in the ionosphere across the earth's magnetic field, but interactions then occur in such a way as to displace the plasma vertically. Where strong wind shears of the appropriate sense exist, the plasma is squeezed into a thin concentrated layer, being moved downwards from above, upwards from below. As the wind pattern descends the layer descends too into a thin concentrated layer; moved downwards from above, upwards from below. As the wind pattern descends the layer descends too into an ever denser atmosphere, until finally at a height of about 100km it is brought to a halt…

"Sporadic-E then owes its transient character to interactions between atmospheric wave, the ionospheric E layer, and magnetic and electric fields. All but the magnetic field are constantly changing, so that the right conditions for layer formation occur – well, sporadically. If the question is asked why the explanation has been so long in coming, I should explain that physicists the world over have contributed to the solution – the answer is that the region concerned, roughly 100-200km above the earth's surface, is inaccessible to satellites and therefore to regular on-the-spot measurements.

"One final point. Were the sporadic-E layers to be composed simply of ionized atmospheric gases, they wouldn't persist. They are, in fact, composed of ionized metallic atoms, mainly magnesium, silicon and iron, probably the remains of burned-up meteorites. The descending wind shears sweep up the metallic ions and bring them down as sporadic-E layers out of the thermosphere into the lower regions where atmospheric turbulence then churns them away into oblivion. Sporadic-E layers seem to be the product of Nature's vacuum cleaning!"

Dr Brian Austin, G0GSF, has been investigating the effects of sporadic-E on HF, especially on 7MHz. He writes: "The sporadic-E (Es) season is once again upon us. As I indicated last year we frequently find that Es clouds, blobs or patches [in the form of intense thin layers of varying extent and height] are much in evidence above the UK and propagation responds accordingly.

"What surprises me is the fact that Es seems only to be associated in the radio amateur literature with openings in VHF propagation when its effect on HF propagation (at least locally in the region of influence of the Es cloud) is highly significant. This is particularly noticeable at this point in the sunspot cycle because the critical frequency of the Es layer often exceeds that of the F region by a very significant amount. In addition, it often happens that the Es layer is so highly ionised that all layers above it are frequently completely obscured or 'blanketed' to use the term of the ionospheric community. Under such circumstances any upwardly [or downwardly – G3VA] propagating EM wave will be reflected or refracted by the Es layer and not by those other layers regardless of their features since they are, essentially, invisible to the EM signal.

"**Figure 3** shows a graph I have plotted of the ionosphere above the ionosonde at Chilton throughout the 24-hour period of July 17, 2006. The solid line shows the critical frequency (foEs) of the sporadic-E layer, while the data points are the critical frequencies of the F region (foF2 during daylight, foF at night) where these could be read off the ionograms. Note that foEs exceeded foF2 throughout most of the 24-hour period. For those hours around midday, blanketing was total; in fact there was no sign at all of the higher layers. During that period, HF propagation would have been dominated by sporadic-E, at least where the Es cloud was at the point of reflection.

"Note that, at times, the foEs reached as high as 11MHz, therefore NVIS propagation even on the 10MHz band, would have been dominant. It is also likely that inter-G NVIS working on the 7MHz band would have been much in evidence but I cannot confirm this as I was not operating that day – my wedding anniversary!"

In some countries, including India and parts of the USA, sporadic-E conditions are virtually a daily event in some seasons. Perhaps we may see more Es events in the UK in the future climate changes.

**FUNNELLING THE ANTENNA.** A new concept of implementing a broadband HF antenna for recently built naval and civilian ships is outlined in a paper "Naval Structure Antenna Systems for Broadband HF Communications" by Italian engineers G Marrocco and L Mattioni (*IEEE Trans Ant & Prop, April 2006*, pp1065–1073).

To quote from the introduction: "Recently manufactured ships for both civil and tactical purposes are equipped with the Software Defined Radio system having the capability to simultaneously receive and transmit multiple waveforms, of the same or different type, within the 2MHz to 2GHz range. To host *multi-channel* communications, actual [current] naval systems employ either a small set of broadband antennas together with combining networks or a plurality of narrow-band antennas operating in different sub-bands. In the first case, because of the losses introduced by the power combiners, the resulting system efficiency can be highly reduced. In the other case, the coexistence of several radiators requires large space and produces undesired interactions among antennas and the ship superstructure. These issues are particularly critical in the HF range because of the large wavelengths involved.

"The problem of efficient antennas in the available space has already been considered in the context of

**FIGURE 3**

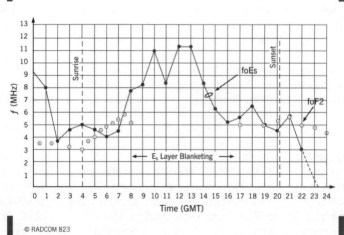

FIGURE 3: PLOT PREPARED BY G0GSF OF THE IONOSPHERE THROUGHOUT THE 24-HOURS OF JULY 17, 2006 SHOWING THE foES EXCEEDING AND OFTEN BLANKETING THE F LAYER (foF2) AS SEEN ABOVE THE CHILTON IONOSONDE.

# TECHNICAL TOPICS

avionic environments where since the late 1950s, some structural solutions have been proposed. In particular, the whole aircraft or some parts of it can be used as a radiating element in the HF range, when properly fed (notch or towel bar antennas), since the aircraft size is comparable with the radiated wavelength."

The paper shows that broadband shipboard structural systems could use structures such as the funnel or a big mast in combination with properly shaped and loaded wire radiators. It is shown that it is possible to achieve a broadband HF radiating system which permits handling both sea-wave and near vertical incidence sky-wave communications.

While I don't suppose that there are many amateurs owning steam yachts with large funnels, the idea of using "bulky" structures rather than conventional masts and towers, in order to improve bandwidth etc, is not without interest. A number of papers recently stress that it is the total *volume* of an antenna that has an important bearing on the efficiency of a short vertical antenna. A scaffold tower, for example, might be used as a temporary antenna, although on lower HF bands it would need a good ground-plane – not a problem on a ship!

**SECRET LISTENERS OF WW2.** Many of us in the UK tend to associate secret listening during World War II with the Voluntary Interceptors of the Radio Security Service (MI8c) who monitored the German Military Intelligence networks and forwarded their logs to Box 25, Barnet. The traffic ended up at Bletchley Park where it was deciphered into ISOS or ISK. A number of recent books, such as the Guy Liddell Diaries Vols I and II, have underlined the importance of ISOS for the successful deception operations of the double-agents etc, etc. The annual VI/SCU3 reunion at BP organised by Bob ("Nos") King, G3ASE, still brings a few of us (plus Whaddon SCU1 personnel etc), each April to BP. As for others, secret listening refers to those in Occupied Europe who defied Nazi edicts banning the listening to BBC or 'Black' broadcasting stations, and also to the secret receivers constructed in the Prisoner-of-War camps.

An excellent, well-illustrated article by Dick Rollema, PA0SE, on secret listening in Holland appears in *Radio Bygones* (August/September, 2006, pp14-21) drawing on his own experiences as a schoolboy participant in his family's listening to the BBC and the London-based *Radio Oranje*. The sets illustrated range

**FIGURE 4**

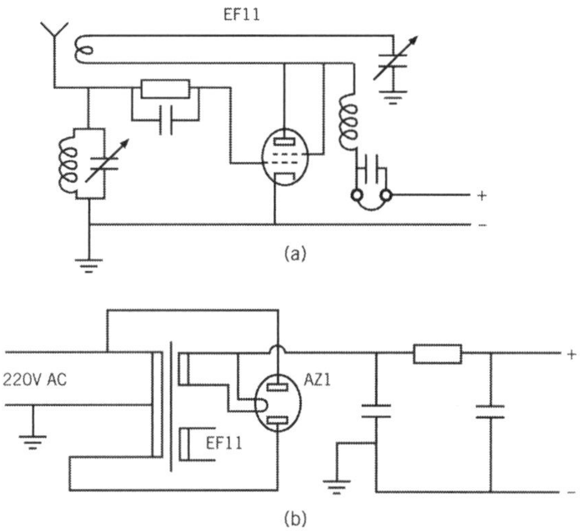

FIGURE 4: CIRCUIT OF P-O-W RECEIVER FOR THE 31-METRE BROADCAST BAND CONSTRUCTED BY A SOUTH AFRICAN AMATEUR AT STALAG 8C. THE LATEST NEWS FROM THE BBC WAS DISTRIBUTED AS RUMOURS FROM A NEW PRISONER "JOHNNY HIGGINS" AND THE RECEIVER BECAME KNOWN BY THIS NAME.

from crystal sets built in a matchbox or concealed in a space cut out of a dictionary up to a LW receiver with three acorn-type valves in a tobacco tin. PA0SE also discusses techniques used to overcome jamming.

I recall during the period in early 1945 when I was on loan to the Netherlands Intelligence Department (Bureau Inlichtingen or BI) as an operator for the clandestine Dutch Inland Radio Service being shown by a member of the Dutch Underground a mains-operated receiver with a single acorn-type valve built into a very small tobacco tin (2oz size?) that had survived a search while he was on a train. He took it out of his pocket as though preparing to refill his pipe and held it in his hand while subject to a personal search.

'TT' in January 1986 (see *Technical Topics Scrapbook 185-89,* p77) told briefly the stories of PoW camp receivers built in Germany by Captain Ernest Shackleton, G6SN; in Hong Kong by Herb Dixon, ZL2BO; and by Tom Douglas, G3BA in Malaya. This despite the difficulty of finding, scrounging, stealing or obtaining corruptly from the guards, those components that could not be fashioned from available bits and pieces. MI9 sought to smuggle radio components and instructional notes into the camps although none of these three constructors seem to have had access to such sources. Communications from many of the camps to the UK were established by means of "letter codes" in which short messages were concealed in long letters. There is some evidence that construction of transmitters was undertaken for use in an emergency, though no contacts were made.

Brian Austin, G0GSF, has drawn my attention to a 24-page booklet by a South African amateur, Ken Wade-Lehman, ZS5KWL, "Clandestine Communique" that tells the story of his years as a PoW following his capture, along with some 32,000 others, on June 21, 1942 by Rommel's Africa Korps at Tobruk: "My total assets on that unhappy day amounted to the vital water container and chromium plated cigarette case filled with 'Abdulla' – the cigarette vaunted to be of 'Imperial Preference'".

No wonder that at some stage, most PoWs started purloining odd bits and pieces from their captors, salving their consciences by the thought that any damage done to the enemy assisted the Allied war effort! In December 1943, at Stalag 8c, the Germans brought in a talkie cinema projector to show propaganda films, when the audio failed the operator discarded two 'dud' valves: "Picking these up and pocketing them was a natural reaction for any trained scrounger ... The idea of making a clandestine radio receiver did not, at the time, enter my mind."

When later a different projectionist brought in a similar equipment, Ken was able to swap the two duds for two working AF valves. Over the following months, bits of wire, bribing a guard with chocolate and then applying pressure procured a pair of headphones. A fellow prisoner used his Red Cross watch repair kit to manufacture a variable tuning capacitor. A coil was wound on an empty Bakelite shaving stick container. Cigarettes were traded for bits and pieces including a pair of $32\mu F$ electrolytic capacitors joined by a 1K 10-watt resistor, a full-wave rectifier valve. Construction of a crude mains transformer supplied heater current at the two appropriate voltages. To avoid detection, sets were buried under hut floorboards. Despite many searches, the secret listeners and their receiver (**Figure 4**) were never caught.

# Technical topics

*Pat Hawker sheds new light on the origin of the Zepp antenna, protecting the DC output sockets of radios, a new portable energy source and delves further into LC balun matching networks and PDM transmitters.*

**DJ5IJ ON THE 'TELEFUNKEN ZEPP'.** In 'TT' March and June, 2006, I attempted to trace the origins and the pros and cons of the Zeppelin antenna, which has been widely used by amateurs since the 1920s. I adopted the belief of DT0TR/OE8AK, first published in 1985, that its origins were to be found in the early patent of Dr Hans Beggerow. This hypothesis has now been undermined by a most informative contribution from Karl Fischer, DL5IJ. He writes:

"Fig 5(a) of the March 'TT' shows the schematic principle of German Patent "DE 225204 – "aerial structure for airships" – issued to Dr Hans Beggerow in 1909. I translate from the original document:

'Up to now, the surfaces of the balloon and the basket were used for attachment of or directly as the electrical counterpoise for the aerial. That was a risky measure considering the danger of explosion caused by high voltages present at locations where oxyhydrogen could develop. ... This invention removes that trouble and allows keeping the devices within the basket without exposing the airship to high voltages. According to the invention, the aerial consists of two dangling wires of different length which form a Lecher system in the proximity of the of the airship. ... In that way the voltage antinodes are moved away from the airship.'

"So your original assessment [March] of the point and purpose of Beggerow's patent is absolutely correct. In fact, a non-radiating 'feeder' was clearly not his intention, but to avoid high voltages close to the hydrogen-filled airship, which might cause sparks and thus have disastrous consequences. But in the June 'TT' you revised your opinion ; concluding from the fact that a Lecher-line is part of the antenna that '... it was thus intended from the beginning as an end-fed twin-wire resonant feeder system'. However this revised conclusion is wrong!

"Note that in the patent, the lengths of the twin wires were not specified by the inventor; it is said only that they must have

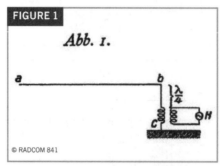

FIGURE 1: (ABB 1) DIAGRAM FROM THE 1926 TELEFUNKEN PATENT. RESONANT HORIZONTAL ELEMENT FED BY QUARTER-WAVE VERTICAL FEEDER (THUS FORMING AN INVERTED L ANTENNA FED AGAINST EARTH)

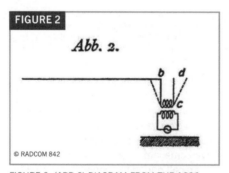

FIGURE 2: (ABB 2) DIAGRAM FROM THE 1926 TELEFUNKEN DIAGRAM, CLEARLY REPRESENTING THE CLASSIC ZEPP ANTENNA AS USED BY MANY AMATEURS FOR 80 YEARS. DJ5IJ SHOWS HOW THE FEEDER RADIATION CAN BE INSIGNIFICANT WHEN THE HORIZONTAL ELEMENT IS TRULY RESONANT

different lengths and this causes the whole system including the Lecher line to radiate [as suggested in March]. In that way neither part of the balloon structure itself belongs to the antenna, carries currents nor acts as a return path or 'counterpoise' any of which conditions might lead to dangerous voltages close to the hydrogen. Instead, each wire can be regarded as being a radiating counterpoise to the other wire. The whole system can be described as a non-resonant (or not necessarily resonant) off-centre-fed radiator, folded at the feedpoint.

Seventeen years after the Beggerow patent, a German patent DE 436462 was issued to the Telefunken corporation, described as a 'method and arrangement for the excitation of horizontal antennas'.

**Figures 1 and 2** and the following quotation are (translated) from the original patent:

*"According to the invention a technically valuable solution for the excitation of such antennas ... is to excite horizontal antennas from one side so that the transmitter is connected to one end of the horizontal antenna by means of an approximately ¼-wave long feeder. Such an example is shown in Figure 1 where a, b stands for the horizontal antenna. The transmitter H is located at a distance of λ/4 from the horizontal antenna and inductively coupled to the antenna. The vertical feeder from end b of the antenna is grounded at point c. Another idea of the invention is to cancel radiation of the vertical feeder by the application of a parallel wire which is excited in 1/4 wavelength by the same generator with the oscillations on this wire being in antiphase. An example is shown in Figure 2 where, as evident, the radiation by b, c is cancelled by the additional vertical part c, d (the voltage distribution is depicted by the dotted line) ...."*

"Though this description says nothing about the length of the radiator, the voltage distribution on the feeder in Figure 2 implies that the radiator is voltage fed and thus must be an integer multiple of a half-wave. This is the classic Zeppelin or Zepp antenna as widely used by radio amateurs, with a half-wave radiating element connected to one terminal of a ¼-wave feeder and the other terminal left open. Clearly the classic Zepp is not Beggerow's 1909 airship aerial but an invention of Telefunken patented in 1926. So much for history; now for the technical aspects.

"The critical comments by Dud Charman, G6CJ and others [as noted in 'TT'] are unjustified. The Zepp concept is sound. The problems caused by feeder radiation (pattern distortion, RFI and a 'hot' shack) are not inherent but arise from misunderstanding the principle of operation. If simple guidelines are followed such problems need hardly exist. I would claim that most centre-fed dipoles exhibit more feeder radiation than a correctly installed Zepp antenna.

"An efficient unbalanced-to-balanced (balun) between a modern transmitter and the twin-wire Zepp feeder is a necessary prerequisite. Current balancing means being transparent to balanced currents while separating the potentials of the ports so that no common-mode currents can flow. An ideal transformer would make an ideal balun, but unfortunately in reality the ports are coupled by stray capacitance between the windings that opens up a path for common-mode currents and hence can cause feeder radiation. Great care should be taken to minimise common-mode currents. In practice even a good balun is usually no guarantee of a non-radiating feeder. Even if current balance is forced at the feedpoint this may not be maintained everywhere along the feeder line.

"A feeder is non-radiating only if the currents remain accurately balanced throughout. This will be the case if the terminating load is either perfectly balanced (equal impedances to ground) or floating (without any current-path to ground). Clearly a Zepp-antenna does not fully meet this requirement, since one wire of the feeder is left open while the other is connected to the radiating element. But at resonance, the radiator is voltage fed and very little current flows since an end-fed radiator has nearly the same high impedance as the feeder. But as one moves away from resonance the real part of the impedance quickly decreases and the reactive part increases, making the radiator impedance increasingly different from that of the open feeder end; thus giving rise to unbalanced currents and a radiating feeder.

"NEC simulations show that feed-line radiation can be expected to be typically about –25dB below that of a resonant radiator; 10% deviation [of the element] from resonance yields about ten times more feed-line radiation or –15dB; 15% deviation yields about 50 times more feeder radiation, a mere –8dB. Feeder radiation caused by a non-resonant radiator cannot be cancelled by insertion of reactive components in the feed-line, thus

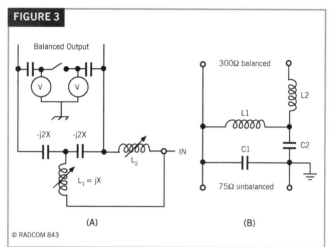

FIGURE 3: (A) THE ALFORD CIRCUIT AS DESCRIBED BY G4LU IN THE SEPTEMBER 'TT'. (B) BASIC CIRCUIT OF THE MATCHING BALUN NETWORK DESCRIBED BY G3KGN IN 1958.

the 'All-Wave Zepp' described in 1931 by G6LI ('TT' March) is an inherently dysfunctional design.

"The final key to success is the correct alignment procedure. The very first step is to adjust the radiator for resonance; that is for minimum feeder radiation at the fundamental frequency. This is best done by coupling a current probe (eg as described by the late L A Moxon, G6XN, in 'HF Antennas for all locations') to the feed line as far as possible from the balun network (this is because directly at the balun, the currents on the feed-line are [at least theoretically] always balanced!). Then, one commences trimming the radiator length for minimum meter deviation. Only after that is accomplished should the feeder length be adjusted for resonance of the whole system as well as the feeder spacing or alternatively the coupling network, for optimum impedance match to the transmitter – but without any further adjustment of the radiator. Although it is seductively simple to resonate the whole system without first tuning the radiator itself, such a procedure ends up with excessive feeder radiation and its associated problems as sure as night follows day!"

Perhaps I should add that it has long been argued that a resonant Zepp element will be slightly longer than a dipole element owing to reduced "end effect".

It is clear from DJ5IJ's explanation that a Zepp with little or no feed-line radiation requires careful adjustment at its fundamental frequency and is basically a monoband antenna. But I would argue, as I did in the March 'TT', that the same antenna could also be used as a multiband Beggerow-type antenna in conditions where the problems of feeder radiation/pattern distortion etc are of minor significance. The whole system could be brought into resonance by an ASTU at the transmitter. For example a 7MHz Zepp should prove reasonably effective as an inverted-L with the open ended feeder wire forming a counterpoise on 3.5, 5, 10MHz and even to some extent on 1.8MHz and the higher HF bands.

PROTECTING EXTERNAL POWER SOCKETS. The August 2006 'TT' included a contribution from LA6PB describing the construction of a fast circuit-breaker used to protect the internal 500mA fuse in the 13.5V DC output jack for powering external units, as found on many transceivers. After dismantling his transceiver for the nth time, he decided that a quick acting circuit breaker would prove a better solution than dealing with yet another blown fuse.

Harry Leeming, G3LLL, writes: "LA6PB notes the inconvenience, if the 13.5V outlet socket is overloaded, of replacing the internal fuse. He should think himself lucky that he does not have a Yaesu rig such as the FT757 or FT747.

"The 13.5V sockets on these rigs do not have fuses. Should a short-circuit occur, all the 20A (or whatever is available from the PSU or battery) goes through the internal wiring. I have had quite a few of these rigs brought to me for servicing with melted wiring and fused printed-circuit panels, some so badly damaged they were hardly worth repairing.

"I hate to see good engineering ruined. So please, if you take power from the back of your rig, either fit a LA6PB-type circuit breaker or a fuse. Alternatively, if you want only a few milliamps, put a 20Ω resistor in series with the load so that a large current cannot be drawn."

**NEW PORTABLE ENERGY SOURCE.** The June 2006 'TT' included an item "Non-mains power sources" that included a brief survey of the various means of powering equipment in the field or for emergency operation or simply as a means of keeping on air in the event of a local power cut. It was noted that for 100-watt transceivers, the traditional petrol-electric generator remains supreme (although recently challenged by the hydrogen fuel-cell, see the September 'TT'). But mention was also made of the hand- or pedal-operated generator/alternator that can provide up to about 40 Watts or in some circumstances rather more from an energetic user.

As someone with an ex-military unit that can be either hand cranked or fitted with pedals, I can vouch that this is not only a heavy unit but also not one that I would really like to use for long in the field or for that matter in the home, although capable of providing 110V AC at up to about 40 watts. I did try to fit it to a bicycle-type exercise machine but this proved impractical and I have to admit it has remained virtually unused for many years.

It was therefore with considerable interest that I read a "short take" product review by John Marino, KR1O, of ARRL (*QST,* August, 2006, p47) of the Freecharge WEZA Portable Energy Source, manufactured by the London, UK firm of Freeplay Energy. WB8IBY concludes: "The WEZA is ideally suited for emergency operations. It can provide unlimited operation of electronic devices that draw medium current."

Basically, the WEZA appears to be unique in using human power to recharge a self-contained, sealed 12V, 7Ah lead-acid battery by using a foot pedal with an efficient flywheel for charging. A LED meter display shows the amount of battery charge available. The unit incorporates facilities for charging the battery also from AC mains supplies or from a 10.7 to 21V DC sources such as solar or wind generators. The foot pedal is primarily intended for emergency situations.

The WEZA represents a further product in the Freeplay Energy range, claimed: "To make energy available to everybody all of the time." Dimensions are 290mm height, 220mm width and 463mm length. Its weight, including battery, is 8.06kg so it is portable and could be carried as a backpack.

KR1O reports using the WEZA during a 12-hour power cut to power a Yaesu FT-857 set at 20W for over an hour before recharging by foot was necessary, and also used an ICOM IC-703 10W low-power HF radio on SSB for over three hours before a recharge was needed. Use of the pedal mechanism for about 15 minutes brought the battery from a low state to fully charged. He notes that additionally the WEZA will recharge mobile phones, GPS units and laptop computers. A pair of heavy-duty jumper cables enables its use to jump start an automobile or marine battery. He adds "The WEZA appears practically indestructible and built to withstand harsh rugged outdoor use ... in situations where communications must be maintained under adverse conditions, the WEZA offers a very practical solution for unlimited time operation."

KR1O reports that the average selling price in the USA is $277 (about £150 at the time of writing) but on enquiry in the UK I was quoted a launch price of £299. This seems a large differential, even taking VAT etc into account, for what is claimed as a UK product. It would seem a crazy situation if a UK product costs less if imported from the States and perhaps the UK price may have been adjusted by the time these notes appear. The

THE FREECHARGE WEZA PORTABLE ENERGY SOURCE

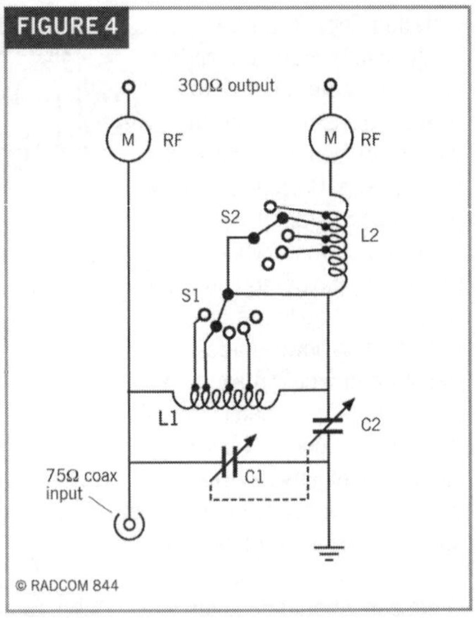

FIGURE 4: THE PRACTICAL MULTIBAND TUNER PROVIDING 300Ω BALANCED OUTPUT AS DESCRIBED IN 1958 BY G3KGN. FOR INDIVIDUAL BANDS COMPONENT VALUES ARE: 3.5MHz C1, C2 300pF, L1, L2 14µH; 7MHz 150pF AND 7µH; 14MHz 75pF AND 3.5µH; 21MHz 50pF AND 2.2µH; 28MHz 37.5pF AND 1.8µH. L1, L2 ARE 16 SWG ENAM FOR MULTIBAND UNITS C1-C2 IS A 500pF GANG VARIABLE CAPACITOR AND L1,L2 ARE EACH 24 TURNS OF18SWG WIRE WOUND 10TURNS PER INCH ON A 1¾- INCH FORMER TAPPED AT 4, 6, 9,AND 14 TURNS RESPECTIVELY

unit is available to buy online at www.freeplayenergy.com or by calling 020 7851 2600.

**MORE LC BALUN MATCHING NETWORKS.** The description by Stan Brown, G4LU, of the Alford circuit providing a balanced output at a desired impedance from an unbalanced 50Ω input ('TT'. September 2006, p69) has brought in information on several alternative circuits. Don Nappin, G5MLS, recalls an article "All-Band Balun Unit – 75Ω unbalanced into 300Ω balanced" by A C Edwards, G3GKN, (*The Short Wave Magazine,* October 1958, pp436-437). This used a variation of the Alford circuit, interchanging the positions of the capacitors and inductors: **Figure 3**. G3GKN included details of a switched unit for 3.5, 7. 14, 21 and 28MHz, using two identical tapped coils and a two-gang 500pF receiving type variable capacitor. **Figure 4** shows the circuit diagram but I hope to include further details of this unit together with at least some of G5MLS's comments in a future issue.

# TECHNICAL TOPICS

Colin Brock, G3ISB/DJ0OK, draws attention to a "Broadband universal balun transformer" published in *Radio Handbook,* edited by the late William I Orr, W6SAI. G3ISB gives the reference as "20th edition, p26.9" but I assume this is the same as in my 22nd edition (1981) p26.11. To quote from the Handbook:

*"A Broadband LC Balun –* The derivation of a broadband lumped constant balun is given in **Figure 5**. Illustration (A) shows two pi-network circuits with the inputs connected in parallel and the outputs series connected. For this example the balun is assumed to match a 50Ω unbalanced line to a 20Ω balanced load, a common condition for a Yagi beam antenna. One network is the conjugate of the other. The circuit can be redrawn, as in (B), omitting C1 and L2, as they form a resonant circuit at the design frequency. The final revision is redrawing the circuit as a bridge, as shown in (C). There is no coupling between the coils and they should be mounted at right angles to each other. The bandwidth of the balun is inversely proportional to the transformation ratio, and a balun having a transformation ratio of unity has a theoretically infinite bandwidth."

G3ISB comments: "The network, when redrawn as in (C) can seen to be a conventional bridge circuit comprising two inductances and two capacitances, all of equal reactance. It is configured into a ring of alternate L and C units, the inputs and outputs being across a pair of them. The values of the components are found from $X_L = X_C = \sqrt{Z \times Z_{ant}}$
where Z = input coaxial cable impedance, $Z_{ant}$ = balanced output (antenna) impedance.

"I experimented with this circuit using resistors as loads, fed from a 50Ω source (spectrum analyser with tracking generator) and was surprised at both the simplicity and the efficacy. Subsequently it was put to work to feed a 7MHz folded dipole cut to 7100kHz and it was found that the whole 200kHz-wide European band could be covered with a return loss of under –15.5dB.

"This means that almost any transceiver, including my Elecraft K2, can be plugged straight into a 50Ω coax feeder, with the network itself located directly under the centre of the antenna – where it should be!

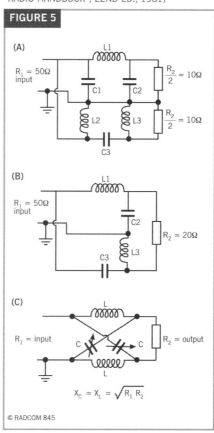

FIGURE 5: DERIVATION OF THE "BROADBAND UNIVERSAL BALUN TRANSFORMER" OR, AS PREFERRED BY G3ISB/DJ0OK, THE "BRIDGE BALUN". FOR EXPLANATION SEE TEXT. (SOURCE: "RADIO HANDBOOK", 22ND ED., 1981)

"Bill Orr called the circuit a 'broadband universal balun transformer' but I prefer the term 'Bridge Balun'."

**G4ENA's QRP PDM TRANSMITTER.** The September 'TT' p70-71 included the advice by G3LNP that those interested in AM operation should experiment with PWM (PDM) techniques if they wished to get anywhere close to SSB audibility (and to avoid the use of bulky iron-cored components). After outlining the use of PWM/PDM for modern high-power broadcast transmitters, I added: "I wonder if any reader could start the ball rolling with a simple practical design for QRP."

I have to confess, as Jack Ponton, GM0RWU, has reminded me, that I entirely overlooked the detailed article on the design, construction and adjustment of such a transmitter in "DF Transmitter for 160 metres" by Peter Asquith, G4ENI (*Radio Communication,* May 1993, pp36-37). This accompanied his associated article "DF Receiver for 160 metres" (pp34-35). The 12V battery-operated transmitter employed, as G4ENI put it, "the modern technique of generating AM by pulse width modulation, thus permitting the application of a high efficiency and very compact switching VMOS power amplifier stage. The PCB outline is only 100 x 50mm."

The design used, apart from the RFP 2N10L PA, three ICs – TL072, a 74HC04 and a 78L05 regulator – plus a few discrete components.

G4ENI added: "The standing carrier power is about two watts. To achieve more output power, the supply voltage can be increased to a maximum of 24 volts. The VMOS PA would need additional heat-sinking at these levels, however. … The range will depend on many factors, with a good aerial system up to ten miles is possible. … Power supply can be any form of wet or dry battery. A capacity of about 1Ah is sufficient to last the duration of the longest DF exercise."

GM0RWU comments: "This is a very neat design that could easily be adapted for other powers and bands. With modern chips it should be capable of running at least up to 70 MHz with a class C PA. With a Class D/E PA the upper frequency would depend on how far technology for switching transistors has moved on since the early 1990s.

"Although I never got round to building the G4ENI design, I did think through a simple adaptation of it for 3.5MHz:

Replace the crystal with a 3.69 MHz ceramic resonator and variable capacitor, giving coverage of roughly 3.6 to 3.7MHz.

Use another G4ENI design described in "QRP + QSK – a Novel CW transceiver with Full Break-in" (*Radio Communication,* May 1992, pp33-35) for an IRDF110 switch-mode PA.

The switch-mode PA is so efficient that it doesn't even need a heat-sink."

GM0RWU has made further comments on AM and SSB transmitters that I hope to refer to in a future issue.

**HERE & THERE.** May I take this opportunity of thanking all those 'TT' readers who have sent me congratulations on my appearance in the Queen's 80th Birthday Honours List? Too many to reply to individually but all most appreciated. Thank you!

# TECHNICAL TOPICS

PAT HAWKER, MBE
37 DOVERCOURT ROAD,
DULWICH, LONDON SE22 8SS

G3VA

# Technical topics

*How do we get 'inactive' amateurs back on the air?* ♦ *More interesting chips* ♦ *Mixer improvements* ♦ *Propagation beacons* ♦ *Doherty AM revisited* ♦ *GU5ZC SK*

**INACTIVE AMATEURS.** During the current period of low sunspot activity there has been a most noticeable lack of HF activity on 14MHz other than contest and special (expedition etc) events. At such times any "open bands" fill with stations exchanging 599 or 59 reports, with each contact over in a few seconds. The traditional contact conducted at a leisurely pace has become almost a rarity on the higher bands, although still flourishing on the lower frequency "nets". One gets the impression that, particularly in the USA, there exist hundreds, possibly thousands, of high power stations with effective beam antennas that remain dormant during weekdays and contest-free weekends. Only by listening to the beacons, or consulting the web, does it seem possible to assess whether a particular band is open for DX.

More disturbing are the reports of recent licensees who never seem to appear on the HF bands, and confine their activities (if any) to the use of hand-helds working through local repeaters. A recent *QST* guest-editorial by Dan Romanchik, KB6NU, ("No Ham Left Behind", September 2006, p88) notes that the number of licensed US radio amateurs in April 2005, at 667,318, showed a net loss of 20,542 from the peak in April 2005. The 2005 total comprised 28,869 Novices, 318,221 Technicians, 137,093 Generals, 76,706 Advanceds, and 106,238 Amateur Extras.

These are impressive figures (bettered only by Japan). But KB6NU accepts that there may be between 25% and 50% of all licensees who are [completely] inactive: "For whatever reason, these folks lost interest and are Amateur Radio operators in name only. … I think it is more important to have active, engaged amateurs than to have a large number of licensees. Inactive hams don't work CW or experiment with circuits. The question then is how to encourage Amateur Radio operators to be more active. The rule changes over the last 10 to 15 years have enabled many to obtain licences, and, I think overall, that's a good thing. But getting that first licence is only a start, not an end in itself. Let's face it: if all an amateur knows is what the Technician class licence manual covers, all that he or she is really prepared to do is to buy a handheld and talk on a repeater. That's fun for a while, but the novelty quickly wears off."

KB6NU believes that the Amateur Radio community must do more to help new hams to develop basic skills: how to solder; how to make voltage and current measurements; how to make a dipole antenna; how to choose an HF radio; how to read specifications and evaluate what is on the market. In other words, he is advocating much of what in the UK is already part and parcel of the introductory practical course that is taken by those seeking a Foundation licence. Whether this goes far enough or whether it needs some follow-up to encourage further study of basic theory and design of radio and electronics is a pertinent question that could affect the future of the hobby. Newcomers need to be welcomed and, where appropriate, provided with practical advice and guidance, and encouraged to extend their technical knowledge and, I would argue, shown the advantages of acquiring the ability to work CW.

**MORE POTENTIALLY USEFUL CHIPS.** With the progressive development of new generations of solid-state devices intended for mass-production applications, the radio amateur can play a significant role in identifying applications to Amateur Radio not foreseen by the chip manufacturers.

For example, Nyall Davies, G8IBR, writes: "In looking for some MMIC amplifiers on ebay I came across the HM174MSS. This is a low-cost ($2.50) MMIC SPDT switch made by Hittite with a switching time of 10ns. The insertion loss is specified as 0.5dB but the IP3 as typically 60dBm (minimum 55dBm). The IP3 spec is much higher than my experiments with the FST3125 suggest.

"I have not yet investigated the HM174MSS but it would appear to have obvious application to high-performance amateur radio equipment for front-end bandpass-filter switching and for wide-dynamic-range HF mixers. Being a change-over switch would also halve the component count. I put in a bid and got ten of these devices for $5.50 and am looking forward to trying them.

"The circuits in **Figures 1** and **2** outline their potential use as a ring mixer with a single transformer and as an H-mode mixer. They are shown without any consideration of the biasing arrangements but demonstrate the low component count possible with a changeover switch."

Colin Horrabin, G3SBI, comments that the 10ns switching time of the HM174MSS may impair its performance as an HF/VHF mixer but agrees that potentially it should be an excellent device for RF front-end switching of bandpass filters, replacing

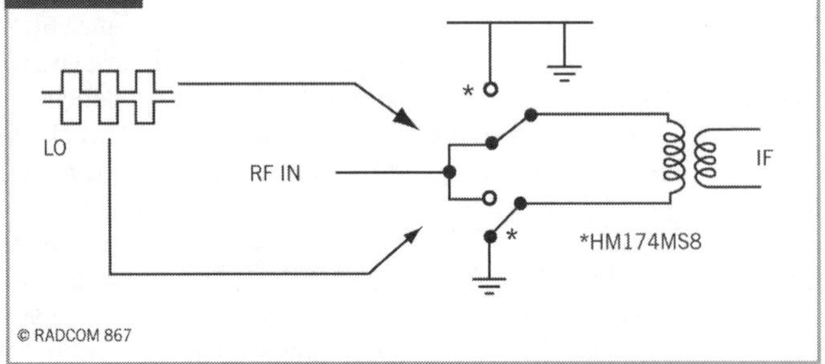

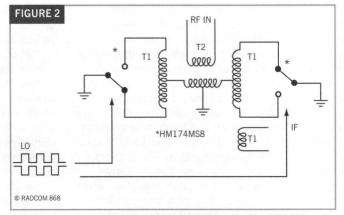

FIGURE 1: OUTLINE OF A POSSIBLE RING-BRIDGE SWITCHING MIXER BASED ON THE HITTITE HM174MS8 QUAD SPDT SWITCH, AS SUGGESTED BY G8IBR.

FIGURE 2: OUTLINE OF TWO-TRANSFORMER H-MODE MIXER USING THE HM174MS8 QUAD SPDT SWITCH, AS SUGGESTED BY G8IBR.

# TECHNICAL TOPICS

electro-mechanical switches, etc. In turn, he draws attention to the recent Fairchild FSAV332. This is described as a quad video switch with individual 'enables' for use as a high-speed video switch. The switching time is specified as only 1ns. Initial checks by PA0AKE also suggest it has better switching symmetry than the FST3125 and thus gives promise of excellent performance as a switching mixer.

Progress also continues to be reported in RF power MOSFETs suitable for use in HF transmitters. Dave Williams, G3CCO, draws attention to a design of a 1.5kW HF amplifier described in the Sept/Oct 2006 issue of *QEX* by JE1BLI and JA1DJW, based on the use of two ARF1500 MOSFETs in push-pull, eliminating the need for the (lossy) combiner networks used with multiple amplifiers. Each ARF1500 is capable of providing an output of 750 watts (efficiency about 50%) at up to 40MHz with a maximum working HT of 125V (100V in the *QEX* design) and breakdown voltage 500V. Clearly, care was needed in the *QEX* design to cope with the high input capacitance and the heat dissipation, not to mention the provision of some 3kW of DC at 100V. Nevertheless, the possibility of 750W HF output from a single solid-state device marks a significant development.

The ARF1500 is manufactured by Advanced Power Technology of Bend, Oregon (www.advancedpower.com) for use in industrial, scientific, medical (ISM) equipment and for HF communications.

**G3SBI's IMPROVED VHF SQUARER.** The performance of any switching-type mixer depends on the accuracy and balance of the square-wave switching waveform. In modern switchers this may depend on such factors as the 'skew' of the IC logic switches; with the skew assuming increased significance as the frequency is increased. In recent months, Colin Horrabin, G3SBI, has been investigating and then developing an improved squarer suitable, for example, to adapting the down conversion H-mode mixer of the CDG2000 transceiver (fully described in a series of articles in *RadCom* during 2002) for operation at 50MHz and; potentially, for other applications requiring an accurate VHF switcher.

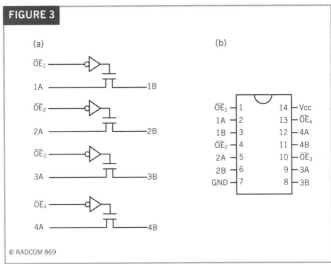

FIGURE 3: (A) LOGIC DIAGRAM OF THE FAIRCHILD FSAV332 QUAD VIDEO SWITCH WITH INDIVIDUAL ENABLES. IT FEATURES –84dB NON-ADJACENT CHANNEL CROSSTALK AT 10MHz. –49dB OFF-ISOLATION AT 10MHz. 1NS SWITCHING SPEED WITH APPARENTLY GOOD SYMMETRY. IT MAY PROVE SUPERIOR TO THE FST3125 IN HF MIXER APPLICATIONS.

G3SBI writes: "A number of people have asked me if the basic CDG2000 configuration is suitable for operation at 50MHz. It is, but not with a divide-by-two squarer as used with the existing VCO which runs between 45 and 78MHz. It would thus mean using a fundamental squarer running between 59 and 63MHz to provide the 9MHz IF. The classic way of doing this is to use the 74AC86 exclusive-OR gate. In this arrangement two gates in the chip are used to square-up the VCO sine wave and this output goes to the input of two gates connected together. One input to these gates is held high, the other low. The outputs from these gates provide, at least on paper, a true and inverse square wave to drive the FST3125 switching mixer. This configuration has been previously covered in 'TT'. Note that, with this arrangement, should the sine wave input be removed the AC86 will take a very high current and burn out.

"In an ideal world, the AC86 outputs would, as suggested above, be a symmetrical true and inverse square waveform with no overlap. In practice, this is not the case, and the situation gets worse as the frequency increases.

"Gian Moda, I7SWX, has come up with one solution to this problem and this can be found on his website. In his circuit, a dual 3-volt logic LVC discriminator with alternate inputs connected gives a true and not true output with transitions within 100ps. However, in practice this is not the ideal solution since when an inverting gate is used as a VCO squarer it automatically compensates for any asymmetry in the input sine wave with frequency. The discriminator approach only works correctly with a perfect sine wave.

"In the usual AC logic data sheet no mention is made of 'skew'. Skew is the [small] time differences between different gate outputs when they are all fed with the same input waveform; that is the problem with the AC86. The 3-volt LVC logic is faster than AC logic and skew is actually quoted in the data as 1ns maximum. This is not good enough for our purposes - but there is a solution:

"A Philips LVC86A can be used to generate an edge from each input transition. A high-speed D-type (275MHz maximum clock rate) can then be used to divide the edge clock frequency by two; the maximum skew of the D-type between Q and not-Q outputs is 100ps maximum. There still remains a problem. When power is first applied, the D-type may be locked in a particular state and does not give the LVC86A squarer the necessary correct input logic transition, with the result that the input sine wave will not go through the gate and run the system. This problem can be overcome by using a LVC14 Schmitt trigger to produce an edge from a one-second delay on power-up; this edge is used to send the D-type in the opposite direction to its power-up state, sending the squarer input gate through its input transition and so starting the system running. An additional advantage is that if the input is removed its current consumption drops to near zero and unlike the AC86-squarer does not self-destruct.

"**Figure 4** shows the new squarer. Do not be tempted to omit the 100Ω resistor although it produces a slight time delay. Although the IC devices are all surface-mounted, all the passive components used in the prototype were wire ended. It goes without saying that the PCB uses a ground plane.

"There is at present a potential logistics problem for would-be constructors. It can be difficult to obtain the high-speed D-type device. Surprisingly, most stockists have only the fancy chips in this range, and not the basic logic elements. However Dave Roberts, G8KBB, found me details of a suitable part via the web. Again, as many readers know, AOR UK have supported the CDG2000 project in many ways and I thank Mark Summer of this firm for obtaining sample parts to enable me to test out this project. I hope that once stockists realise there are applications for the basic logic elements they will stock them."

THE DK0WCY/DRA5 PROPAGATION BEACONS. For almost a decade, I have been copying the transmissions from the German Amateur Radio aurora-warning and propagation beacon DK0WCY on 10,144kHz and 3579 kHz (see the brief note in 'TT' September, 2006) including the extended ionospheric data introduced on July 1, 2006. However, particularly this year, reception in the London area of the 10MHz transmissions has proved unreliable with the skip often too long throughout the 24 hours, and the 3579kHz transmissions confined to limited times (see September). I was thus most pleased to receive the following information from L D Davey-Thomas, G3AGA, (Penzance, Cornwall):

"I noted your remarks about DK0WCY in the September issue, but I wonder if you are aware that the same information is available from DRA5 on 5195kHz at slightly different times. The DRA5 schedule is: Hour + 5, + 25, + 35, + 45 CW; Hour + 15 RTTY, Hour +55 PSK31, continuous throughout the 24 hours. The RTTY and PSK51 transmissions from DRA5 and DK0WCY contain some additional information not included in the CW transmissions.

"Some months ago I was finding it very difficult to receive the transmissions on 10.144kHz and contacted DK0WCY by email. In reply I was told about DRA5 and find reception of this station pretty reliable here when 10.144MHz has faded out." Since receiving G3AGA's letter, I have found DRA5 far more reliable than DK0WCY. The following is a typical CW 'datagram' received on 21 September:

VVV DE DRA5 1/17/N/N (long dash) VVV DE DRA5 INFO CONDS 21SEP 1834UT = MAG KIEL K 2 KCUR 1.7 = IONO RUEGEN F0F2 5.8 MAX HOP 2297 MUF 17 MUF 1K 9 = SUN WIND 324 BZ N1 XRAY A1 FLARE NONE = FORECAST 21 SEP = SUN QUIET MAG QUIET = INDEX 20 SEP = R 11 REQ 6 FLUX 71 BOULDER A5 KIEL A6 + VVV DE DRA5 1/17/N/N

The corresponding continuous call from DK0WCY was: DK0WCY BEACON 1/17/N/N (long dash) with identical datagram format to DRA5.

G3AGA also kindly provided me with the three-page extended "DK0WCY Transmission Schedule" (English text), which is also available on the net at www.DK0WCY.de. The transmission schedule provides a detailed explanation of the abbreviations and terms used in the transmissions from DK0WCY (also used by DRA5).

While the meaning of much of the information will be self evident to those familiar with the terms used in propagation forecasts etc (see for example a previous note on DK0WCY, 'TT' March, 1998), there are some items that may puzzle those without access to the transmission schedule. First, the short cycle transmitted continuously between the datagrams. In the above example 1/17/N/N: 1 is the Kiel current disturbance level of the geomagnetic field (Kiel current K); 17 is the current MUF, defined as the maximum frequency just bent back from the ionosphere in MHz for a vertical antenna radiation angle of five degrees as applicable to Juliusruh (Island of Ruegen). N/N represents 'no events'; the first letter refers to 'events' currently under way; the second to 'events' expected to occur soon. The nature of the 'events' is denoted by one or more letters. Most common is N - none; otherwise: A, 'aurora'; F, 'fadeout' (SID/SWF); X, 'solar shockwave' (this disturbs the geomagnetic field and increases the K index); P, 'polar cap absorption' (may last several days); Z, 'beacon on maintenance, expect reduced power and/or interrupted operation'. During an aurora the long dash changes to a series of dots.

The datagram abbreviations/definitions include: BZ field strength of Z-component of the interplanetary field in nT (P positive, N negative). FLARE number of X-ray flares during past 24 hours. KCUR K-Index (most recent 180 minutes). MAX HOP maximum hop distance via F layer in km. MUF 1K maximum frequency for 1000km hop in MHz. R relative sunspot number (SSN). REQ equivalent sunspot number (a more accurate measure for solar activity compared to R). SUN WIND solar wind speed in km/s. XRAY intensity of x-ray background radiation.

MORE ON SERIES & DOHERTY A.M. Dave Gordon-Smith, G3UUR, strongly disagrees with G3LNP's statement ('TT' September 2006) that series modulation is an 'efficiency' form of amplitude modulation. "Series modulation is most definitely high-level modulation with the same RF efficiency at the quiescent carrier level as 100% modulation. In essence, the standard series modulator is no different from a class A modulator coupled by a modulation transformer to a class C RF power amplifier, except that the HT is twice the normal voltage and no transformer is required. The overall power consumption is the same in both cases, with little difference in power consumption between full and no modulation because of the class A modulator. As far as the DC arrangement is concerned, series modulation is just a class A modulator stacked on top of a class C RF amplifier, or vice versa. The difficulty with series modulation is that the more efficient classes of AF amplification, such as class B, can't be incorporated into the modulator without transformers or AF chokes, and then the great advantage of the original simple arrangement would be lost.

"With regard to AM produced by PWM techniques, over the years there have been a number of amateur designs for producing AM in this way, both high and low power, and one low-power design was by G4ENA in 1993 [see note in the November 'TT' – G3VA]. Other designs from the past, mainly American, are now available on the Internet. It's just a matter of tracking them down with a good search engine and suitably qualified key words."

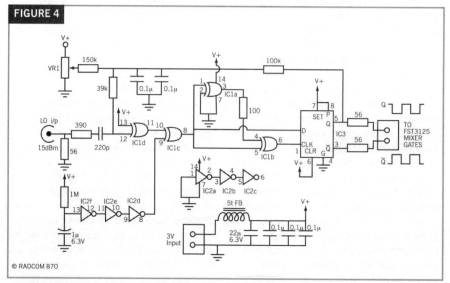

FIGURE 4: G3SBI'S FUNDAMENTAL FREQUENCY SQUARER FOR USE AT VHF, PERMITTING FOR EXAMPLE THE MODIFICATION OF THE CDG2000 TRANSCEIVER FOR 50MHz. IC1 LCV86, IC2 LCV14, IC3 74AUC1G74. VR1 IS A 20K MULTI-TURN TRIMPOT USED TO ADJUST MARK/SPACE RATIO OF OUTPUT.

# TECHNICAL TOPICS

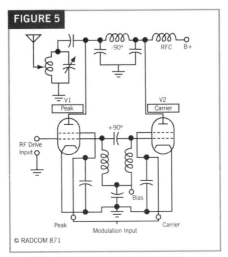

FIGURE 5: SIMPLIFIED DIAGRAM OF HIGH EFFICIENCY SCREEN-MODULATED DOHERTY MF BROADCAST AMPLIFIER AS DESCRIBED IN THE 1965 ARTICLE BY JOSEPH B SATBTON OF CONTINENTAL ELECTRONICS.

Carl Thomas, G3PEM, long-time in the broadcast radio industry, writes: "In the September 'TT' you mentioned the Doherty amplifier and the IBC papers presented by New Street [Marconi] engineers.

"To my knowledge the first commercial use of the [modern form] of the Doherty amplifier was by Continental Electronics, USA in the late 1950s/early 1960s. On Radio Caroline South (MI Amigo) we fitted a CE 50 kW Doherty amplifier transmitter. It was selected for its high efficiency, thus requiring lower generator power; important in view of the small size of the Amigo's machine room. Additionally, the transmitter had smaller building door requirements, enabling us to lower the cabinets into the transmitter hold. The then-current Marconi transmitter was too big!

"Prior to our fitting the CE transmitter, Joseph Sainton had an article "A 500 Kilowatt Medium Frequency Standard Broadcast Transmitter" (*Cathode Press*, house journal of the Machlett valve company, Vol 22, No 4, 1965, pp23-29). I met Joe (I believe he held an American amateur licence] when he came to the UK and found him a most interesting character."

G3PEM enclosed a copy of Joe's 1965 article from which I have extracted a few relevant notes:

"The high efficiency screen modulated amplifier, conceived two years ago, combines the principles of simple screen grid modulation with the Doherty linear amplifier. ... It has been said that screen modulation is incapable of producing 100% negative modulation (complete carrier cut-off). This misconception has probably arisen by comparing screen modulation to plate modulation where carrier cut-off is achieved by modulating the plate voltage to zero volts. By direct comparison a screen-modulated amplifier will not modulate 100% by reduction of screen voltage to zero volts, but by swinging the screen slightly negative 100% modulation is easily achieved. This negative excursion is generally about 10 or 15% of the peak-to-peak audio screen voltage required for 100% modulation. Moreover, the RF plate swing varies in linear fashion down to within about 5% of peak-to-peak screen voltage near cut-off. This means that up to 95% modulation the stage is linear but will have a slight rounding of the negative peak from 95 to 100% modulation. This rounding increases harmonic distortion by 0.5% at 100% modulation and is completely eliminated by overall feedback. ...

"The recent development of tetrode tubes having plate dissipation in excess of 10kW led to the feasible application of a circuit combining screen modulation with the Doherty at high power levels. In this circuit, the Doherty amplifier is screen modulated rather than operated as a linear amplifier of a modulated wave. ... The drive power required is a small fraction of that required for triodes, and the driver stage can be operated class C also. All of this adds up to high overall transmitter efficiency. ... Our new 50kW transmitter requires only 82kW of transmitter input power at carrier level. Other types, including plate modulated and phase to amplitude, require from 94 to 98kW input power. Basically, the amplifier consists of two ML-8545 tetrode tubes with a 90-deg lagging network connected between the anodes and a 90-deg advance network between the grids; **Figure 5**."

The basic advantages of screen modulation include the very low power modulator; the elimination of the modulation transformer which allows the use of AC rectified feedback around the audio stages; and the reduction to about one-half the peak RF and DC voltage applied to the modulated amplifier compared to plate modulated amplifiers of the same power output.

**Lt Col HENN-COLLINS, ex-G5ZC/GU5ZC SK.** *The Times* (27 September, 2006) obituary of Christopher Henn-Collins who died on 6 August aged 91, highlighted his adventures in Poland when he and his detachment of 12 signallers were sent to establish radio communications for the British Military Mission. Arriving after Warsaw had fallen, they became among the first British troops to come under fire. The obit mentioned also his later work in North Africa. 'TT' readers may recall his technically pioneering work as a senior military signaller, professional telecommunications engineer – and radio amateur for over 60 years. Licenced as G5ZC in the 1930s, he relinquished his call GU5ZC only a few years ago.

"Early Over-the-Horizon VHF" ('TT' March 1995 and *Technical Topics Scrapbook, 1995-1999*, pp21-22) described how in 1943 while in charge of the Radio Division of AFHQ he was concerned with the very first VHF radio-relay system (Algiers to Tunis). He observed that often RTTY signals spanned the 600km directly without the use of the intermediate relays, pioneering VHF ducting and tropospheric scatter.

"Pre-history of Amateur SSB" ('TT' March 1989. *TTS, 1985-1989,* p290) described how, after seeing the effective use of SSB/RTTY by the Americans between North Africa and the USA, he proposed that the Army should develop mobile SSB. His proposal was turned down by SRDE on the grounds of the high frequency-stability requirement. Undaunted, he obtained permission and encouragement from Sir Archibald Gill, GPO EiC, to try SSB on the amateur bands. Using an ex-German frequency generator and a Canadian C43 transmitter, as the basis of a linear amplifier, during leaves in 1947, he built an SSB rig and then tried it out from his parent's home on the Channel Islands. To quote: "However, contacts proved disappointing since nobody seemed to know how to tune an SSB signal. ... The disappointment stayed with me for about ten years. Then on a visit to my mother, I put the equipment on the air again. By then SSB had become established on the amateur bands ... [My signals created] one of the biggest pile-ups on 14MHz. Later I gave the whole equipment to the local radio club."

G5ZC was thus the first UK amateur to build and use an SSB rig – and one of the first in the world. In 'TT', July 1988, I had noted the pioneering work of Mike Villard, W6QYT at Stanford University resulting in a first SSB 3.5MHz contact between W6YX and W6VQD on September 21, 1947. This is usually considered the birth of amateur SSB with full suppression of the carrier, although as noted in 'TT' March 1989, two American groups tried unsuccessfully to establish SSB on the amateur bands during the 1930s.

Christopher resigned his commission in 1947, forming Henn-Collins Associates, undertaking project and consultancy work for government agencies and commercial firms worldwide. I recall meeting him for lunch in the 1960s while he was advising Cable & Wireless on their entry into satellite communications. He provided me with a front-page scoop for *Electronics Weekly* and subsequently with details of his pioneering over-the-horizon and SSB transmissions.

# TECHNICAL TOPICS

PAT HAWKER, MBE
37 DOVERCOURT ROAD,
DULWICH, LONDON SE22 8SS

G3VA

# Technical topics

*The triode is a hundred years old this year but its birth was not without controversy.*

**CENTENNIAL OF THE AUDION TRIODE.**
'TT', January 2004, noted the approaching centennial of the 1904 invention by Professor (Sir) Ambrose Fleming of the thermionic diode, based on the 1883 Edison Effect. This event marked the entry of thermionic valves into the then emerging field of radio communication. But the Fleming diode was developed as a new form of RF detector; indeed it marked only one, and not necessarily the most sensitive, of a number of approaches to the detection of weak radio signals. It was another two years before Lee de Forest (1873-1961) created the "Audion" three-electrode (triode) vacuum tube (valve). After later improvements this proved its ability to amplify and oscillate and soon became the unquestioned building block of the communications and electronics industries, dominating them for over half a century. Even today, after another half-century, thermionic devices compete in several applications, including RF power amplification, with the later solid-state technology.

De Forest's reputation as a pioneer inventor of radio and talking pictures has had its ups and down. His bitter patent struggles with the Marconi Company over the alleged infringement of the Fleming patents and with Howard Armstrong over the invention of 'oscillation', the financial vicissitudes, including charges of commercial fraud, the string of failed companies, his failed campaigns to achieve a Nobel Prize or a US Medal of Merit, the persistent suggestions that he put a "grid" into a Fleming diode in order to circumvent Patent Law – all are still sometimes held against him. But there is no doubt that with more than 300 patents, de Forest deserves to be remembered as a truly gifted and prolific American inventor.

The development of the Audion triode is detailed in the book "Lee de Forest and the Fatherhood of Radio" by James A Hijiya (Associated Universities Presses Inc, 1992). To quote briefly: "Fleming was looking for a rectifier that would enable him to use a mirror galvanometer; when he discovered the valve, he had reached his goal and had little incentive to continuing developing the instrument. De Forest's two-electrode Audion (1904) was much less successful, never working well enough as a detector to be practical; only one receiving set using the de Forest diode was ever sold. Therefore, de Forest needed to continue improving his Audion, and he ultimately did so by adding a third electrode.

"On 25 November 1906, de Forest ordered from manufacturer McCandless a new kind of tube, one with a third electrode interposed between the filament and the anode. To prevent the third electrode from blocking the passage of 'particles' between the filament and the anode, de Forest specified that it consist of a wire instead of a solid plate. John Grogan, an assistant to McCandless, suggested that to create a greater surface, drawing electrons from the filament, the wire be bent back and forth, and de Forest named this innovation the 'grid'. He then made this triode (Audion) into a detector by connecting the anode to an earphone and connecting the grid to an antenna for receiving wireless signals. … De Forest began testing the new tube on 31 December 1906, filed for a patent on 23 January 1907, and received it (No. 879,532) on 19 February 1908."

Although the triode Audion worked, de Forest did not accept the 'electronic' theory that Fleming was by then expounding, and continued to believe that some residual gas was necessary – that is as a 'soft valve' rather than with a hard vacuum. It was not until 1912 when engineers Harold Aitken and Irving Langmuir, at AT&T and General Electric (GE) respectively, realised the need to evacuate the tube thoroughly, that the Audion became suitable for practical use as an amplifier. The World War of 1914-18 gave enormous impetus to valve design and the manufacture of valves with consistency of characteristics. De Forest was destined to die virtually penniless on 30 June, 1961. Twelve years later, the hundredth anniversary of his birth in 1873, the important contribution made by the de Forest Audion was recognised by the issue of an 11c Airmail postage stamp.

AUDION TRIODES FEATURED ON A US POSTAGE STAMP ISSUED IN 1973 TO MARK THE HUNDREDTH ANNIVERSARY OF LEE DE FOREST'S BIRTH IN 1873.

**PROTECTING POWER SOURCES.** Steve Boden, G4XCK writes: "The 'TT' item (August 2006) on LA6PB's fast circuit breaker brought to mind problems that I had as a test technician in the 1960s. All too frequently, units would come from the line with short-circuits on the low voltage input, causing the bench power supply to trip. Resetting involved turning off the mains power and waiting 30 seconds. This occurred often, up to twenty times a morning!

My first solution was a low voltage bulb in series with the output. This worked well; the non-linearity of the bulb meant very little voltage drop at normal currents, whereas a short-circuit lit the bulb to full brilliance. However, for some units under test even the low resistance of the bulb at the normal low current affected the test results, so I developed a circuit not unlike the one described by LA6PB.

"More than twenty years later a similar problem arose. After having the pass transistors fail (short-circuit) on a power supply unit, my friend John, G4WZK built an over-voltage protection device using a thyristor to short-circuit the supply if the voltage rose above a safe level. The aim was to blow the fuse in the PSU, but we found that a 30A fuse took quite a long time to blow

FIGURE 1: G4WZK'S CIRCUIT-BREAKER PROVIDING OVER-VOLTAGE PROTECTION WITH A FAST RESPONSE TIME TO RISING CURRENT.

# TECHNICAL TOPICS

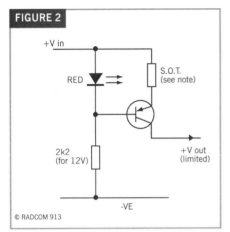

FIGURE 2: G4ANA'S CHEAP AND SIMPLE OVER-VOLTAGE PROTECTION FOR LOW-POWER APPLICATIONS.
SOT RESISTOR = $V_{LED} - V_{BE}/I_{LIMIT}$, BEST SELECTED BY TRIAL AND ERROR.

since the transformer limited the current due to the core saturating. We reduced the fuse rating to 10A, but it still took ten seconds to blow; conventional circuit breakers took even longer and always the thyristor was destroyed. It was time to bring my old circuit breaker out of retirement!

**Figure 1** shows the final design. With the type of reed relay used, only two turns are required around the sense reed, resulting in virtually no voltage drop. Some points:

- The larger type of reed relay seemed to give more stable results.
- Note the use of bulb filaments as current limiters (still my favourite method).
- There is an incredibly fast response to rising current. A 5A electric drill always tripped the circuit even though trip current was set to 40A, whereas a resistive load of 0.35Ω (39A) dropped straight across the output did not.
- A screwdriver dropped across the output trips it without so much as a spark.
- To adjust the actual trip current, provision is made to move the reed relay relative to the coil.
- Most relay coils work well with the 12V 5W filament in series but very low resistance coils may require a different type of bulb.
- Should the unit trip but the short-circuit remain, pressing the reset switch causes the 21W filament to light brilliantly, with less than 2A flowing.

"Nowadays whenever I build a heavy duty PSU, I include this trip circuit as a matter of course; for supplies for lower current outputs I replace the fuse with an appropriate festoon bulb, providing an easy means of protecting the supply."

Steve Cook, G4ANA was interested to see the 'TT' comments by Harry Leeming, G3LLL (November, 2006) as a follow-up to the item by LA6PB (August, 2006) on the need to protect the 12V auxiliary power outputs from modern equipment. He writes:

"I had an unfortunate experience recently when the power-on mini toggle switch on my (homebrew) 455-to-12kHz IF converter, built to allow me to receive DRM transmission on my AOR AR7030+ receiver, chose to collapse and short-circuit the incoming power from the receiver to the metalwork. The AR7030 has G3LLL's recommended 22Ω series resistor between its power input and output, but this is only a ½W component and fried, resulting in changing to roughly 1.5Ω resistance in series with the output socket. It was a hair-raising experience to find only a few volts on my converter after replacing the switch. Clearly, for high currents, a fast circuit-breaker is the practical solution, but for low power outputs I propose an alternative simple solution.

"I use the well-known constant-current circuit in which an LED is used to bias the base of the series transistor (**Figure 2**). The usual advantage given for this circuit is that the temperature coefficients of transistor and LED track, allowing the current to remain effectively constant over a wide range of temperature. However, if it used for protection, a more useful feature is that the LED remains 'off' so long as the current is below the limit, and turns 'on' when this is exceeded, providing a visual indication that something is amiss. At currents below the limit, the transistor is simply 'on', giving minimal extra voltage-drop. The transistor can be any type capable of

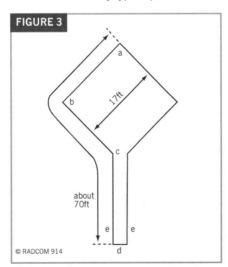

FIGURE 3: MULTIBAND LOOP AND MATCHING STUB AS ADVOCATED BY G6XN OVER 40 YEARS AGO, AND USED RECENTLY BY ZS6AAA. THE LOOP RESONATES AT APPROXIMATELY 7, 14, 21 AND 28MHz AND CAN ALSO BE USED AS A BEAM ELEMENT AT 14 AND 28MHz. CAN BE FED FROM BALANCED LOW-IMPEDANCE LINE AT POINT D, OR BY 600Ω OPEN-LINE AT ABOUT POINT E.

dissipating the power given by ($V_{in} \times I_{LIMIT}$). My usual choice is any generic PNP TO220 device out of the junk box capable of handling 100mA or thereabouts without a heat sink. Result, for a negligible cost, no more fried resistors!"

**MULTI-BAND HF QUAD-LOOPS.** The undoubted pioneer of multi-band loop and quad elements was the late Les Moxon, G6XN. He first began using a 14/21MHz quad without the usual 'nest' of elements as early as 1955, mentioned his ideas in *Wireless World*, March 1955 and in more detail in *CQ* (November, 1962). He was disappointed that his ideas were overlooked by the amateur radio handbooks for several years until they were outlined again by him in a contribution to 'TT', February, 1973 (see *Antenna Topics*, pp72-74). He later covered various multi-band loop elements in his classic *HF Antennas for all Locations* (RSGB). In particular, G6XN brought to notice the use of a loop plus stub configuration comprising about 144ft of wire that resonates at 7, 14, 21 and 28 MHz comprising a square loop with 17ft sides with some 36ft of open twin-wire matching stub: **Figure 3**. This forms a 1λ loop at 14MHz, λ/2 at 7MHz, 3/2λ at 21MHz and 2λ at 29MHz; for current distributions see **Figure 4**.

Andrew Roos, ZS6AAA in "The Compact Quad Multiband HF Antenna" (*QST*, August 2006, pp38-40) presents full details and SWR measurements etc of a similar arrangement, described as "An easy to build loop with a simple matching section offers four-band HF coverage." In effect, the antenna comprises a 1λ loop on 14MHz fed through a matching ¼λ stub with characteristic impedance of some 800Ω and then, at ZS6AAA, with some 100ft of RG213 coax. As the loop impedance is about 120-140Ω, the SWR on the coax is about 3:1 – well within the range of most auto or manual ATUs. On 7MHz, the high feed-point impedance of the ¼λ wave loop is transformed by the ¼λ stub and then directly into the coax. Similarly on 21MHz the 1½λ loop high impedance is transformed down by the ¾λ stub. On 28MHz, the 1λ stub reflects the low impedance of the 2λ loop. ZS1AAA recommends the use of a 1:1 current (choke) balun to connect the coax to the 800Ω matching section with both loop and stub formed from a continuous 140ft length of wire. The configuration is reported to provide an acceptable SWR of 5:1 or less on all four bands, with good low-angle radiation. The loop can be square or more conveniently diamond

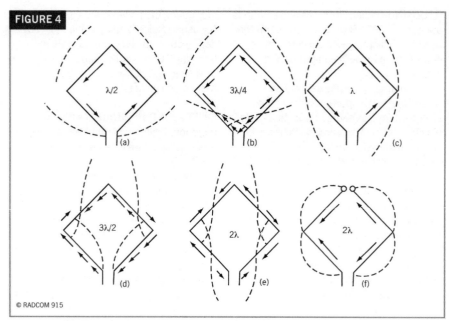

FIGURE 4: CURRENT DISTRIBUTION ON LOOP ELEMENTS. THE ARROW LENGTHS REPRESENT DIFFERENT FIELD STRENGTHS.

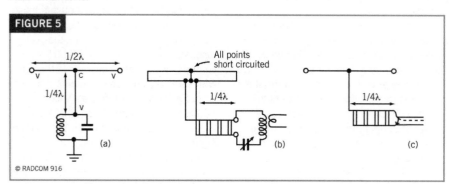

FIGURE 5: THE VERTICAL-TEE (INVERTED GROUND-PLANE) ANTENNA AS DESCRIBED IN 'TT' JULY 1970, SHOWING VARIOUS VOLTAGE FEED SYSTEMS.

shaped but not too squashed. For the open-wire stub he uses 8in spreaders with the wire spaced about 6¾in apart. For full constructional details see the *QST* article.

**VARIATIONS ON THE CLASSIC 'T' ANTENNA.** As a result of the recent notes on Zepp antennas with twin open-wire resonant feeders, several reminders have come my way recently of various forms of relatively simple 'T' antennas that in the past have proved capable of good performance. They can be designed for either monoband or multi-band applications, but never seem to have attracted much attention.

On the "Box 25" net, David White, G3ZPA reminded me that over thirty-five years ago I described in ' TT', July 1970 (*Antenna Topics*, pp34-35) "The vertical-Tee or Inverted-Ground Plane" antenna based on a simplification of the 'Bobtail array'. At the time I believed it to be original: **Figure 5**. It later proved once again that there is little really new in the field of antennas, with several readers pointing out that similar designs had appeared in pre-war American journals. The design was later endorsed by G6XN who described some further variations. The July 1970 item noted the earlier antenna described by G3JR in the *T&R Bulletin*, March 1936 (see below).

Later ('TT' September, 1972 or *Antenna Topics*, pp 67-68), in an item "Polarisation switching made easy" I drew attention to how a Telefunken MF broadcast antenna could switch polarisation at dusk to reduce interference. This technique was used also by Ken Glanzer, K7GCO, (*CQ*, July 1972) who showed how a horizontally-polarised half-wave dipole fed by a quarter-wave open-wire feeder could be simply switched at the base into a vertically-polarised inverted ground-plane: **Figure 6**.

Don Maclennan, G3GKM, has drawn attention to the use of three-wire Zepp-type feeders in combination antennas. He enclosed a cutting from the ARRL 'Hints & Kinks' Volume 3, 1945 in which VK6MO showed how such feeders could be used for multi-band end-fed antennas. He also enclosed a photocopy of "Design for Multi-Band Aerial" by G P Morgan, G8DV (*Short Wave Magazine*, June, 1947, pp215-218) that provided a "Combination System for All-Band Operation".

G8DV used the three-wire system to adapt the 14MHz monoband DX antenna described by C A Heathcote, G3JR, as "An Omni-Directional Low-Angle Aerial" (*T&R Bulletin,* March, 1939, pp502-505). In this antenna the horizontal section was a full-wave and the vertical section a half-wave, permitting the vertical section to be centre fed with 80Ω twin-wire balanced feeder: **Figure 7** (this was before coaxial cable became widely used by amateurs).

G8DV's version used the same basic dimensions as G3JR but with a three-wire feed system that allowed it to be used as an effective DX antenna on 14 and 28MHz and also by switching the feeder connections on 3.5 and 7MHz bands (this was before the release of 21MHz and long before the WARC bands). But there would seem no reason why, with a suitable wide range ASTU, it should not radiate reasonably well on all bands. The G8DV 'combination' antenna is shown in **Figure 8**. For 14 and 28MHz the transmitter output is fed via an ASTU to A and B, with C left floating. For use as a 'T' antenna, A and B are joined together with the third wire C forming a counterpoise; the ASTU output is fed to A/B and C.

Clearly the G8DV/G3JR antenna requires more space and higher supports than the 'inverted ground-plane' but would probable provide better DX performance on the main band for which it is designed. In both antennas there is high RF voltage at the lower end of the vertical element and also as a T antenna on other bands. This could provide a safety hazard but also means that there will be greater induced

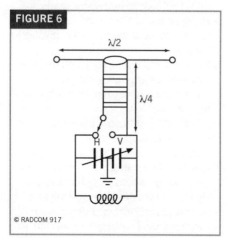

FIGURE 6: K7GCO IN 1972 SHOWED HOW A HALF-WAVE DIPOLE FED WITH QUARTER-WAVE RESONANT LINE COULD BE SWITCHED AT THE BASE TO PROVIDE A VERTICAL-T ANTENNA WITH VERTICAL POLARIZATION.

# TECHNICAL TOPICS

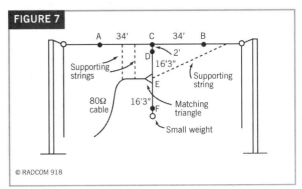

FIGURE 7: G3JR'S OMNI-DIRECTIONAL LOW-ANGLE ANTENNA AS DESCRIBED IN 1939. THE DELTA MATCHING TRIANGLE HAS 4" SIDES.

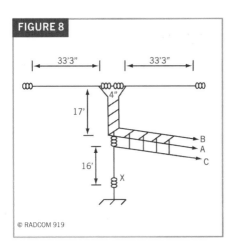

FIGURE 8: THE "G8DV COMBINATION ANTENNA", BASED ON THE G3JR ANTENNA BUT USING THREE WIRES TO CHOOSE BETWEEN ITS USE AS A FULL-WAVE HORIZONTAL DIPOLE ON 14MHz OR AS VERTICALLY POLARISED T ANTENNA ON OTHER BANDS.

ground loss than with a conventional monopole or ground-plane antenna. The vertical element must be carefully trimmed for resonance, as should the horizontal element (preferably before being joined). G8DV gave the lower section of his G3JR 14MHz dipole element as 16ft 6in, compared with 18ft for the top half; for the G8DV he gave 16ft and 17ft as shown in Figure 8.

All told, this family of resonant T-type antennas seems worthy of revival and further investigation.

LARGE HORIZONTAL LOOP ANTENNAS. To the best of my knowledge, one of the first suggestions of the value of large loops as multiband HF antennas was a contribution by S M de Wet, ZS6AKA, in an item "Experiments with multiband loop aerials" in 'TT', June 1972 (see *Antenna Topics*, pp 63-64). ZS6AKA pointed out that "the large 'loop' configuration is consistently omitted from the standard engineering textbook, possibly because the mathematical analysis of practical harmonic loops would present a particularly sticky problem". This was of course before the widespread use of computer modelling based on the NEC exploitation of the "Method of Moments". ZS6AKA presented advice on calculation of impedances etc resulting from experiments with rectangular, triangular, rhombic, vertical triangular (Delta) loops that have all now become quite widely used: **Figure 9.**

A recent article "The Horizontal Loop – An Effective Multipurpose Antenna" by Scott M Harwood Sr, K4VWK, (*QST* November, 2006, pp 42-44) carries the sub-heading "The horizontal loop need not be resonant and can work well in a number of ways". K4VWK sets the scene as follows: "It's been well documented that a large horizontal loop will perform well as an amateur radio antenna. It may also be one of the most misunderstood of antennas. Many hams believe a loop must be resonant on the lowest operating frequency to work well at the design and higher frequencies. The fact is, as I will show later, a loop need not be resonant at all to perform well. One purpose of this article is to demonstrate how to use computer modelling to perfect a loop for one's needed location. This paper is not an antenna-modelling tutorial. program such as *EZNEC* and *NEC Win-Plus* are relatively inexpensive and easy to use."

K4VWK suggests that those wishing to learn more about antenna modelling in general should look at the *ARRL Antenna Modeling Course* (for information see www.arrl.org/cce/courses.html) authored by L B Cebik, W4RNL, or at least look at his web site www.cebik.com which includes several papers on loops (e.g. "Horizontally Oriented, Horizontally Polarised, Large Wire Loop Antennas" and "Horizontal Loops": How Big? How High? What Shape?".

K4VWK notes that W4RNL points out two general misconception: (1) the longer the loop, the more the gain; (2) a loop gives an omnidirectional pattern on all HF bands. It is stressed that low angle gain is proportional to height – the higher the antenna, the higher the gain at low angles. At higher than design frequencies the loop is not omnidirectional; the radiation patterns and performance are affected by shape and location of the feed point.

K4VKW notes that for 1.8 and 3.5MHz DX a loop under 30 to 50ft high may not be the best choice (although possibly excellent for NVIS). As a rule of thumb, a good length for a loop is about 5λ at the highest desired DX operating frequency.

In his three-page article, K4VWK discusses whether a loop is the right choice in particular circumstances, presenting in tabular form the performance as governed by shape and feed point of circular, square and triangular loops; optimising a loop antenna "for your location"; resonant versus non-resonant loops; tuner considerations; and loop construction and erection. Among his "final thoughts" he stresses that "modelling software is a great aid in discovering new high performance antennas, but it is not absolute gospel ... its results need to be tested and verified." He echoes the view of W4RNL that "The advantage of the horizontal loop will not show itself in any one contact or in a short period. Satisfaction with the antenna grows with time and changes in the propagation paths, as successful communication with almost everywhere shows up in the log."

CORRECTION. My apologies to Karl Fischer, DJ5IL, for wrongly giving his call as DJ5IJ and then DL5IJ in his contribution on the Telefunken Zepp antenna ('TT', November 2006, PP69-70).

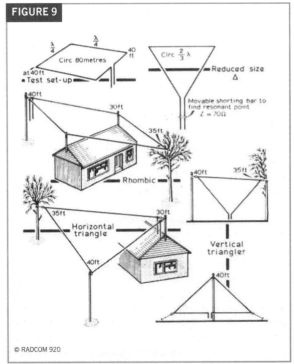

FIGURE 9: SOME OF THE HARMONIC LOOP AERIALS TRIED BY ZS6AKA.

# TECHNICAL TOPICS

PAT HAWKER, MBE
37 DOVERCOURT ROAD,
DULWICH, LONDON SE22 8SS

G3VA

# Technical topics

*Pat Hawker considers the benefits of simple equipment.*

**KK7B's MICRO R2 & HOME CONSTRUCTION.** Many years ago in discussing the pros and cons of building relatively simple communications receivers for amateur radio, I wrote (for many editions of *A Guide to Amateur Radio*): "Having read so far, the newcomer may be thinking 'Well I can see that a good communication receiver is a fairly complex piece of equipment, but does this mean there is no alternative to buying a ready-made set, and forgetting about constructional work?' The answer is no. Emphatically there is no better way than constructing a receiver – even a comparatively simple one – for finding out infinitely more about receivers than you can ever learn from reading theory. Even a straight regenerative [or direct-conversion] receiver will teach you a good deal about the finer points to look for in tuning capacitors, slow motion dials, careful adjustment of regeneration controls and the like... ."

Although component priorities may have changed, something of the same thoughts appear in the introductory notes to "The Micro R2 – An Easy to Build 'Single Signal' SSB or CW Receiver" by noted designer Rick Campbell, KK7B (*QST*, October 2006, pp26-31) (**Figure 1**). To quote selectively: "The block diagram of a basic amateur radio station [transmitter, Morse key, microphone, receiver, power supply, antenna] has not changed for more than a century. It could be a spark station or Elecraft KX-1.

"In earlier times, the amateur radio operator built and maintained the various pieces – or at least obtained a working receiver and transmitter and then assembled them into a working station. These days that enjoyable task has often been delegated to the radio designer, so that even a kit transceiver like the KX-1 leaves few decisions to the owner-operator. ... Why should kit designers and a few lucky radio engineers at radio manufacturers have all the fun? There are good reasons. Modern radios have several thousand parts [components]. ... The practical difficulties of parts procurement and modern constructional techniques, to name a couple, conspire against the successful completion of an all-homebrew station.

"So most of us don't build complete stations, even though we often think about it. We repair and restore old boat anchor receivers, wire up the stereo speakers, and maybe build a simple QRP kit or a transmitter like Wes Hayward's Mk 2 (*QST*, April 2006). Occasionally we dig through the junk box and stare off into space while holding that old National Velvet Vernier dial, daydreaming about the receiver we imagined building with it. Oh, for the good old days.

"We can recapture the joy of assembling a unique station by dividing the effort up into smaller tasks and working on them one at a time... . Start with a simple [QRP] transmitter ... . Make a few contacts using the receiver section of a [factory] transceiver. Then spend some time thinking about an appropriate companion receiver. I like the simplicity of a three-transistor receiver as described in Chapter 8 [Direct Conversion Receivers] of *Experimental Methods in RF Design (EMRFI)* (ARRL, 2003), page 8.5, but I prefer the performance of the classic Drake 2B receiver. Thus I designed a relatively simple single PC board receiver that performs like a 2B receiver over a 100kHz segment of a single band.

"A major advantage of a station with a separate receiver and transmitter is that each can be used to test the other. There is a particularly elegant simplicity to the combination of a crystal-controlled transmitter and simple receiver. The receiver doesn't need a well-calibrated dial because the transmitter is used to spot the frequency. The transmitter doesn't need sidetone, VFO offset or other circuitry. ... Even the receiver frequency stability may be relaxed as there is no on-the-air penalty for touching up the receiver tuning during a contact."

In his *QST* article, KK7B provides a full description, including constructional details, of his Micro R2 receiver. It is an advanced direct-conversion design (**Figure 1**), using all-pass networks to reduce the audio image by up to 37dB (SSB) and, for a CW-only version with alternative resistor values in the all-pass filters, by up to 50dB. He claims a performance equal to that of the classic receivers of the 1960s. While considerably more complex than the simple three-transistor design, construction should prove reasonably straightforward, using standard sized components etc. Active devices are two J310 FETs, two 2N3804 bipolars, five NE5532 or equivalent dual low-noise high-output op-amps, an LM7806 or equivalent, a 6V three-terminal regulator and two Mini-Circuits TUF-3 diode ring mixers: **Figure 2** shows the front-end only, providing AF I and Q outputs for processing.

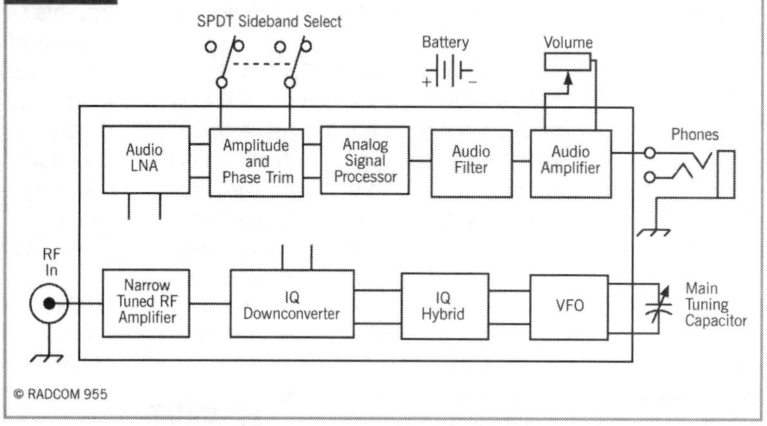

**Figure 1**: Block diagram of the Micro R2 high-performance 'single signal' SSB/CW direct-conversion receiver described by KK7B in *QST*, October 2006

To build KK7B's complete Micro R2, it would be necessary to consult his four-page article, However an experimenter could try alternative methods of processing the audio; for example the use of a polyphase filter rather than all-pass filters as discussed by the late Harold Wilson, G3OGW, ('TT', April, May, 2001) and by PA2PIM ('TT', August, 2006).

The increasing popularity of direct conversion and image-rejection mixers as now being used, for example, in software defined transceivers, is a reminder that even the simple regenerative detector in a two or three-valve 'blooper', preferably supported by a peaked audio filter can still perform well as a CW receiver, although usually much less satisfactory for AM or SSB phone. My own memory was stirred recently by working F6CIA on 7MHz CW. He was using a 1942 Paraset (Whaddon Mk VII), one of the simplest possible effective transmitter-receiver designs:

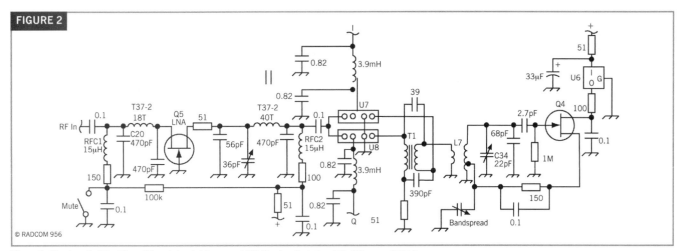

**Figure 2**: 'Front-end' of the Micro R2 as intended to cover 100kHz of a single band (in this case 7MHz) delivering I-Q audio signals to the audio low-noise amplifier ahead of the analogue signal processor. both fets J310; U7,U8 mini-circuits TUF-3 diode ring mixer. T1 17 turns two colours no 28 bifilar T32-2 L7 36t 28AWG enam on T50-6, tap at 8t with 2t LINK. For further details see KK7B's article.

**Figure 3, Photo 1 & 2.** His signals in London were weak but fully readable and he seemed to copy my 70W signals without difficulty. It all reminded me of the months in 1943-44 taking French cipher traffic from Mk VIIs and other Whaddon transmitter-receivers for real!

HIGH VOLTAGE & OVERLOAD SHUT-DOWN. Colin J Brock, G3ISB/DJ0OK, writes: "LA6PD's rapid action circuit breaker ('TT' August 2006) reminded me of a very simple, but effective, means of both high-voltage control and overload shut-down. It was extensively used in the control circuits of the high-power transmitters at both Criggion and Rugby radio stations. The circuit can be used without modification to control amateur radio power supplies.

"As can be seen from **Figure 4**, the basis of the system is a relay, the contacts of which are placed in the circuit that it is required to control. The relay coil is controlled by two series push-buttons; one with a NO contact (red) and the other with a NO contact (black).

"When the black button is depressed, a second contact of the relay closes across the button, thus ensuring a constant supply to the relay, and thus switching the HV (high-voltage) supply 'on'. If the red button is depressed, the HV- supply will be switched 'off'. As indicated in the diagram, all sorts of useful contacts can be placed in series with the red 'off' button to interrupt the HV should anything untoward occur, such as interlock, HV-supply overload, the non-appearance of essential, auxiliary supplies such as failure of bias supplies, etc."

FAST 'HUFF & PUFF' STABILISER IN A PIC. Ever since the original publication in 'TT' in 1973 of the effective VFO stabiliser developed by the late Klass Spaargaren, PA0KSB, that I subsequently named 'Huff & Puff' in view of its action resembling that of an old gas engine, the system has continued to attract attention and further advances. These include a fast acting version first described by Peter Lawton, G7IXH in 'TT' December 1997 and subsequently in more detail in QEX, December 1998, that represented very significant advances over the previous stabilisers. Now Ron Taylor, G4GXO, writes:

Photo 1: Whaddon MK VII 'Paraset' with headphones and AC PSU for 110, 125, 150, 210, 225 and 240V 50 cycle AC. (Source: PA0SE)

Photo 2: A rare glimpse of the business end of the Paraset (see also figure 3). (Source: PA0SE)

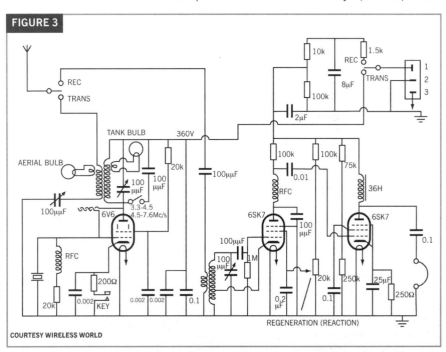

**Figure 3**: Circuit diagram of the attaché-case VII/2 'paraset' simple transmitter receiver built by MI8//SIS section VIII at SCU1 at Whaddon and Little Horwood. as used for many clandestine radio links during WW2, alternative psus were available for AC mains or 6V battery supplies.

# TECHNICAL TOPICS

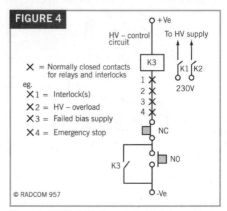

Figure 4: G3ISB/DJ0OK recalls that this form of both high-voltage and over load shut-down protection has been widely used at GPO high-power transmitting stations.

"I was approached by G7IXH who, having seen our counter-type X-lock VFO stabiliser kit, asked if it would be possible to implement a Fast Huff-Puff scheme within a PIC (programmable IC). G7IXH's original design and other derivatives which followed, such as the 'Slow-Tuning Fast Stabiliser' by Chas Fletcher, G3DXZ ('TT') and the ingenious designs of Hans Summers (http://www.hanssummers.com/radio/huff puff/) were built from discrete logic but as an overall process. The fast stabiliser seemed to lend itself to software and hardware architecture of many types of PIC. Peter asked and encouraged me to 'have a go'.

"The PIC16F628A was chosen mainly because it was to hand but also because in common with many similar PICs, its architecture seemed to be well suited to the task. Unfortunately, the smaller 8-pin 12F devices were unsuitable as the clock crystal shared the input that would be used for the VFO signal.

"After several false starts a working design was produced with a 5kHz sample rate and an 80Hz lock point spacing – not ideal but it proved the technique was viable. Having got this far, I drew an equivalent circuit of the software and hardware processes to better understand the stabilising action. My first reaction was one of surprise, it reminded me of some 'TT' items of many years ago (August, 1972 and July 1975) and a project in *VHF Communications* describing VFO stabilisation systems using PAL colour TV delay lines. A phase-shifting circuit, possibly developed during WW2 for D/F applications, called a 'Goniometer', was used to shift the VFO phase to provide continuous tuning. (Goniometers were used in Bellini-Tosi D/F systems almost a hundred years ago – G3VA). These designs were entirely analogue and whilst probably could not be claimed as forebears of 'Fast Huff Puff' they were clearly on a parallel evolutionary path.

"**Figure 5** shows a simplified block diagram of the PIC stabiliser. The input signal is squared and divided by two (as a consequence of the PIC hardware) after which it is sampled by a clock derived from the processor crystal clock. It is this that sets the long-term stability of the stabilisation process. Note that this sampling frequency is much lower than the VFO frequency and the duration the sampling will span many cycles of VFO. Importantly however the process is consistent in time and produces a 1 or 0 depending upon the phase of the VFO at the instant the sampling gate shuts. The sampled signal branches into two paths, one goes direct to the phase detector (a software XOR function), the other enters a delay line formed by a shift register, again clocked by the sampling clock. The shift register delay is the number of stages multiplied by the sampling clock period. The output of the shift register which carries the delayed samples, is passed to the other input of the XOR for comparison with the current samples. After that the output of the XOR is presented externally to the PIC for averaging by the time constant circuit before being applied to the correction varactor in the VFO.

"The XOR phase detector is worth further mention since its action in this type of stabiliser has puzzled many. The following analogy may go some way to giving an impression of how it works. As a

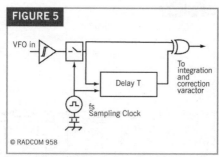

Figure 5: Simplified block diagram of G4GXO's fast huff-puff stabiliser using a PIC microchip.

Junior Officer at RMAS, one of the key leadership messages we had drummed into us was: "Always make a decision, don't hesitate. It doesn't matter if it is the wrong decision; it can be put right later. Just decide...". The XOR works on a similar principle. It always produces an output irrespective of whether one is needed. Sometimes it is right; other times it is wrong, but when wrong the phase relationships at its input cause it to correct itself on the next 'decision'. As Peter, G7IXH, described, over many samples the XOR will statistically always produce an output that will drive the VFO towards the nearest lock point. The phase detector displays a 'zero crossing' at VFO frequencies corresponding to harmonics of the sampling frequency.

"Further development of the software has now produced an improved version with a 10kHz sample rate and 1000 step shift register, again using the 16F628A. This combination produces a 20Hz lock point spacing, although by lowering the sample rate to 5kHz, a 10Hz lock point spacing will be possible. Lock point spacing (Hz) = $2f_s/N$, where $f_s$ is the sampling frequency and N is the number of shift register stages.

"When used in connection with a roughly built 12MHz VFO, the results proved remarkable. The VFO settled down very quickly after switch-on and remained on frequency for many hours. The tuning range of the varactor (a red LED) used in our test VFO is about 20kHz for a 0 to 5V control range. Current PIC performance should allow operation at input frequencies up to 50MHz, although this has yet to be tested. Higher input frequencies at VHF and UHF could also be accommodated by either mixing down or dividing the oscillator source before applying it to the PIC input. If division is used this will increase the lock point spacing as a factor of the divisor value.

"Whilst using a microprocessor

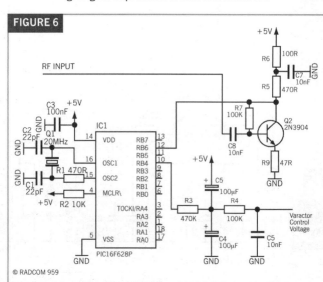

Figure 6: Frequency stabiliser using a programmed T8CK1/RA4 microchip permits significant advantages in circuit simplicity and cost.

to stabilise a VFO may seem like overkill, there are significant advantages in circuit simplicity and cost. **Figure 6** shows the minimal component count. In some applications, the addition of an op-amp integrator may be desirable to provide adjustment of the varactor tuning range. There is also plenty of scope for future development with newer PIC families, offering potential for larger shift registers and lower lock point spacing at higher sampling rates. Controls could also be easily added to tailor the delay and sampling rate to suit a particular application. Software for the circuit in Figure 6 can be downloaded from Ron's web site at http://www.cumbriadesigns.co.uk (select "resources" > "downloads" from the home page). There is a very useful graphical simulation to aid the development of Huff & Puff schemes at http://tinyurl.com/y5hk67.

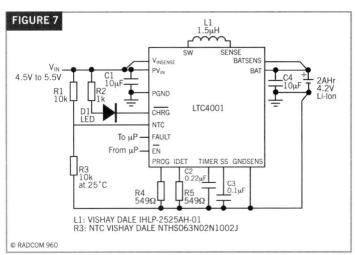

Figure 7: Li-Ion 2A battery charger with 3-hour timer. Temperature qualification, soft-start, remote sensing and C/10 indication.

**SIMPLE 2A HANDHELD BATTERY CHARGER.** No. 393 in the series of Linear Technology Design Notes by Mark Gurries, published as a two-page advertisement feature in *EDN* (July 2006), features two 2A Li-ion battery chargers for handheld devices, both based on the LTC4001 device.

These are introduced as follows: "As the performance of many handheld devices approaches that of laptop computers, design complexity also increases. Chief among them is thermal management – how do you meet increasing performance demands while keeping a compact and small product cool in the user's hand?

"For instance, as battery capacities inevitably increase, charge currents will also increase to maintain or improve their charge times. Traditional linear regulator-based battery chargers will not be able to meet the charge current and efficiency demands necessary to allow a product to run cool. What is needed is a switching-based charger that takes just about the same amount of space as a linear solution – but without the heat."

The Design Note describes two 2A chargers based around the recently introduced LTC4001. **Figure 7** shows a feature-laden charger. It requires just the LTC4001, a small 1.5mH inductor, two small 1206-size 10mF ceramic capacitors, and a few other miniature components. It provides a monolithic 2A, 1.5MHz synchronous PWM (pulse width modulated) standalone battery charger that can be packaged in a 4mm by 4mm 16-pin OFN package. It provides such features as a three-hour timer, temperature qualification, soft-start, remote sensing and C/10 indication. Even simpler solutions are possible without these special features. Data sheet download for the LTC4001 is at www.linear.com.

**MUGGING UP ON 'ANCIENT MODULATION'.** The recent items (May, July, September, November and December, 2006) on amplitude modulation systems – including series, PWM, Doherty, anode (plate) etc – made free use of such terms as 'Class C', 'efficiency modulation, 'high-level modulation' etc. It was also made clear that while SSB is generally the more acceptable mode on the DX bands (at least when crowded) on account of the reduced spectrum requirement, there remains a genuine case that can be made, not only on grounds of simplicity, of the pleasure of restoring one of those heavy old transmitters brought down from the attic or the garage, and the ease of construction of new equipment for AM. As G3UUR put it, last July, "It is useful to have a carrier to squash the noise, particularly on 80m and 150m and it is so much nicer to listen to AM than it is to SSB. Personally I find it a strain listening to most SSB signals. ... I'll continue to use my restored vintage and valve equipment on the amateur bands. ... I learn more from playing about with these older rigs than I ever would shelling out several thousand pounds for a 'black box'."

It has become clear that some readers have either forgotten or have never learned the basic use and/or terminology of AM. Not surprising since it is now some forty years or so since AM. was largely replaced by SSB. It seems appropriate therefore to quote briefly from an old edition of The "Radio" Handbook (Editors and Engineers, 9th Edition, 1942). An introductory note "Systems of Amplitude Modulation" explains: "There are many different systems and methods for amplitude modulating a carrier, but they may all be grouped under two general classifications: *variable efficiency* [often called simply *efficiency systems* – G3VA] in which the average input to the stage remains constant with and without modulation and the variations in the efficiency of the stage in accordance with the modulating voltage accomplish the modulation; and *constant efficiency* [notably high-level anode modulation – G3VA] systems in which the input to the stage is varied by one means or another to accomplish the modulation... ." In both cases the RF power output should increase when the stage is modulated.

The various systems of variable efficiency modulation include (for valves): grid-leak modulation, class BC grid modulation, class C grid modulation, screen-grid modulation, suppressor-grid modulation, and (a special case) cathode modulation. The class B linear amplifier also falls in this classification. The limiting factor is the anode (plate) dissipation of the valve. The efficiency of the stage is doubled when going from the unmodulated state to the peak of the modulation cycle. Hence, the unmodulated efficiency must always be less than 45%.

"Constant efficiency, variable-input modulation systems operate by virtue of the addition of external power to the modulated stage to effect the modulation. There are two main classifications: systems in which the additional power is supplied as audio-frequency energy from a modulator, usually called [high-level] plate modulation, and those systems in which the extra power is supplied as DC from the plate supply. The former includes plate and series modulation; the latter includes the Doherty linear amplifier and the Terman-Woodward high-efficiency grid modulator – both operate by virtue of a carrier amplifier and a peak amplifier connected together by electrical quarter-wave lines."

**CORRECTION.** 'TT December, 2005, p72, G3PEM is Colin Thomson not, as given, Colin Thomas and, in the caption to Figure 5, the surname of Joseph Sainton was wrongly given as 'Satbton'. Apologies!

# Technical topics

*'When I use a control, it should do exactly what I expect it to do' says Pat Hawker (with apologies to Lewis Carroll)*

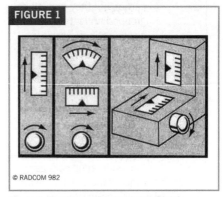

Figure 1: People expect a certain relationship between the movement of a tuning knob and their associated display. Some examples of acceptable control/display relationships. Pointers should avoid ambiguity and parallax errors.

**HUMAN ENGINEERING IN THE DIGITAL AGE.** The basic rules of good human engineering – ergonomics – had, by the mid-20th Century, become well established, if not always adhered to in practice. For the amateur operator these involve primarily the design and layout of front panels, the use of suitable control knobs and switches whose status can be read at a glance and which exercise their control in a manner that would be expected by a human operator. Controls placed in logical positions with sufficient space between them, well suited to the physical size of a large human hand and with a clear association between displays and associated controls. There are also matters of the internal mechanical design to permit the repair or replacement of faulty components, the provision of slow motion tuning drives free from backlash, etc.

As I noted in the February 2005, 'TT' page 73, quoting from a 1972 article in *Electronics* by Roy Udolf and Irving Gilbert, and an old DSIR booklet *Ergonomics for Industry 2- Instruments and People,* neglect of human factors leads to inefficiency and errors. Receivers, transceivers, control consoles and the like need to *communicate* with the operator, and the communication is two-way since users generally have to do something, like turning a knob or selecting a switch, to obtain the information or to respond to it. A system that takes full account of human engineering can be much easier and more pleasant to operate than one that does not – and less liable to incorrect and potentially damaging or dangerous settings.

As I put it in 2005, "Such common symbols as 'red for danger/stop' or 'green for OK/go' should be used, but their meanings must never be reversed. People expect certain relationships between the movement of controls and their associated displays (see **Figure 1** for some examples). Pointers and knobs should be designed to avoid ambiguity and preferably give some simple '1-9' calibration that allows an operator to see at a glance whether, for example, a rotary gain control is almost fully advanced or nearly minimum. Controls should always operate in the expected manner: fully clockwise for maximum effect; a toggle switch always turned down for 'on' etc. **Figure 2** shows some of the types of rotary control knobs that were popular on amateur gear over many years (before the era of digital push-button switching controls) that provided an instant analogue visual indication of their setting.

"Panels should not be cluttered, unnecessary labelling (legends) should be avoided to eliminate operator confusion due to sensory overload. With panels viewed from the front, labels should preferably be placed *above* the controls to which they refer, and large enough to be read comfortably at the normal operating distance, never less than about 20in (50cm)."

Many of the better old boat-anchor receivers and transmitters of the analogue valve era came near to meeting most if not all of these guidelines. But is it only a sign of the generation gap and the technofear of advancing old age that makes me question the ergonomics of modern receivers and transceivers? Increasingly, these are menu-driven units with progression towards the keyboard operated software-defined radios with tens of small multi-function push-button switches. Display screens can lack contrast and be difficult to 'read' in full daylight. As someone who finds even a cordless digital telephone difficult to use without reference to the instruction booklet, I find myself hesitating to splash out on a sorely needed new rig.

That I am not completely alone in regretting that equipment manufacturers seem to have largely thrown away the guidelines to good human engineering is shown in comments by receiver-enthusiast Michael O'Beirne, G8MOB in connection with the recent FTdx9000D HF transceiver from Yaesu. In fairness to the company I would stress that his comments are directed at the ergonomics and not the overall performance which was well covered in the review by Peter Hart, G3SJX ("Top of the Range", *RadCom,* December, 2006, pp 19 – 22). Again, this equipment is only a notable and rather extreme example of trends evident in other recent equipment, and by many other manufacturers.

G8MOB writes: "Ergonomically [the FTdx9000D] is not brilliant. There are so many knobs and buttons that you are lost within seconds. [Peter Hart, G3SJX gives the number as 'over 130 controls on the front panel as dual concentric rotaries and push-buttons' adding 'The most frequently used are large and assessable but some are quite small and a bit inconspicuous.' – G3VA]. Think of Hampton Court Maze and multiply that by a big factor. I suspect the panel designer's brief was to acquire some graph paper and fill every square with a knob or a button or meter. If so, he has been highly successful.

"As noted in 'TT' and elsewhere good ergonomics involves turning a knob clockwise to increase effect, make the important function controls big and immediately visible; confining the flimflam

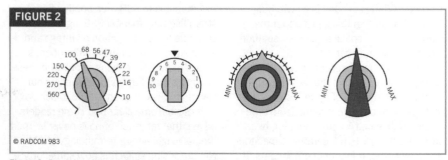

Figure 2: Examples of types of rotary control knobs popular over many years in the era of all-analogue receivers. Gain-control knobs should preferably be provided with some simple form of 'calibration' or at least enable the operator to see at a glance whether fully advanced or near minimum setting. Controls should always operate in the expected manner: fully clockwise for maximum effect.

to the periphery or to easily accessible pre-sets inside. To that I would add be sure to label the panel with decent sized contrasting print and assume that the operator has some brains and can work out what he needs to do.

"The FTdx9000D brochure describes a 'My Bands' feature as follows: 'When changing bands using the multi-function knob on the right side of the front panel, you can customise the band selection by omitting bands you do not need at the moment, thus reducing clutter and confusion when changing bands in a contest.'

"This seems unfortunate. There is a much simpler way of omitting bands – use a band knob as some of us have been doing for 30 years or more and skip over the click position of the unwanted band, not by creating a complicated way of selecting the band and then an even more complicated way of way of not selecting it!

"Looking at the panel in the shop, I could not see any obvious way of changing bands. Only when home, examining the brochure cover picture carefully, did I spot the band buttons on two of the edges of the right-hand TFT – a poor place to locate them, and they really are tiny! If you look carefully the buttons are concave and catch the light by reflection, obliterating most of the writing on the button. Big flat-matt-buttons would have been more ergonomic or the designer could have binned the small buttons and created room for a decent-sized keypad as found on most modern rigs. It is surely not asking too much of an operator to punch in '7' and 'MHz' to get to 40 metres!

"There are two independent receivers built in, but their controls are all over the place in no logical order…. If you do need two receivers, locate all the controls for one on one side of the panel, and those for the other on the opposite side, as in the IC-7800.

"For a panel already overpopulated, the designer insists on inserting a massive 75mm tuning knob, taking up useful panel space, awkward to use, and reminding me of the control knob on large, old Variacs. Again, he has placed the LSB mode button *above* the USB button! … I was most disappointed with the panscope display on the right-hand TFT screen. It is desperately slow, at least a second behind real time, no doubt the result of delays in the DSP. The delay is most frustrating – old fashioned analogue systems using a CRT are far superior, typically the Racal MA2313E and, I believe, the old Kenwood spectrum displays."

I would emphasise that G8MOB's criticisms are primarily of the ergonomics and not the technical performance of this transceiver. As he puts it: "The irony is that the technical specifications are very good even if seemingly impaired by its ergonomics."

While G3SJX concluded: "The FTdx9000D is a most impressive and eye-catching radio with an awesome armoury of features to assist the serious DX and contest operator. The overall performance is excellent in most areas but it is a large and heavy radio and needs plenty of desk space…."

Although G8MOB shares my liking for some of the classic receivers of the past, both valve and solid state, and in his case several of the Racal professional HF communications receivers, he is by no means adverse to modern receivers utilising up-conversion, key-pads, menus and firmware-defined systems. Provided, always, that the panels are reasonably uncluttered and follow the traditional guidelines for good human engineering while taking advantage of the improved analogue RF engineering of the front ends that is now possible. One must agree that any equipment or instrument that presents over 100 front-panel controls is hardly likely to be user-friendly.

He writes: "The Racal RA3701 receiver produced in the 1990s shows what can be done when gimmicks are avoided and brain applied to function. Superficially, the circuitry looks straightforward but I am told that the software is very sophisticated. Yet all controls are instantly visible, standing in neat straight lines like soldiers on parade; there is a minimum of visual clutter. One large button per function except for extracting deeper layers of software; the right-hand alpha-numeric display takes you through the options in a helpful and intuitive way; even a software-ignoramus such as I can navigate through six layers of software with confidence and at speed.

"With a third-order, out-of-band intermodulation (IP3) of +33dBm at 20kHz off tune (RF amplifier off), it has little need for complex front-end tuning even in a high signal strength environment. Racal produced a sub-octave tuner as an optional module but I have seen only one in the UK and I don't believe the military ever required it. As further example, Tony Hibbart, G8AQN of AJH Electronics has told me that he used a much earlier Racal RA1772 within spitting distance of the massive BT antennas at Rugby, yet never needed to switch in the manual preselector despite the very strong local signals.

"This order of performance has been possible professionally for some 30 years, yet the factory-made amateur rigs are only now getting there, with many still having a long way to go."

"The average ham would expect a few more facilities than on these professional receivers; typically a variable notch filter and a variable passband control, possibly an antenna-select option, but that's about all. All the other controls and gimmicks found on many current designs are not needed *because the circuitry is of excellent design and construction.*"

WHAT FUTURE FOR HF R&D? As radio amateurs we need to face up to the fact that HF has largely fallen into disfavour for professional and military communications, and increasingly for external broadcasting. One wonders to what extent this can be revived by *Digital Radio Mondiale.* This seems far from certain in view of the increasing use of satellite distribution, downloading from the web etc). One result is that the research effort into improving HF receiver performance, HF antennas, HF propagation research studies etc have all diminished rapidly during the past two or three decades. The ending of the Cold War, with its emphasis on Signals Intelligence (SigInt) derived from weak manual or automatic Morse or RTTY signals has meant that few governments and/or firms are now investing much money in HF research and development. Other, perhaps, than for complex digital modes and spread-spectrum signals. Increasing it is the semi-professional amateur experimenter who has become the driving force for future improvements. One has only to think of the efforts of some of the core contributors to 'TT' and to the feature pages of *RadCom.*

One of the very few factory-built receivers designed in the UK specifically for SWLs and radio amateurs that took full

**Photo 1:** AOR's renowned AR7030 receiver

## TECHNICAL TOPICS

advantage of improved front-end technology and pioneered the use of menu-driven operation is the compact AOR-UK Model AR7030 (**Photo 1**). Designed and first marketed in the mid-1990s it must be one of the few models that have remained on the market for over ten years. In a detailed tabulation of some 59 receivers and transceivers measured by Robert Sheerwood, WB0JGP and George Heidelman, K8RRH of Sheerwood Engineering (see 'TT', May 1997, pp153-4) the AR7030 came second only to the specially modified Drake R4C/CF 600/6 receiver (the stock R4C came near the bottom of the table).

I can personally vouch for the excellent performance of this small, lightweight receiver but I have to agree with G8MOB and others that ergonomically it leaves something to be desired; it requires careful study of the operating manual and even then is visually challenging in operation. The small black push buttons are almost lost against the compact black enclosure while the small white labels are difficult to read by those of us who no longer have 20/20 vision. The tuning knob is of convenient size and smooth in action, but the rotary volume control and memory selector (which is also used to switch bands by setting the memories) are small for those of us with large clumsy fingers. It certainly takes time and practice to use the AR-7030 effectively without reference to the user's manual, yet undoubtedly it was/is a truly pioneering iconic design. Its linearity throughout all stages and low-phase-noise oscillator also makes it ideal for use with external analogue or DSP audio filtering for CW/digital-modes operation without the optional narrow-band CW filter. MF/HF AM broadcast station reception is improved by the ability to use synchronous demodulation of both sidebands. If the size of the front panel had been larger the receiver might have been less portable but more of the human-engineering guidelines would have been possible. Size matters!

Peter Waters, G3OJV insists in the Waters & Stanton advertisements that "Software Defined Radio is the future. I am sold on it and so will you be. No, it is not like controlling your XYZ radio by computer. SDR transfers most of your radio circuitry into the PC – even SSB generation – offering ultra linear processing and unprecedented circuit stability.... This transceiver [marketed version of the SDR-

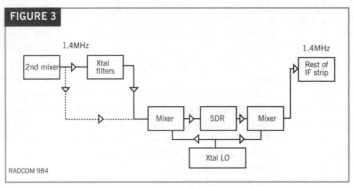

Figure 3: Suggested use of a SDR board to 'smarten up' the 1.4MHz IF strip of a RA3701 receiver.

1000] outperforms hardware designs that cost three times the price."

Certainly the SDR rigs that I have worked on SSB on the 3.5MHz band have notably pleasant audio and seem to perform well. But the overall performance in terms of dynamic range etc of an SDR rig, particularly on the higher HF bands, must inevitably at present still depend on good RF analogue engineering of the front-end and the suppression of internal switching noise. I believe that the stand-alone transceiver, even if increasingly dependent on digital techniques and built-in software, still has an assured future. Price is not everything, particularly since costs can be reduced by home-assembly, for those willing to construct or assemble some recent designs and kits such as the Elecraft K2 that can provide a performance as good as or better than the latest factory built models, at a significantly lower cost.

Michael O'Beirne, G8MOB writes in respect of SDR generally: "I may be an old fogey, but I am not totally convinced that a PC and a radio make the best of bedmates. Quite apart from QRN (noise) issues, who wants a radio tuned by a computer? The answer is the professionals because it is vastly cheaper than paying a trained radio operator plus all the add-on costs – NIC, pension, sickroom benefit etc, etc. [SDR also appeals to the military because of its flexibility for coping with ever-more complex new digital modes and signal-concealment techniques that are of only marginal interest to the radio amateur – G3VA].

"No, I want a 'proper' radio, preferably in a 19-inch cabinet or rack with a proper 2-inch tuning knob and proper manipulable knobs/controls, not a black-box with an on/off switch and a red LED indicator.

"However SDR does have a good part to play ... the SDR module can replace or be additional to the IF strip. I know someone who has an SDR module with which he plans to smarten up his Racal RA3701 as in **Figure 3**. The common LO is applied before and after the SDR unit so that there is no error in the return to the main IF of the receiver. An alternative is to feed the IF output to an external SDR module and extract the audio from the module.

"This seems to me a sensible approach

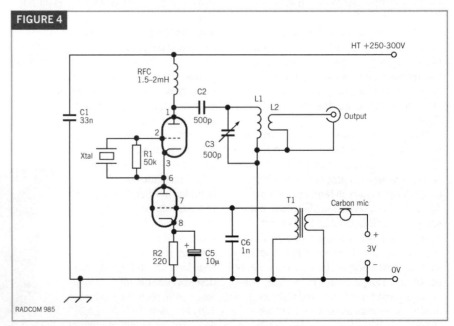

Figure 4: Circuit diagram of G3BDQ's "single-valve" (12AT7 double-triode) series-modulated A.M. transmitter. Note in the diagram the heaters are wired for 6V AC operation. When using a 12V-heater supply as in Figure 6, the heater centre tap (pin 9) is not used. Full constructional details etc in the January 2007 Practical Wireless. For 1.8MHz, L1 (34[micro]H) is close wound with 24 SWG enamelled copper wire and has 42 turns on a 35mm plastic film container, with a two-turn link winding at the 'earthy' end of L1.

to SDR rather than by boxing the SDR into the guts of a PC."

**G3BDQ's "ONE-VALVE" QRP A.M. TRANSMITTER.** In the May 2006 'TT' item "'Ancient' Series Modulation (pp102-3) I drew attention to an article "Receiving FM programmes in the Medium Wave Band" by Peter Lankshear (writing from New Zealand) that appeared with full constructional details in *Radio Bygones*, No 18, August/September, 1992 (*not, as given, 1982*). I included the circuit diagram of his unit as an example of a low power series modulated 'transmitter' which used a double-triode valve (12AU7, ECC82, B329, 6SN7, B65 or ECC32) plus a triode audio amplifier. The unit was intended to produce a few milliwatts of RF with one section of the double-triode acting as a self-excited MF tuneable oscillator intended to provide coverage of his house and was intended to distribute programme material from an FM receiver to AM sets within the house. I added: "For this application, such a device breaches UK regulations. A few quick tests showed that the system worked well on MW and could, almost certainly with a few modifications and a good antenna, be shifted into the 1.8MHz band. It would then become, with microphone input, a simple 1.8MHz QRP transmitter, or used to provide drive for a high-gain grid-driven linear amplifier, although I never got round to trying this out in practice."

I have no idea whether it was this item (or possibly the 1992 *RB* article) that may have inspired John Heys, G3BDQ to develop "The Rother – 1.8MHz amplitude modulated transmitter" described with full constructional details in *Practical Wireless*, January 2007, pp52-53. As shown in **Figure 4,** his low-power transmitter features crystal-control rather than a self excited oscillator and dispenses with the audio amplifier by using a high-output carbon-insert microphone (old telephone-type insert). He limits the HT supply to 250-300V in order not to exceed the cathode-heater insulation rating. The PSU (**Figure 5**) uses two of the readily available 230/12V transformers back-to-back to provide 12V heater and 250V HT supplies isolated from the mains supply.

He finds that (presumably with a good antenna) and an RF output of some 250-500 mW, ranges of up to about 20 miles can be achieved with this

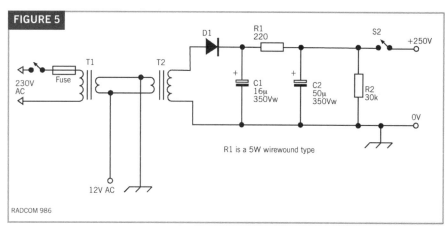

Figure 5: PSU for G3BDQ's single-valve QRP series modulated AM transmitter as described in Practical Wireless

simple "one-valve AM transmitter." He also suggests that it could work on 3.5MHz with a suitable crystal. By adding a simple AF amplifier as in Peter Lankshear's original distribution unit, it would be possible to use a higher quality, lower output microphone. The May 2006 'TT' item also showed a basically similar solid-state series-modulated arrangement.

**KK7B's MICRO-T2 SIGLE-BAND SSB TRANSMITTER.** The February 'TT' outlined KK7B's design of the Micro-R2 – an easy-to-build, single-band solid-state high-performance SSB/CW direct-conversion receiver with I/Q processing that he fully described in the October, 2006 *QST*. In the December, 2006 issue (pp28-33), Rick Campbell, KK7B describes, with full constructional details, "The Micro-T2 – A Compact Single-Band SSB QRP transmitter" to go with the Micro-R2. It uses a crystal-controlled VXO to cover about 100kHz of the 7MHz band with an output of about 1mW but he also provides details of a 0.5W amplifier. He points out that the design can be adapted to provide USB or LSB output on any single band from 1.8MHz to 50MHz. The design uses I/Q signal processing based on all-pass op–amps and two Mini-Circuits TUF-3 diode ring mixers. **Figure 6** shows the block diagram of the circuitry on the PC board.

**HERE & THERE.** Dr Brian Austin, G0GSF (ex ZS6BKW), reports the death last December at the age of 91 years of Horace Dainty, ZS5HT, later ZS5C, a noted designer of military radios. Together with David Larsen, ZS5DN later ZS6DN, he was responsible in the early 1960s for the development in South Africa of the first relatively low-cost HF/SSB pack set (RT14) which was soon to form the basis of the Racal fully transistorised "Squadcal" (5W pep output) pack set. The Squadcal was marketed by Racal in 1965 as the first to put a complete SSB station on a man's back for a cost of about £300. Although Racal sold some 25,000 Squadcals to more than 50 countries, it was never adopted by the British Army due to niggles by SRDE. The full story of the work of Horace Dainty and David Larsen is told in G0GSF's two-part article "The SSB Manpack and its Pioneers in Southern Africa" (*Radio Bygones*, Issue No 93, February/March 2005 and Issue No 94, April/May 2005). Earlier, It was Dainty's company, then known as SMD, that built the first six HF receivers utilising Trevor Whadley's triple-loop system; one of which was demonstrated to Racal led that firm to produce the classic RA17 receiver.

**CORRECTION:** 'TT', January, 2007, p64, 2nd column. In the item on protecting the 12V auxiliary output from G4ANA's AR7030, it was wrongly stated that the 22Ω series resistor when fried changed in value to 1.5Ω. This should have read 1.5 KΩ.

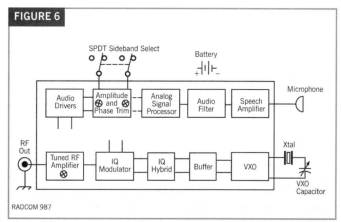

Figure 6: Block diagram of the PCB circuitry of KK7B's MicroT2 "A compact single-band SSB transmitter"

# Technical topics

*Pat Hawker's wide ranging column includes a radical new suggestion for spectrum licensing, the start of the new solar cycle and more on the debate over Marconi's S S S.*

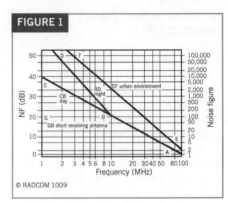

**Figure 1**: Relationship between maximum usable sensitivity, minimum acceptable sensitivity and frequency showing the effects of the high site noise levels in urban sites existing in the mid-20th Century. It was seldom necessary for the receiver to have a noise factor less than 15 for frequencies below 21MHz unless a very short, inefficient antenna is used. The lines BC and BD represent reception on half-wave dipoles about a half-wave above ground. Minimum acceptable sensitivity is represented by the line EF, reflecting a typical urban environment some 50 years ago.

**WHAT LIES AHEAD?** As 'TT' enters its 50th year of publication (first appearing in the RSGB Bulletin, April 1958) there are clearly significant challenges facing Amateur Radio. As someone who tries, perhaps unsuccessfully, to avoid becoming a POT (pessimistic old timer) it is difficult to maintain an entirely optimistic outlook. Increasingly, the world takes the computer world as the future of global telecommunications in a digital era; and links this with the regulators' current belief that the radio spectrum exists primarily as a financial asset to be exploited by governments. Rather, that is, than as a means of communicating over long and medium-distances and a scientific-research medium that requires strict national and international regulation to ensure that noise levels do not rise to the extent that extremely weak signal reception becomes significantly impaired.

In the UK, it would seem there has always been an agenda on the part of the regulator in respect of amateur "experimental" licences. In the early days, there was the desire to protect the Admiralty, maritime, aeronautical and other official W/T services from interference, later extended to medium-wave broadcasting. Throughout the 'thirties, the Postmaster General sought to protect Post Office telephone and telegram services by enforcing the "no third-party" rules. Post-WW2, the protection of the fledging VHF television services became a priority along with the desire of the Services to keep available a pool of experienced radio operators.

When the Home Office, and then its offshoot, the Radio Communications Agency, took over the role of regulator, the agenda became less clear; some shift towards deregulation with the target of balancing out administrative costs against licence revenues, including the phasing out of the former free investigation service for broadcast licence holders. Then an initial auction of UHF spectrum showed the potential for the spectrum to become a major financial asset.

Today, with the role of regulator taken over by Ofcom, the agenda for commercial licensing has been set out recently by William Webb, Head of Research and Development at Ofcom, in an article "A licence to do (almost) anything you want" in *IET Communications Engineer* (December/January, 2006/07, pp13-16). If you are puzzled by the 'IET' remember that the former 'Institution of Electrical Engineers' (IEE) is now part of the 'Institution of Engineering & Technology' (IET).

I should make it clear that the proposed flexible approach to spectrum licensing advocated in his article is not concerned with amateur licences, which are not mentioned – neither is there any reference to International Regulations or European Union directives. ITU regulations seem likely to come under increasing pressure to be radically revised to permit more flexible licensing.

William Webb stresses: "Research suggests that if licences were more flexible this could increase the value the UK generates from the radio spectrum by nearly 1bn Euros per year".

Spectrum is thus seen primarily as a marketable commodity. But it is not, I hope, too pessimistic to suggest that the plans to introduce more flexible licences for commercial licensees may have unintended consequences for amateurs and some of the other minority services that depend on the reception of very weak signals. All DX-minded amateurs (from LF to GHz) seek to use signals at noise-levels as near as possible to the unavoidable solar noise level which drops dramatically as the frequency increases, permitting the use of receivers with great sensitivity (low noise figure). The basic noise floor is already seriously degraded, particularly at MF and HF, in most urban areas and increasingly in many rural areas by the mass of local electrical noise coming from switching-mode power supplies, low-energy electric-light bulbs, microwave ovens, corona discharge from high-voltage distribution lines, etc etc and not forgetting interference, at HF, from broadband over telephone lines and, potentially, from overhead power lines. EU EMC directives are valuable but set limits well above the natural noise floor.

There is little doubt that the characteristics indicated in **Figure 1** are seldom appropriate today at most typical locations. The basic external noise includes atmospheric noise, which is dependent on frequency (decreasing with rising frequency), time of day, diurnal and seasonal variations (that is, ionospheric conditions) plus the local noise generated by a myriad of electrical appliances which traditionally is greater at lower frequencies but now spreading to higher frequencies. The operational 'minimum detectable signal' of a receiver depends on its bandwidth and its antenna as well as the noise factor of the receiver. The threat to amateurs from the recent US ruling on broadband data transmission over power lines is being challenged in the courts by ARRL. The UK amateur licence has traditionally permitted mixed mode operation but this led, about 1950, to voluntary band-planning; found necessary to protect narrow-band modes.

The IET article, addressed to professionals, opens as follows (selective quotation): "If you want to transmit radio signals then, unless you are operating licence-exempt equipment, you need a licence to do so. Up until now, radio licences (or, in the rather antiquated terms used by the relevant law 'wireless telegraphy licences') have been more or less specific to a particular technology or application. ... A licence that was flexible, allowing a range of technologies and uses would be greatly superior, enabling rapid innovation and unlocking substantial additional value from the use of radio. ... The key reason for managing spectrum is to avoid interference between different users. ... The role of the spectrum manager is to ensure that neighbours both in geography and in spectrum are given compatible licences – that they do not cause *excessive* (italics added) interference to each other."

This may seem at first sight to offer no danger to amateur operators using bands still allotted primarily to them (but remember that in many bands we are *secondary* users). But there could be unexpected consequences even in the primary bands. In the past, amateurs have not been protected against interference unless this also interferes with other licensed services, such as television broadcasting, etc. The position may now be changing for the better with the implementation of the new EU EMC Directive 2004/106/EC.

As a result of lobbying by the IARU, the preamble to the Directive now reads: "Member states are responsible for ensuring that radio communication, including radio broadcasting and the amateur radio service, operating in accordance with International Telecommunication Union (ITU) radio regulations, electrical supply networks and telecommunications networks, as well as equipment connected thereto, are protected against electromagnetic disturbance." It will be interesting to see how Ofcom interprets this preamble in practice.

In a section on "How licence holders interfere with each other" William Webb considers that there are broadly three ways that one licence holder (A) can interfere with another (B). *Geographical interference* with A and B using the same frequency too close to each other. *Out-of-band interference* with A and B close to each other using separate but nearby frequencies with A's transmissions spilling into neighbouring bands. *In-band interference* with A and B close to one another using separate nearby frequencies but with B's receivers unable to reject A's transmissions. **Figure 2** shows the typical radiation for GSM where the assigned bands extend 100kHz each side of the zero point but emissions continue well beyond this and could interfere with neighbouring assignments.

Amateurs have always suffered from all these three forms of mutual interference, so what would be new? The answer, I suggest, is that, for the first time, the spectrum will be auctioned or sold off in blocks to a new form of middleman trader who would then sell-on or lease channels to the highest bidder or at 'retail' market prices. His clients could then use any technology or application provided it did not interfere with other 'purchasers'. It would seem that no defined limits would be placed on noise generated in-band or

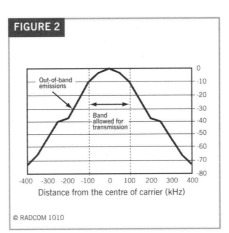

**Figure 2**: When a transmission is made on a specific frequency, the energy emitted extends across a much broader band. For example the diagram shows typical radiation of energy by a GSM UHF cellular transmitter where the assigned band extends 100kHz each side of the zero point, but emissions continue well beyond this, causing out-of-band interference into adjacent channels. Source: IET Communications Engineer

out-of-band unless it prevented other users operating their services. As William Webb puts it: "The market will determine the use of radio spectrum, and the role of the regulator will diminish to that of a spectrum policeman."

I gather that the proposals do not find favour with all current professional licensees since, in effect, the new spectrum traders, rather than Ofcom, will, it seems, have near monopoly rights over their portions of the spectrum and be free to charge 'purchasers' whatever the market will bear. For amateurs it seems inevitable that we will be faced with still further raised spectrum noise floors, mainly from electrical appliances but also some from in-band and out-of-band mixed mode transmissions making reception of extremely weak signals more difficult. It is difficult to be altogether optimistic in spite of the new EU EMC Directive.

**SUNSPOT CYCLE 24 SOON?** There seems to be every reason to expect conditions on the higher HF bands to begin to improve before long. Sunspot Cycle 24 is expected to kick off this year, possibly even by the time this 'TT' is published, but it is more likely that its official start will be around mid-summer. Hopefully Cycle 24 will follow the usual pattern of a steeper rise than the subsequent decline. Curiously, both Cycles 22 and 23 have been marked by double-hump maxima with two distinct peaks, one or two years apart.

More sunspot activity means increased solar flux (as measured in Canada at 2300MHz) with the higher HF bands open for stronger DX signals over longer periods and more readily workable with simpler antennas and lower powers. For many of us, the year 2006 has been most disappointing with many days during which there has been few DX signals on the amateur bands above 10MHz with even the usually reliable 14MHz band often in poor shape. My records of the DK0WCY/DRA5 beacon transmissions show that the solar flux has often been below 75 and occasionally down to 70 which seems to be about as low as it ever gets. The daily sunspot number is a less reliable guide to actual HF conditions than the solar flux and the current geomagnetic K and A ratings.

**Figure 3** shows monthly maximum and minimum figures for solar flux based on my daily logging of the DK0WCY/DRA5 beacons (fairly complete except for those "contest days" on which the beacons do not operate). It also shows monthly Kiel KA figures with anything above KA 20 representing magnetic storm conditions. Notice the extremely severe storm in mid-December, 2006 when signals collapsed on most HF bands.

**FIGURE 3**

|  | 2005 | | 2006 | | | | | | | | | | | | 2007 | |
|---|---|---|---|---|---|---|---|---|---|---|---|---|---|---|---|---|
|  | N | D | J | F | M | A | M | J | J | A | S | O | N | D | J | F |
| Max Flux | 102 | 106 | 94 | 79 | 101 | 101 | 93 | 86 | 87 | 89 | 87 | 80 | 97 | 03 | 92 | 90 |
| Days Flux ≥ 90 | 12 | 17 | 5 | 0 | 12 | 14 | 0 | 0 | 0 | 0 | 0 | 0 | 5 | 5 | 0 | 2 |
| Days Flux ≤75 | 0 | 0 | 0 | 6 | 1 | 1 | 7 | 14 | 19 | 11 | 11 | 18 | 0 | 8 | 0 | 11 |
| Min Flux | 77 | 85 | 77 | 74 | 75 | 75 | 72 | 72 | 70 | 70 | 70 | 70 | 77 | 72 | 76 | 74 |
| Max Kiel kA | 26 | 28 | 32 | 22 | 33 | 34 | 21 | 25 | 22 | 39 | 23 | 23 | 22 | 86 | 35 | 24 |
| Days Kiel kA ≥20 | 1 | 3 | 1 | 1 | 3 | 5 | 2 | 4 | 1 | 4 | 3 | 4 | 1 | 7 | 7 | 2 |

**Figure 3**: Monthly tabulation of selected HF ionospheric data for solar flux as measured on 10cm at Penticon, Canada and geomagnetic field KA rating as measured at the Kiel beacon. Compiled from my logging of the daily K0WCY/DRA5 beacon transmissions over the period November 2005 to February 2007

# TECHNICAL TOPICS

One word of warning, some writers have forecast that we may be entering a new "Maunder Minimum", the long period in the 18th Century when there was virtually little or no observable sunspot activity. This view was expressed last Autumn in a feature article in *New Scientist* but I recall that a similar forecast was made as Cycle 22 was running out, There may be another Maunder-type long minimum period – but hopefully not yet!

**MARCONI AND HIS 1901 'S' CLICKS.** In a December 2001 'TT' item, "12 December 2001: Did Marconi's ears deceive him?" I explored some aspects of the long controversy over whether the 400m (750kHz) or thereabouts signals from his Poldhu, Cornwall transmitter could really have reached windy Signal Hill, Newfoundland, over a daylight path, at sufficient strength to be detected as clicks on the crude untuned receiver that had been pressed into use by Marconi and Kemp. Both were convinced that they heard (and logged reception of) the three clicks several times that day. December 1901 and 2006 were both near the bottom of a solar cycle.

The contrary view was expressed openly by many contemporaneous and later scientists, most recently by Dr John Belrose, VE2CV at the "IEE International Conference on 100 years of Radio". Because of legal action by the cable company, Marconi was prevented from repeating the experiment, and the debate still continues. There is however no doubt whatsoever that in late February 2002, Marconi and C S Franklin, aboard the 'SS Philadelphia', did receive (at night) Poldhu signals at distances far beyond the horizon and thus opened the way to world-wide radio communication, as well as discovering the difference between day and night propagation.

Because of the legal action Marconi was prevented by the cable companies Marconi was prevented from repeating the experiment in Newfoundland and the doubters pointed to the possibility that what was heard was static or local electrical interference. In the 2001 'TT' item, I recalled that Gerald Garratt, G5CS, of the Science Museum told me, some 30 years or so ago, of his belief that Marconi *did* hear clicks from Podhu. But, since the transmitter must have been rich in harmonics up to HF, he received harmonics rather than the fundamental 400m signals.

It was therefore of interest to read in *Radcom* (December 2006, p 11) the News item "Research into famous Marconi reception" of the planned experimental beacon transmissions to be run from November 2006 to the end of January 2007, by the Poldhu Amateur Radio Club in conjunction with the Marconi Radio Club in Canada. The Poldhu Club was to use the 160m beacon GB3SS to make regular two-minute transmissions every quarter of the hour on 1900kHz, with amateurs in Newfoundland aiming to copy and analyse the transmissions.

There were soon reports (*Radcom*, January 2007, p10) of the GB3SS signals being heard across the Atlantic, under the heading "Experiment supports Marconi". But the value of the experiment was called in question by Brian Clows, GW4HBZ, ("The Last Word", January 2007 p97) who pointed out that in 1901 Poldhu was transmitting on about 800-850kHz, not 1960kHz.

Late January 2007, I received a letter from L D Davey-Thomas, G3AGA, one of those concerned with GB3SS, headed "Poldhu Top Band Beacon Experiment" and from which the following notes are extracted:

Rare facsimile signed photo of Guglielmo Marconi (© The Marconi Company Limited, used with permission)

"After three months of operation the Top Band beacon at Poldhu will go QRT. ... The experiment was suggested by Bart Lee, KV6LEE in an attempt to throw some light on the age-old question of what Marconi (and Kemp) did or did not hear on 12th December 1901. Poldhu ARC set up the beacon so that propagation across the [North] Atlantic could be monitored over the period of the (forecast) sunspot minimum for this solar cycle when conditions should resemble closely those of 1901. ...

"The beacon power was 100 watts and the aerial a simple Marconi 'T' – there are constraints on what we can do on the National Trust site; also the site is very exposed to winter gales. Not least, it is difficult to get an effective earth because the soil structure is a deep layer of shale. The 'T' was about 50ft high with the top about twice the length of the vertical, The ground plane consisted of eight 65ft radials, pegged to the surface since the National Trust would not permit us burying them. After trimming, the feed-point impedance came out at 20 +j0Ω. A simple 'L' section was used to match the feed-point to 200ft of 50Ω coax back to the shack. The SWR in the shack varied from 1.1 to 1.3, depending on the rainfall! (Heavy rain = low SWR).

"In spite of the simple aerial and low power, the beacon has been heard regularly in daylight across the Pond, in some cases even before noon UTC. Reports from Newfoundland and Prince Edward Island seem to indicate that the beacon has been audible throughout most of the day. ... The G4JNT transmitter was programmed to send a series of differing power levels – five steps each 6dB down on the previous one (100W to 100mW). It is surprising how many reports claim to have heard all five steps.

"Besides reports from VE, VO and W, there have been a number from all parts of Europe. Best DX is reported by ZL4OL who seems to have heard the beacon on several occasions... to confirm reception he played back (by email) the PSK31 transmission – 100% correct. Many reports are being collected/collated overseas. I can comment only on those seen here at Poldhu.

"In summary, the experiment seems to indicate that a daylight path exists across the Northern Atlantic in the 2MHz region, at least [around the turn of the year and] during periods of sunspot minimum. Some reports suggest the received signal strength can be quite high when conditions peak.

"My personal view is that Marconi *did* hear the famous S S S from Poldhu but not on 366 metres – but that is another story!"

Without pre-empting G3AGA's "another story" I would hazard the guess that he subscribes to the theory expressed many years ago by G5CS, that Marconi received harmonics emitted by his high-power Poldhu transmitter on his untuned simple receiver and kite aerial. It seems worth reminding readers of a 'TT' item ("Daylight propagation on 1.8MHz" ('TT', April 1994, or *Technical Topics Scrapbook 1990 to 1994*, p285-286) in which P Hobson, G4HOJ, noted that signals on a September day over a 125-mile path showed skywave enhancement throughout the 24 hours despite the thirteen hours between sunrise and sunset. There were only 4-5 hours during which the signals were received only at very low levels. My own experience in dealing with complaints of interference to the early ILR co-channel stations on MW convinced me that the D-layer which absorbs signals from distant stations during 'daylight') does not reform immediately at dawn and, during the winter, begins to reappear some time before dusk, both more especially at the upper frequency end of the MW (AM) band.

When one considers that the Great Circle path from Cornwall to Canada takes the signals north-west towards the Arctic Circle (and consequently very short periods of winter daylight) combined with the present sunspot cycle minimum, it is perhaps not surprising that the 1960kHz signals were so often received in Newfoundland. It certainly supports the view, advocated by G5CS over many years, that Marconi really heard, on his untuned receiver, harmonics of his Poldhu high-power transmitter! The experiment *has* served a useful purpose!

**REPAIRING AN AVO TEST METER.** The merits of the long-established AVO multi-testmeters (VOMs), Model 7 (1000 ohms/volt) and Model 8 (20,000 ohms/volt) – with their meter movements protected from overload by effective cut-outs – have been stressed before in 'TT'. But even these bulky, reliable and long-lasting analogue meters occasionally develop faults.

Brian Horsfall, G3GKG writes: "I recently discovered, to my horror, that the AC ranges on my trusty old Model 8 AVO meter were producing readings that were only a fraction of what they should have been. I have other, more modern digital instruments but still reach for my AVO when needing to measure higher voltages or, as in this case, either alternating or direct current greater than a few mA. (I like to see evidence that 'current' is actually capable of doing something – like physically moving a needle!)

"When I found the DC ranges still OK, I immediately suspected the AVO bridge rectifier. I had visions of needing to find an expensive exact replacement. ... Just as an experiment I substituted a small DIL-packaged silicon diode bridge from the junk box. But, having established that the rectifier was having to cope with only about 1.6 volts for full-scale deflection, I felt it would be necessary to resort to at least using germanium diodes to even approach the characteristics of the copper-oxide elements in the original component.

"In the event, I was amazed to discover that the new accuracy and linearity, compared against readings on my Solarton 7040 DVM, were at least up to the original specification with the voltage drop across the instrument's input terminals is several times lower than that quoted therein. Note that the AC scales themselves are the same ones as used for DC measurement and in themselves completely linear.

"Subsequent examination of the AVO circuit diagram revealed that the rather clever design uses a current transformer to ensure that the rectifier is always dealing with the same input range, regardless of the setting of the range switches. Furthermore the characteristics of the rectifier elements themselves have little bearing on the performance. It struck me as reminiscent of the RF current transformer circuits using ferrite toroids that I have employed myself in various applications. Not forgetting the adaptation I contributed to 'TT' (June 2003 and *TTS 2000-2004*, p177) for limiting the power output of modern transceivers to the 10 watts of the M3 licence that led to some controversy."

**HERE & THERE.** Roger Bunney, G8ZMM, since 1971 the contributor of the monthly "DX and Satellite Reception" feature to *Television and Consumer Electronics* tells me that the journal has ceased publication. The magazine started life as *Practical Television,* one of the stable of "Practical" magazines under F J Camm at Newnes. It was later edited for more than 30 years until early last year by my former colleague John Reddihough. Under John, it became for many years a highly profitable magazine directed primarily at TV service engineers. Its demise leaves TV, radio and electronics engineers without a magazine specifically covering their interests as we move towards the all-digital, surface-mounted, plasma and LCD solid-state display era. Roger adds that DX TV is advancing into other fields than Sporadic E reception of analogue VHF TV. Enthusiasts are now busy with DTT, DAB, FM and satellite transmission on the Ku and soon perhaps in the Ka bands – concluding that "Much of the VHF TV activity in Europe is migrating to DTT at UHF so perhaps DXTV as we know it today will by about 2012 have completely disappeared."

Challenges to 'traditional' history seem to be in vogue. Andy Green, EI3HG, draws attention to a 14-page Radio World *'RW-Online'* article by James O'Neal – "Fessenden: World's First Broadcaster?" – in which he challenges the usual account of the Christmas 1906 broadcasts from Brant Rock; traditionally accepted as the world's first broadcasts (see for instance a recent news item in *Radcom*, December, 2006. p10). O'Neal insists that no contemporaneous accounts exist and the story first appeared in 1926 and more fully in 1940 in a biography by his widow.

Dick Rollema, PA0SE draws attention (thanks to William Oorschot, PA0WFO) to an article in *IEEE Spectrum INT* (November 2005 pp 47-51) "How Europe Missed the Transistor" by Michael Riordan who teaches the history of physics at Stanford University, California. Riordan provides a detailed account of how, in late 1948, two physicists from the German wartime radar project, Herbert Mataré and Heinrich Walker, while working at a Westinghouse subsidiary in Paris, developed (seemingly independent of the work at Bell Laboratories) a strikingly similar device to BTL's point-contact transistor. They called it the 'transistron' and had overcome manufacturing problems holding up the Bell transistor.

**CORRECTIONS.** Chris Cory, G3MEV, spotted that the circuit diagram of the Whaddon Mk VII/2 ("Paraset") in the February 'TT' (Figure 3, p83) appeared to show, due to the absence of a crossover half-loop, that the grid of the 6SK7 AF amplifier was connected to the screen-grid. Additionally, in the caption Whaddon was described as Section VIII of MI8/SIS – this should have been MI6/SIS as in the text.

**ONE VALVE QRP AM TX.** John Heys, G3BDQ, notes that C4 (50nF, 350V, mica) was missing from the PW diagram and hence from 'TT', March, Figure 4. It connects from the 'earthy' end of the crystal to ground.

TECHNICAL TOPICS  PAT HAWKER, MBE
37 DOVERCOURT ROAD,
DULWICH, LONDON SE22 8SS

G3VA

# Technical topics

*Pat Hawker delves into antennas, crystal sets and software defined radios.*

**DIRECT DIGITAL RECEIVERS – R&S EM510.** Some time ago, a contributor to "The Last Word" suggested that amateur radio should join forces with the multitude of computer hobbyists and become, in effect, a junior, specialised section of the dominant hobby. While one must agree that computer information technology now plays an increasing and ever more significant role in amateur radio, with its data modes, Echolink, DX cluster, electronic logging and QSLing etc, and, currently, software defined radio, I was glad that the suggestion did not seem to receive the support of other members. I hope that I do not appear too Machiavellian by recalling that in "The Prince" the cynical, but realistic, 15th Century Italian advised "A prince should never join in an aggressive alliance with someone more powerful than himself, unless it is a matter of necessity.... Wise Princes, therefore, have always shunned auxiliaries and made use of their own forces. They have preferred to lose battles with their own forces than win them with others ... no true victory is possible with alien arms."

Amateur radio should *not* become an integral part of the computer hobby but this does not mean it should not take advantage of IT. While, personally, I prefer to use standalone rather than computer-linked equipment, there is no doubt that there will be a significant effect from the true software defined high-performance receivers, with analogue to digital conversion directly at the RF signal frequency, which are now beginning to reach the market. Furthermore, as some have reminded me, the large visual display screen of a PC, combined with a good keyboard, can provide a more user-friendly set-up than some modern stand-alone transceivers with over 100 controls crowded on to their front-panels - as discussed in the March 'TT'– at least for those already well versed in IT.

Adam Farson, VA7OJ/AB4OJ, at the suggestion of Brian Austin, G0GSF, has kindly sent me a copy of the brochure (German text) on the new Rohde & Schwarz EM510 professional direct-sampling LF/MF/HF receiver. This new high-performance digital, wideband receiver from R/S, a long-established German equipment manufacturer of high-class, professional communications equipment and test instruments, has also attracted the interest of other readers, including Gian Moda, I7SWX.

VA7OJ writes: "The most significant aspect of the EM510 is that it uses the direct-sampling technique. The 16-bit analogue-to-digital converter (ADC) is

The new Rohde & Schwarz EM510 high-performance professional direct digital wideband receiver.

clocked at 64MHz and samples a sub-band of the entire LF/MF/HF range (9kHz to 32MHz). The antenna signals feed the ADC via a bank of switched RF high-pass and low-pass filters and a very linear RF pre-amplifier. Thus the receiver samples directly at the incoming RF.

"To my way of thinking, direct sampling is the direction in which HF receiver architecture is heading, The Tayloe QSD (I/Q switching detector/mixer) is but a side-trip.

"Another example of the direct-sampling receiver is the RFSpace SDR-14. Whilst much less capable (and correspondingly less costly) than the EM510, the SDR-14 embodies the same direct-sampling principle. The relevant Internet links are http://tinyurl.com/ywvfj2 for the EM510 and http://www.rfspace.com/sdr14.html for the SDR-14."

The EM510 brochure, even with my limited understanding of the German text, shows clearly that this is a versatile, high-performance receiver used with a dedicated R/S keyboard-controlled computer. It features an extremely high scanning frequency (64MHz); absence of any image or IF frequencies; no phase-noise or spurs from local oscillators or mixers; high linearity despite low current consumption; reliability (high MTBF) and low count of critical components.

To quote selectively from the very detailed technical specification: Frequency coverage 9kHz to 32MHz with 1Hz resolution and stability $\leq 1 \times 10^7$ with input for 10MHz reference; phase noise $\leq -130$dBc (1Hz) at 1kHz offset; IP2 intercept point >70dBm, typically 80dBm, in 'low distortion' mode 1 to 32MHz; and > 50 dBm, typically 65 dBm (normal mode 1MHz to 35MHz; IP3 intercept point $\geq$ 30dBm, typically 35dBm, in low distortion mode, 1 to 32MHz and $\geq$ 20dBm, typically $\geq$25dBm in normal mode. Demodulated bandwidths 30 filters from 100Hz to 10MHz (for optional panoramic display); modes FM, AM, Pulse (AM Pulse), φM (synchronous demod of AM ?), USB, LSB, ISB, CW, I/Q. Audio filtering for notch, noise limiting, bandpass, etc. Plus many other facilities for professional users. etc.

The brochure does not show price (which would presumably be negotiated according to the number ordered under contract by a professional user). For a single EM510 receiver, it would unlikely to be within the range of a normal amateur radio budget. It is presented here to show the capabilities of the latest direct-digital, computer-controlled HF receiver from a leading European firm.

**HIGH-SENSITIVITY CRYSTAL SET.** As a complete contrast to the EM510, Bob Culter, N7FKI in "High Sensitivity Crystal Set" (*QST* January, 2007, pp31-33) advises us to build a "crystal radio" that does not require an outside antenna or ground (or battery) by using a new zero-voltage-threshold MOSFET. As an introduction, he writes: "If you are like me, you may have been bitten while building an AM broadcast band (MW) crystal set. In my case, I was using a galena 'cat's whisker' detector. Such a radio typically required a long outside antenna and ground-rod or water-pipe ground-connection to function – even with the use of a low-threshold (0.3V) germanium detector diode such as the well-known 1N34A. To rekindle the spark of your youth or to interest young people in radio, consider building a radio that doesn't need an outside antenna or ground (or local source of power)".

N7FKI explains that there is a trade-off between sensitivity and selectivity. A conventional diode detector, with appreciable threshold, needs to be connected across virtually the whole of the

# TECHNICAL TOPICS

tuned circuit. Placing the tap across 10% of the coil would mean that the coil would need to produce some 3V to overcome the 0.3V threshold. He claims that it is now possible to have your cake and eat it too by using the newly introduced ALD110900A zero-threshold-voltage MOSFET made by Advanced Linear Devices (www.aldinc.com/pdf/ALD110800.pdf).

**Figure 1** shows the circuit diagram of N7FKI's experimental MW broadcast set using a 7.5-in ferrite-rod antenna/tuning coil. He also provides details of an HF receiver (**Figure 2**) covering 90 to 40 metres (3.3 to 7.5MHz) for use with an external antenna, although admitting that reception is difficult at his West Coast location. I suspect that better results would be possible in the UK with the many very strong 6MHz-band and 7MHz-band broadcast stations.

A 'TT' (April 2000) or *TTS 2000-2004* (pp 16-18) item on "Crystal sets,

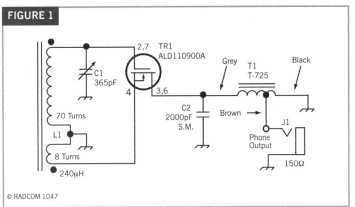

Figure 1: Circuit diagram of N7FKI's high-sensitivity crystal set. L1 240µH, 7.5-in ferrite loop antenna with 8-turn secondary wound on lower end over the 70-turn primary. T1 audio transformer matching 25kΩ to 150Ω headphone impedance. Headphones ex old-style telephone earpiece or equivalent.

headphones & pi-networks" included a contribution by Tony Harwood, G4HHZ, describing the benefits he obtained by using a pi-network to match an external antenna to an OA81 germanium diode: **Figure 3**. This arrangement provided both good selectivity and sensitivity (In the Southampton area, he received six stations on LW and some seven on MW). He used a 1000pF tuning capacitor (paralleled 2 x 500pF/gang from old broadcast receiver) with relatively low-resistance ex-military headphones, Type SGB CLR (50Ω per earpiece). He found these as good as the CHR (2000Ω per earpiece).

The use of a self-powered zero-threshold detector in conjunction with a pi-network/external antenna might prove an interesting project. It would be advisable to include the 1N4152 diode between gate and grid as shown in Figure 2. *[We hope to feature a project based on these designs in a forthcoming RadCom – Ed.]*

## REMOTE ANTENNA CURRENT PROBE.
Ian Braithwaite, G4COL in *Sprat* (Issue Nr 129, Winter 2006) describes a simple device for checking and optimising the RF current fed into a current-fed element (e.g. half-wave dipole, shorter doublet or typical ground-plane) at its element feed point: the result of pondering the question "How does one know the antenna is working properly?" In effect, an RF probe is placed at the feed-point of the antenna element and the rectified DC voltage fed back down the antenna feeder to the shack, where it is measured, much in the way that UHF mast-head pre-amplifiers are powered.

As presented, G4COL's device is suitable for QRP operation with transmitter output powers of up to some 10 watts, and it may need modification for higher powers, for example using a larger toroid in the probe to avoid saturation, and suitably rated components, etc.. Unlike an RF current probe or SWR meter in the shack, this device will detect increasing cable loss due to ingress of moisture etc.

The *Sprat* article provides full constructional details but **Figures 4, 5 & 6** show the basic details.

## FRONT-END H-MODE MIXERS AND FILTERS.
Despite the emergence of direct A/D conversion of HF signals, there is still much interest in optimising the performance of stand-alone, analogue receivers and transceivers. This involves skilled and careful measurements of the various critical components used in the front-ends of high-performance home-built models such as the CDG2000 etc, or for the development of new projects. Such work requires the use of high-grade, often high-cost test instruments but is nevertheless being actively undertaken by dedicated groups and individual amateurs on both sides of the Atlantic. Indeed, it is evident that some of the developments

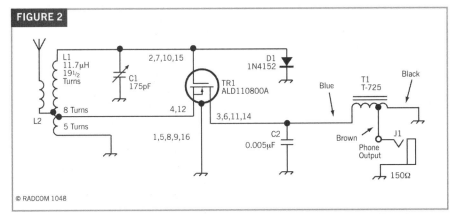

Figure 2: N7FKI's 90m to 40m short-wave version of the high sensitivity crystal set shown in Figure 1. Detector ALD110800A, D1 1N4152. L1 AirDux coil 19.5 turns, 2in dia, 10t/in, tapped at 5 and 8 turns. L2 9 turns on FT-50-41 toroid.

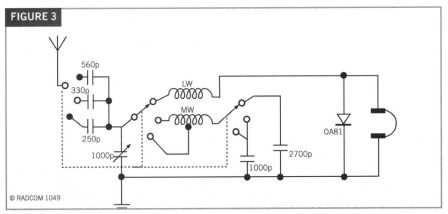

Figure 3: The use of a pi-network matching network for a MW/LW crystal set with conventional OA81 crystal diode as described by G4HHZ in 'TT' (April 2000)

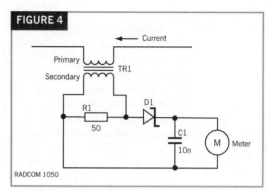

Figure 4: Basic RF current probe

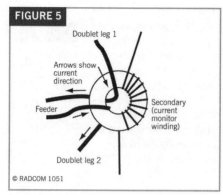

Figure 5: Toroid current transformer as fitted at dipole element feed-point

they have already made are currently attracting the interest of and use by professional designers. For instance, there are reports that the H-mode mixer developed originally by G3SBI and presented in 'TT' in 1993 is being taken up for a new military receiver. One must stress that to take full advantage of the H-mode mixer requires the use of low phase-noise oscillators.

In a report on recent developments in high-performance front-ends, Colin Horrabin, G3SBI, writes: "It is about eight years since Gian Moda, I7SWX, told us to take a look at the Fairchild FST3125 as an H-mode mixer. This device has since been used in the CDG2000 and in a number of other amateur radio projects with excellent results. When using a divide-by-two squarer for the local oscillator drive to the chip good results have been achieved without the use of specialised test equipment to set up the mixer.

"Nothing in technology stands still and Fairchild have recently introduced a number of other bus switches. My friend Martein Bakker, PA3AKE, has been investigating some of these switches with a view to improving [still further] the performance of the CDG2000 front-end design. Of particular interest have been a number of video switches. One of these, the FSAV337, is a drop-in part in the small package for the FST3125; another is a configured single-pole, two-way switch, the FSA3157, for which you do not need to develop Q and not-Q to drive it. The FSA3157 can easily be driven from a simple fundamental frequency-squarer using an AC04 to run the CDG2000 on 50 and 70MHz from the existing VCO (45 to 78MHz).

"Harold Johnson, W4ZCB has, in the past, made H-mode mixers using 1:1 transformers instead of 4:1 transformers. He found the IP3 was about the same; while the bus switch on-resistance was low enough not to increase the insertion loss of the mixer. As a result, PA3AKE has made IP3 tests with a range of Minicircuit transformers in order to identify which provides the best results in H-mode mixers. The Minicircuits data for their later generation of transformers also includes amplitude-unbalance and phase-unbalance information that seems to be important for optimising the H-mode mixer.

"PA3AKE has assembled an impressive range of test equipment for these measurements and I hope his home-made set-up can form the basis of a future article in RadCom. It is important when making IP3 measurements that the two test oscillators do not interact. PA0AKE uses two completely separate and double-shielded AD9951 DDS signal generators (IOCG design, see 'TT' October 2006) with 500MHz external clock and with two-stage J310 grounded gate amplifiers which have reverse isolation better than –75dB, allowing IMD tests up to the +6dBm output level from the home-made hybrid combiner. He displays results using a Wandel & Goltermann SNA-62). The OIP3 of the two generators plus combiner with 20kHz spacing is better than +50dBm between 160m and 12m (+49.2dBM on 10m and +45.7dBm on 6m)."

Because the aim is to measure only the IP3 of the mixer embedded in the CDG2000 PA3AKE has also checked the IP3 of the 9MHz hybridised quartz roofing filter following the mixer at a wide range of signal levels. With post mixer roofing filter (20kHz spacing, the signal level (dBm) / roofing filter IP3 is 4.5/51.0; 2.5/49.0; 0.5/47.8; 47.8/-1.5/45.8; -3.5/44.0; -5.5/ 42.5; -7.5/ 41.0; -9.5/41.0. PA3AKE points out that the roofing filter limits the IP3 of some of the better H-mode mixers, as the IP3 of the filter is around 42.5dBm at the –5dBm level encountered when measuring the mixer at the 0dBm level. However the good news is that it is bottoming out at +41.5dBm allowing for a front-end in the plus forties with modern crystal manufacturing techniques, The reason why crystal filter performance degrades at low signal levels has been unearthed by W4ZCB and will be discussed in a future 'TT'.

PA3AKE has noted that some constructors have experienced disappointing results when using the big Amidon T50 cores for H-mode mixer transformers. He has investigated the performance of the tiny Minicircuits 1:4 and 1:1 transformers, measuring the IP3 of two cascaded transformers while going from 50 to 200 ohms and back to 50 ohms. With +6dBm input on all amateur bands from 180 to 6m, all transformers (TT4A-1A, T4-1, FT37-43, ADTT4-1, ADTT4-1, ADTT1-1 and T1-1T) the IP3 exceeded +40dBm on all bands, with the ADTT4-1 proving the best, varying between +44.2 dBm on 6m, 45.7 on 169m, and +56.4dBm on 40m when tested with the FSA3157. However things proved more complicated when the transformers are used in an H-mode mixer again using the FSA3157 (see below).

G3SBI writes: "It has become evident that one of the problems with the FST3125 is that the switch 'on' and 'off' times overlap. The more recent FSAV332 has equal switch 'on' and 'off' times. The FSA3157 is a SC70 part that can be fitted very close to the transformers."

PA0AKE summarises and tabulates his

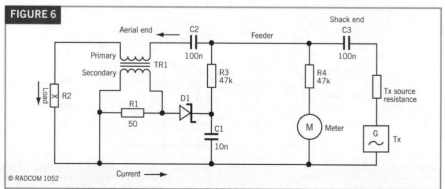

Figure 6: Remote antenna probe system as described by G4COL in 'Sprat' (Journal of the G-QRP Club)

# TECHNICAL TOPICS

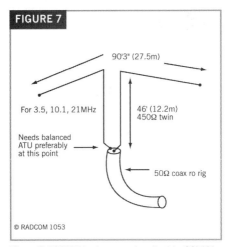

Figure 7: ZS6BKW antenna as described by G3UKV in 'Sprat' with horizontal or inverted-vee layout.

measurements on the FSA3157 + TT4-1A + 74AC04 squarer which allows for tweaking the mark/space ratio for best RF-IF isolation at 9MHz; and also with the TT4-1A transformers and 74AC04 squarer. However he finds the FSAV332 + ADTT1-1 +74AC74A a clear winner. With the ADTT1-1 transformer the IIP3 is > +40dbm on all HF bands and is not bias-point dependent, However spur levels are somewhat problematic and 6m performance considerably less good, probably due to limitations of the PCB layout of the mixer. The ADTT1-1 transformer, first tested with the FSA3157 improves the IP3 especially on the higher bands as it did with the FSA3157 mixer. The price to pay is the increase in conversion loss of roughly 0.6dB.

PA3AKE's full report runs to over 40 pages with many detailed tables. As G3SBI puts it: "A fantastic job carried out excellently in a very short time". Unfortunately there is space here for only these very brief extracts. The full report is available at http://tinyurl.com/2bztdr.

THE ZS6BKW ANTENNA RE-EXAMINED.
It is now some 25 years since attention was first drawn ("Potential of the G5RV Antenna" see 'TT', May 1982 or *Antenna Topics,* p180) to the computerised work of Dr Brian Austin, G0GSF, while still in South Africa as ZS6BKW, in developing his variation of the G5RV antenna that permits its use without an ASTU on more of the HF bands. This work blossomed and was soon presented in a full-length article in *RadCom* (August 1985 pp 614-617) "Computer-aided design of a multiband dipole – based on the G5RV principle" and again as "An HF multiband wire antenna for single-hop point-to-point applications", (*Journal of the IERE*, April, 1987, pp167-173).

'TT' returned to this antenna in January 1983 ("More on the ZS6BKW/G0GSF multiband dipole" (see also *Antenna Topics* pp299-300), again in February, 1993 (*AT,* pp301-2), and in 'TT, June 1994, and *AT,* p310. Nevertheless, I still find many of my contacts give their antenna as a 'G5RV, very few as a 'ZS6AKW'.

Recently, Martyn Vincent, G3UKV in G8PG's "Antennas" section in *Sprat (*Issue Nr 129, Winter 2008, pp32-33) has revisited the ZS6BKW antenna, showing that when implemented as in **Figure 7** it needs no ATU on 7, 14, 18, 21, 24, 28 *and* 50MHz, and can be used with an ATU on 3.5, 10 and 21MHz. His measurements were made with a MFJ Antenna Analyser plus practical use with excellent results on 7 and 14MHz.

As a result, Dr Brian Austin, G0GSF, ex-ZS6BKW, has made fresh computer analyses and is presenting the results as "The Highs and Lows of the ZS6BKW". To quote briefly: "The configuration of the ZS6BKW is shown in **Figure 8.** The dipole radiator is L1, the series section impedance matching transformer (to give its formal name), with characteristic impedance of Z2, is L2 spaced twin-wire. The lower end of L2 presents an impedance, Z3, to the coaxial cable, Z4, (50 ohms as is standard practice in all modern radio systems). A computer-based prediction technique indicated that optimum performance will occur if the antenna system has the following dimensions: L1 = 28.5m, L2 = 13.3m x velocity factor of the line, while Z2 = 400 ohms, None of these dimensions or values is especially critical. Changes to the lengths of L1 and L2 by a percent or so either way will not seriously affect performance, while Z2 between 300 and

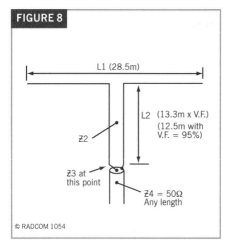

Figure 8: The basic ZS6BKW antenna as recently re-analysed by G0GSF (ex-ZS6BKW)

400Ω will work, though the higher values in this range are preferable. Even 450Ω could be used but it must be appreciated that Z2 is a key element in the matching process and the optimum match occurs with a value rather lower than that.

"With the dimensions shown in Figure 8, the ZS6BKW will produce a better than 2:1 VSWR when measured on the 50Ω cable over significant portions of five of the eight HF amateur bands *without the use of any additional form of impedance matching or antenna tuning*. They are the 40, 20, 17, 12 and 10m bands. It does not match with a VSWR less than 2:1 [without the use of an ATU] on the 80, 30 and 15m bands. "[G0GSF did not consider the 6m band but now believes it should work well]."

G0GSF has produced a series of tables showing impedance, losses on the various sections, efficiencies, and comparisons with the losses of matched half-wave dipoles etc, assuming a height of 10m above rural ground (5mS/m. dielectric constant 13) with L2 semi-air spaced coax with a loss of 0.125dB/30m at 7MHz, velocity factor 0.95) and fed by 15m of 50Ω RG213 or RG58 coax cable. The greater part of the losses occurs in the coax cable section. A loss of –1dB represents an efficiency of about 79%.

**Table 1** (adapted from his Tables 3 and 4) shows the total antenna system loss and efficiency (η) without an ATU across the HF bands for the ZS6BKW fed with either RG213 or RG58 cable and comparison with a series of single-band resonant half-wave matched dipoles at the same height and fed with 50-om cable.

TABLE 1: Calculated total system loss and efficiency of ZS6BKW fed directly with 15m coax, 10m above rural ground and loss & VSWR of half-wave matched dipoles fed with 15m coax 10m above rural ground.

| Freq (MHz) | With RG213 L (dB) | η (%) | With RG58 L (dB) | η (%) | Matched half-wave RG213 | RG58 | VSWR |
|---|---|---|---|---|---|---|---|
| 3.53 | -2.1 | 62 | -2.5 | 56 | -0.26 | -0.46 | 1.04 |
| 7.15 | -0.45 | 90 | -0.75 | 85 | -0.37 | -0.70 | 1.6 |
| 10.14 | -6.4 | 23 | -9.2 | 12 | -0.46 | -0.85 | 1.7 |
| 14.10 | -0.84 | 82 | -1.34 | 73 | -0.50 | -0.97 | 1.5 |
| 18.13 | -0.83 | 83 | -1.33 | 74 | -0.58 | -1.14 | 1.3 |
| 21.20 | -7.9 | 16 | -10.5 | 9 | -0.61 | -1.20 | 1.4 |
| 24.80 | -1.02 | 79 | -1.6 | 69 | -0.68 | -1.34 | 1.6 |
| 28.50 | -0.97 | 80 | -1.7 | 68 | -0.71 | -1.41 | 1.5 |

# Technical topics

*Frequency stability and ergonomics are the main staples this month.*

**QUARTZ CRYSTAL CONTAMINATION.** The quartz crystal resonator (QCR) has always seemed a fascinating, if rather mysterious, device, falling somewhere between an active and a passive component. It has played a key role in amateur radio for over 80 years, first as a means of controlling oscillators and later also for selective IF filters, originally as a single-QCR CW filter, subsequently, with two or more crystals, for bandpass lattice or ladder MF/HF filters. It has progressed along the years in many ways, moving from natural quartz to synthetic quartz, from individual crystal blanks to the monolithic crystal filters (MCF) used for VHF roofing filters, from the early temperature-conscious X and Y-cuts, to the 'zero-temperature coefficient' AT and later improved cuts. An outline of some of these developments can be found in "Quartz Resonators – History and Progress" (September, 1994 and *Technical Topics Scrapbook 1990 to 1994*, pp287-289).

Peter Chadwick, G3RZP, in 'TT' January 1994 drew our attention to the dynamic range limitations of crystal filters. He pointed out that this (and oscillator phase noise/reciprocal-mixing) could set a limit to improving the receiver performance possible even with such mixers as the DMOS Si8901quad FET mixer used by N6NWP in his "Superlinear HF front-end" ('TT' September, 1993) or G3SBI's original H-mode mixer first described in 'TT' November 1993. He stressed that "as the intercept point of the mixer is raised, a limitation to overall performance is likely to be set by the linearity of the crystal filters available on the amateur market", adding "This has been recognised professionally. Incidentally SAW (surface acoustic wave) filters are very good because they don't stress the quartz."

The recent measurements made by PA0AKE (see the May 'TT') showed that while current crystal filters are unlikely to degrade the performance of an H-mode mixer, he found that the intermodulation characteristics of crystal filters degrades at low signal input levels as well as at excessive signal levels. I commented that Harold Johnson, W4ZCB had unearthed a professional paper reporting on this phenomenon and that this topic would be discussed in a future 'TT'.

The paper is "A Simple Single Model for Quartz Crystal Resonator Low Level Drive Sensitivity and Monolithic Filter Intermodulation" by L Dworsky and R G Kinsman (both of Motorola) in *IEEE Transactions on Ultrasonics, Ferroelectrics and Frequency Control,* Vol 41, No 2, March 1994, pp261-268. To quote briefly from their introduction: "The drive power level sensitivity of quartz resonators (QCRs) has been recognised for several decades. It has been described by many names Starting Resistance, Sleeping Sickness, and Second Level of Drive being just a few of them. The problem relates to the fact that the parameters of a linear device should not vary with excitation, or drive, level. Instead it is often found that, at drive levels too low to normally produce nonlinear elastic phenomena, the motional resistance and series resonance frequency of a QCR are a function of the drive level.

"The application problem caused by starting resistance are well known to oscillator engineers. ... Typically, they found that an oscillator which had more than adequate gain to operate with the expected QCR resistance did not start or 'turn on'. The problem occurred both after repeated use of a QCR and with a new QCR which exhibited low enough resistance at operating drive levels. ... Marginal QCRs were particularly frustrating in that the oscillator might or might not start at any given time. ... The problems caused by QCR starting resistance were a common experience to early radio users. ... Pressure-mount, unsealed, QCRs manufactured before and during WW2 were often found to be 'dead', particularly after a storage period. It was routine practice periodically to open the package and clean the quartz wafer with soap and water or a solvent. ... In 1954 Gerber showed that the problem existed in plated as well as pressure mount QCRs."

The paper lists further experiments by various researchers from which it was concluded that the low-level characteristics of QCRs are caused by trapped particles on the surface of the quartz and/or (pieces of) electrodes. In 1973, Horton and Symthe presented test data that showed monolithic QCR filters exhibit both low drive-level and high drive-level intermodulation characteristics which were not correlated. The paper presents further experimental work by the Motorola authors outlining various scenarios and possible cures for scratched surfaces, etc.

I noted that one of the paper's references was to a March, 1981 contribution to the IEE's *Electronics Letters* by D Gordon Smith and D P Almond, "Anomalous nonlinearity in quartz crystal filters". Recognising that 'D Gordon-Smith' must be G3UUR, I drew his attention to the Motorola paper and also the recent measurements made by PA0AKE. He has kindly replied in detail as follows:

"I had forgotten about the note that Darryl Almond and I sent to *Electronics Letters* back in 1981, and am surprised that anomalous non-linearity is still a big problem in modern crystal filters. Manufacturers have the knowledge and means to keep the surfaces of their resonators clean and damage free these days, but the cost of doing so is not economical for them. Twenty-five years ago, synthetic oil was used extensively in the diffusion pumps that provided the vacuum for the plating chamber. This was one of the main surface contaminants that caused microscopic solid debris, stirred up in the plating chamber by turbulence at the time of vacuum release, to adhere to the electrode surface. Microscopic particles from electrode scratches and the excess bonding material used to connect the leads to the electrode panhandles can also remain attached to the surface through the adhesion provided by the oil layer. Once the surface was contaminated with this oil, it was very difficult to remove. Even the most aggressive solvents had little effect on it, and plasma washing with an argon plasma was of limited use in breaking up the oil molecules on the surface because the great multitude of contaminant particles shielded reservoirs of oil beneath them and the process only succeeded in thinning down the surface oil layer, not removing it completely. The result of this thinning only changed the anomalous non-linearity, it didn't really improve the small-signal intermodulation problem at all, and could sometimes make it worse at lower signal levels.

"'Sleeping sickness' and anomalous non-linearity in QCRs are both caused by surface contamination, but 'sleeping sickness' generally occurs only when there is a very high degree of contamination. Anomalous non-linearity, which results in IMD at low signal levels, can occur at very low levels of contamination. Diffusion pump oils and lubricating agents are often known as non-Newtonian fluids. That is, they are either rheopectic or thixotropic, and the shearing action of the piezoelectric

# TECHNICAL TOPICS

plate causes their properties to change if they are sandwiched between the surface of the electrode and microscopic particles that act as inertial anchors.

"The shear thickening or thinning properties of the contaminating fluids causes non-linear loading of the resonator and hence intermodulation. The properties of these non-Newtonion fluids are dependent on temperature, so intermodulation products at low signal levels can vary considerably with temperature. Heavy surface contamination with particular matter and a shear-thinning non-Newtonian fluid can result in very heavy initial damping of the QCR, stopping it from oscillating freely. But, once some shearing action has thinned the fluid, the mass loading damping is reduced and the resonator appears more active – this is usually the behaviour of quartz crystals exhibiting 'sleeping sickness'. Anomalous non-linearity and 'sleeping sickness' require the presence of both a non-Newtonian fluid and particular surface contamination for them to occur.

"Today, turbo-molecular vacuum pumps are commonly used for pumping down plating chambers, and cryo-pumps are often used in the final adjustment stage, so I would have thought that the problem would be much reduced. Obviously it isn't. Could it be that lubricating fluids from the lapping and polishing stages are not being completely removed from the surface of the quartz plates prior to electrode plating, and these are providing a non-Newtonian adhesive film that survives the plating process? In mass-production there will always be a tendency not to clean out the excess plating material from inside the plating chambers after each session, because this causes down-time with its consequent financial penalty. However, manufacturers who are providing crystals for filters ought to be paying more attention to these matters than those producing masses of standard frequency crystals for use with microprocessors and digital integrated circuits. After all they charge much more per crystal than the mass producers.

"Generally, ignoring other gross defects, those resonators with high Q and low ESR at a particular frequency have fewer surface contaminants than those with lower values of Q. Resonators from the HF range, where the highest values of Q are physically possible, are more susceptible to the effects of surface contaminants than those outside this range. As a general rule, however, the non-linearity caused

**Photo 1**: The classic Racal RA17 valve receiver introduced in the 1950s with a front panel layout that grouped the operational adjustment controls on the left-hand side leaving the operators right hand free to copy down incoming traffic.

by microscopic surface contamination will be worse at higher frequencies and narrower bandwidths. This is because for a given set of contaminant particles the anomalous loading effect will be greater for thinner resonator plates, and the non-linear change in ESR will be a greater proportion of the mesh resistance for the lower termination resistance of narrow filters.

"Since the tendency these days is for higher IFs and narrower bandwidths, the designers of modern high-performance receivers are moving more and more into the region where the problem is most pronounced. It seems to me, if you're building your own crystal filter for a high-performance receiver, the only answer is to select crystals for the best intermodulation performance as well as the correct motional parameters and resonant frequency. Alternatively, you could change to overtone crystals where the effect of surface loading is reduced, but that would probably not solve the anomalous problem totally, only shift it. It would also bring with it another set of problems relating to the spurious responses at the fundamental and possibly closer and more active higher overtones ones. What do they say about pain and gain?"

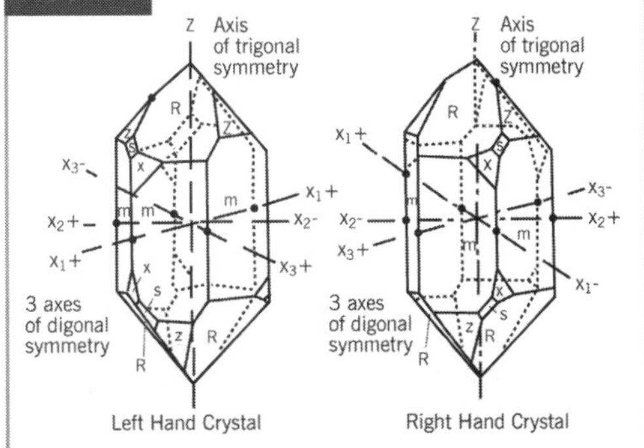

**ERGONOMIC CONTROLS.** The human engineering of Amateur Radio equipment, as discussed in 'TT' March 2007 pp82-84, continues to attract attention. For example, Alistair Dunlop, G7IET, feels that some of the comments rather missed a point in regard to placement of controls. He writes:

"The fundamental aim of ergonomics is to ease and simplify the use of the equipment; this may not best be served just by providing a button for everything. Placement of controls is critical, and ease of use may be simplified by fewer, well-placed controls. With regard to control positioning, I always consider the old Racal 17 (**Photo 1**) as an excellent example. The frequency was controlled by two large knobs, the Magacycles on the *right* and the Kilocycles on the *left;* in other words in the *opposite* order to that expected from seeing a frequency written down (Megacycles then Kilocycles). However, when the receiver was used in anger, you realised that all of the controls you needed to adjust during a contact were positioned on the left side within easy reach of your left hand; leaving your right hand free for writing messages and logging [or using a Morse key – G3VA].

"Thus, you could refine the tuning, change filter bandwidth, adjust the BFO tone and tune the pre-selector all without stopping writing – assuming you were right-handed. I don't think Racal ever made models especially for left-handed operators!

"Compare that to the majority of modern transceivers, particularly such smaller models as the IC-706 or FT-817 and similar. The tuning knob is on the right-hand side so you cannot fine-tune the radio with your left hand – or if you do, you can't see the display. Many functions require the hand to hop all over the place, and almost inevitably draw your attention away from the act of communicating. Maybe radios don't drift like they used to, but there is still often the need for fine-tuning when working in a pile-up; and the same need to record parts of call-signs – I know I still need to scribble notes and hasty logs. The original 'ergonome' was often pictured as a three-foot tall dwarf with six-foot long arms since this was the control positioning on a capstan lathe of the period; perhaps we should have a competition to draw the equivalent physique needed to use a modern radio?

"A related point – digital radios have

**Figure 1**: Characteristic form of large quartz crystals before cutting. Although quartz is one of the commonest natural substances in the earth's crust, large crystals are quite rare with Brazil almost the only commercial source until the development of synthetically grown crystals. This illustration shows the form of perfectly shaped crystals although these seldom occur in nature. For convenience in discussing the properties of quartz crystals, crystallographers make use of a system of axes, with the principal Z axis here shown vertically and three X axes, all perpendicular to Z and at 120° to each other.

many faults as rightly pointed out, particularly their (usually) catastrophic overload characteristics. However, they also seem to improve significantly the ability, in certain circumstances, to communicate; and hence could be considered a major ergonomic advance. Nothing to do with knobs, but the affordable introduction of panoramic reception coupled to instantaneous tuning. Steve Ireland's 'Software Defined Radio' column in the February, 2007 RadCom (pp53-54) shows how this facility can significantly improve the ability to find and work very weak signals. Note that this is not the spectrum analyser or scanning approach used for band scopes; rather, it is equivalent to a bank of receivers covering the frequencies of interest and provides a continuous time-history of all activity on each frequency. Professionals may have had this facility for years, but it is the first time that it has been affordable by the average amateur. [The first detailed explanation of the benefits to amateurs of panoramic reception, together with full details of a home-built panoramic converter for the 144-146MHz band, appeared as a two-part article by B H Briggs, G2FJD in the *RSGB Bulletin*, September and October, 1950 – *G3VA*].

"Of course, it relies on high-speed digital computers and high-end sound cards, but here the usual mad over-specification of current hi-fi audio components actually works to our advantage. This is thus another example of amateur radio riding piggy-back on advances in other fields. In addition, you can't get much simpler than a SoftRock receiver as sold by a well-known retailer. Although the necessary accompanying computer and soundcard are hardly home-construction, you might class them as 'normal modern accessories' in the same way that motorised transport is accepted as part of National Field Day!"

AN ALTERNATIVE PIC STABILISER. Chas Fletcher, G3DXZ, writes: "TT has again spurred me into action. The February review of G4GXO's work in implementing G7DXF's idea of building a 'Fast Huff & Puff' frequency stabiliser using software led me to think it was time to review the stabilisers in my home-brew station. I took advantage of Ron Webb's web site (given in the February 'TT') to look at his software methods, but soon realised that the only PIC I had on hand, an old 4MHz 16F84, was not up to running his software routines. Not to be beaten by a minor problem and itching to get started, I decided to re-write the software and see if the F84 would work. My target was a VFO

**Photo 2:** The later Racal RA1772 solid state receiver of the 1970s carried on the Racal regard for a clean and attractive front panel with due regard to operability.

running in the 3.5 to 7MHz region.

"What followed was an enjoyable four weeks of design, correspondence with G7IXH and G4GXO, and experiment – Ham Radio at its best! I find the combination of digital and analogue techniques most interesting and resulted in the fast-stabilising VFO of **Figure 4**.

"Software-wise, the VFO uses the PIC differently from the G4GXO method. The sample rate is determined by dividing the PIC crystal clock using an onboard counter. The VFO input is sampled directly on a clock edge rather than feeding the incoming VFO to an outboard counter and using a counter bit as a sample. The shift register and control XOR function is provided by an interrupt driven routine associated with the sampling operation. Provided that the code for the interrupt service routine (ISR) will run within the sampling period, its exact duration is of no consequence. Tests have shown that sampling rates around 1 to 2 kHz work very well which leaves 0.5-1.0 ms for processing; in computer terms that is quite a long time. Further, this obviates the need for high PIC clock speeds.

"A good VFO needs to be smoothly tuneable. The power of Peter Lawton's design results in the stabiliser having a tenacious hold, especially if the step size between lockup points exceeds 10 Hz. Even at 10 Hz, the pull of the stabiliser is just felt when small changes in frequency are intended. Going below 10 Hz, although easing the pulling problem, puts more demands on intrinsic VFO stability. Faced with these conflicting demands, I reviewed some of my previous ideas of using CMOS gates as a VFO. In the past such VFOs have proved to drift steadily, but I found this problem can be easily remedied by restricting the amplitude of the oscillator gate's output swing to less than saturation (ie rail to rail). This is done simply with a pre-set resistor in the gates feedback circuit; this can be adjusted to reduce the oscillatory swing to 4V or less with a 5V supply. The effect is magical.

"One other facet of PIC control annoyed me. Whereas it is easy to produce a lock-up step of around 5Hz using cheap mass-produced crystals, I wanted an exact 5Hz step. The ability to use lower-frequency crystals brought the solution, since a crystal intended for controlling serial asynchronous communications (Baud rate) can be divided to exactly 5Hz. Problem solved! Interestingly, the step, once established, is independent of the actual VFO frequency – unlike my previous one where there has been some inter-relationship. Changing from 3.5 to 7MHz needs no other changes than tuning the VFO itself. Shifting frequency beyond a 2:1 ratio can, however, be troublesome, due to the need to control the varicap sensitivity.

"The 16F84 PIC can only support a 256-bit shift register due to its limited RAM. When my samples of 16F627/8 arrived, I adapted the code and expanded the shift register to 512 bits. This made a marginal improvement and is the preferred option.

"The end result is shown in Figure 4. Using a 2.4576MHz crystal as the clock, a counter dividing by 1024 and a 480-bit shift register results before in a 2400Hz sample rate and an exact 5Hz step. Lock-up is virtually from switch-on.

"One other minor annoyance appeared in the form of an occasional initial control voltage of other than half Vdd. The good idea in ('TT' many years ago) of using a pair of capacitors in series across the supply rails would normally be expected to divide the voltage equally at power up, but sometimes they did not. However, as most of the PIC's I/O capacity is unused, the software runs a routine at start-up to force the capacitor voltage to the correct value. It would be nice to be able to detect fast tuning, ie when the PIC is not controlling the frequency, and run this routine during normal operation, but I confess that so far this subtlety has evaded me.

A test point (TP) is provided to monitor the software operation and outputs a pulse of duration equal to the ISR and occurs at the sample rate. Construction is easy, but

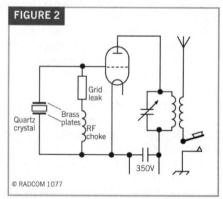

**Figure 2:** The first crystal-controlled transmitter (3MHz) was built by H S Shaw, (W)1XAO of General Radio Co in April 1924, following the publication in 1923 of the paper "The Piezo-Electric Resonator" by Dr W G Cady, and the development later that year of simple crystal-controlled oscillators by Dr G W Pierce and Dr J M Miller. Shaw used a 5-watt triode valve in a Miller-type oscillator. By 1926 a significant number of American amateurs were using crystal-controlled transmitters with UK amateurs soon following suit. (Source QST)

# TECHNICAL TOPICS

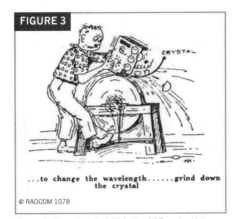

**Figure 3**: How, in July 1924, the QST cartoonist visualised how Amateurs might tackle the problem of shifting crystal frequencies upward by grinding the crystal thinner. In reality, the art of crystal grinding, generally using a little toothpaste as a mild abrasive spread on a flat glass plate, became common practice until the sealed plated crystals were introduced post-WW2.

for queries, comment and software on this topic, please email, if possible, to: chas@g3dxz.go-plus.net.

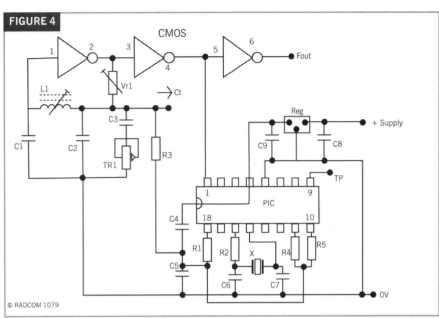

**Figure 4**: Circuit diagram of G3DXF's PIC-controlled Huff and Puff stabilised 3.5/7MHz VFO. Component values: R1 2M2; R2 1K5; R3 100K; R4, R5 2K2; VR1 10K pre-set; L1 6µH TOKO 10E; C1 1nF polystyrene; C2 470pF polystyrene; C3 68pF ceramic; C4, C5 47µF 16V; C6, C7 22pF ceramic; C8, C9 100nF ceramic; X 2.4576MHz; PIC 16F627; Reg 78L05; CMOS 74HCU04; TR1 BC169; Ct tuning capacitor about 150pF max.

**BRIGHT LEDS & CONDUCTIVE POLYMERS.** An item "Bright Future for LEDs" ('TT', November 2003) discussed the future role of the then recently-developed ultra bright (20mA) gallium-nitrate (GaN) LEDs including their use in torches, vehicle headlights and (in detail) as replacement for the 6.3V pilot lights in vintage valve equipment. Also quoted was the 1992 forecast by Glenn Zorpette in *Spectrum* the LED was seen as a possible replacement for the household incandescent lamp when used as a mixture of white and amber LEDs. (The amber LEDs are used to remove the slight blue tint of the 'white' LEDs). *Time* magazine has recently reported that such household lamps are now entering production in the USA and claimed as superior in energy-saving to the current low-energy florescent lamps. Presumably LED lamps may present fewer RF interference problems than many of the present low-energy types, some of which do not seem to fulfil the claims for efficiency in terms of light output or reliability over many years. The use of LED arrays as TV display screens seems likely also to become commonplace before long.

"A Cheaper Chip" by Laura Blue (*Time*, February 19, 2007) points out that, for years, plastic has been electronics' next big step and asks "Is it finally here?" She quotes Hermann Hauser of Amadeus Capital Partners as saying "We believe there is nothing silicon transistors can do that polymer transistors won't be able to do eventually." Meanwhile, the US firm Plastic Logic announced in January that the company intends to build a factory in Dresden, Germany to create its flexible, portable text display – a device that would let you carry your whole library on a sheet of plastic.

Laura Blue writes: "Ever since conductive polymers were developed in the 1970s, researchers and entrepreneurs have wondered whether they could make commercially viable plastic electronics. Unlike microchips made of amorphous silicon and glass, polymer chips are light, hard to break and – perhaps best of all – as cheap as plastic. Although plastic transistors don't perform well enough to make the polymer PC a realistic goal for many years, they are quickly becoming suitable for applications where fragile silicon chips are impractical. Imagine electronics so cheap you could put them in disposable packaging, or so light and flexible you could put them in your clothes. Processes similar to ink-jet printing can literally print circuitry onto materials. ... On January 24, a spin-off from Philips – Polymer Vision – unveiled plans for its own mass-production facility in Southampton, UK. The firm will make a 5in screen that can be rolled up to the thickness of a cell phone ... the Dutch company plans to produce at commercial volume this year."

**HERE & THERE.** One hears rumours that an Amateur Radio QRO rig with its antenna running close to overhead telephone wires has twice 'taken out' the amateur's ADSL (broadband) card at the local Exchange. I wonder if this is exceptional or whether it could become a common problem?

The April 'TT' (p85) noted the closure of the *Television* monthly magazine, published by Nexus to whom it was sold, together with *Electronics World* (former *Wireless World),* by the giant Reed-Elsevier company. This made me wonder what had happened to *Electronics World,* a magazine that started as the *Marconigram*, almost a century ago. As *Wireless World* in the early 1920s it was for several years the official journal of the *Wireless Society of London* (renamed the RSGB in 1923). Its offer of funds to support the struggle by the RSGB to overcome the ban on overseas contacts by UK amateurs, at a time when the Marconi Company was negotiating a large Government contract, brought about its sale to Iliffe where it stayed until the early 1970s when the company became part of the large International Publishing Company (IPC) that in turn became part of Reed and then Reed-Elsevier. Then a few years ago it was sold to the relatively small firm of Nexus. As someone who had contributed to the magazine on Amateur Radio and (later) Broadcasting topics from 1941, at first occasionally and then, from 1969 to 1994, regularly, I watched with some concern the changes in editorial policy introduced by Nexus. The new editor seemed to concentrate on articles directed at professional design engineers rather than catering also for hobbyists. I now understand that, while *Television* has ceased publication, *Electronics World* was sold to another independent publishing firm – St John Patrick – where it continues publication under the same editor, Svetlana Josifovska, as at Nexus.

# Technical topics

*Whither the future? And what can we learn from the past? Pat Hawker investigates.*

**THE PATH AHEAD – SUNNY OR STORMY?** Undoubtedly, Amateur Radio world-wide is facing an era of profound change – in its technology, its regulation and its role in radio communications. The regulatory changes, the increasing dominance of digital over analogue, the availability of an enormous variety of factory built equipment that requires little technical expertise by the user with the consequent increasing commercialisation of the hobby. At the same time, there is access to relatively low cost international audio and image telecommunications systems based on the Internet, using wideband coaxial ocean cables, fibre optics and commercial satellites. There also appears to be a growing reluctance at schools to study the science, physics and mathematics required in practical engineering, coupled with the increasing scepticism of technological innovation and positive 'technofear' among some of us elderly individuals. And the public generally believes that Morse is obsolete. No wonder that some of at times question the future of Amateur Radio as a lifelong hobby, capable of still contributing to propagation knowledge, education and emergency communications.

But are these misgivings justified? Change there may be – but change can also bring new opportunities. After all, *plus ça change plus c'est la même chose*. Some of us feared that the ending in the UK of the Morse test as the sole means of gaining entry to HF would spell a rapid and terminal decline in CW operation – yet this has not proved the case. There is some evidence that many of the new licensees are keener than ever to become proficient CW operators and to construct at least some equipment.

American amateurs are now facing such challenges as the FCC finally follows many administrations in deleting compulsory Morse as the entry to HF and generally relaxing the licensing exams. The correspondence columns of *QST* currently reflect the misgivings expressed a couple of years ago in the UK and elsewhere – a mixture of optimism and pessimism.

For example, Michael Clarke, KG7AN (*QST*, April, 2007, p24) writes, *inter alia*: "Amateur Radio as it existed in the 1960s and before is dead. What we have today are the vestiges of a once great hobby. ... Amateur Radio does not hold the interest of the younger generation as it once did. The magic is gone. .... Even I, a veteran ham, rely on the Internet and cellphone. ... What good is Amateur Radio? This question lies at the very heart of the future existence of our hobby. Amateur Radio will only survive through the tenacity and fortitude of a responsible membership. Isn't this the purpose of the licensing exams, to ascertain the level of competence and responsibility of a person applying for a licence? We should be careful about dumbing down the written examinations.... I believe Morse code proficiency has reached an end as a necessary requirement ... the best course for the future will probably be through education, public service and the *esprit de corps* we hold among ourselves with all our varied interests and mutual organisations. As for Morse code, my primary mode of operation, CW will live forever."

The May 2007 issue of *QST* (p56) includes "Why is it that CW Works so Well?" by Joel R Hallas, W1ZR, *QST* Technical Editor, sub-titled "Just because you no longer have to know how to use it doesn't mean you might not want to!" The main theme of the article is to reiterate what has long been recognised by most experienced operators: that CW gets through much better than voice. This, of course, is due to its reduced bandwidth (permitting a lower noise bandwidth in the receiver) and the lower signal-to-noise requirement for satisfactory reception – with a good CW operator able to copy code at or below the noise level. W1ZR notes that a 100-watt CW transmitter will have much the same range as a 1kW SSB transmitter.

There are other factors, applying also to some of the other digital modes (Morse is a non-return-to-zero digital mode): the transmitter operates under key down conditions at its full output power. An SSB transmitter, without heavy speech-processing, has a constantly varying output, mostly far below peak, dependent on the actual voice characteristics of the operator. Even with heavy processing (which degrades speech quality) it will have an average output significantly below that of an equivalent CW transmission with its narrow bandwidth governed by the keying speed. Then again consider the problem of foreign operators or even national dialects. For most of us, the use of phonetic alphabets is essential to copy correctly call-signs, locations, names even for strong SSB signals.

SSB and AM phone are excellent, enjoyable and effective modes when signals are strong and audio quality reasonable, but are greedy of spectrum space by a factor of five (AM ten) or more times compared to manual CW. One can only echo W1ZR's advice to beginners to try to start using CW at an early stage, even if this means using a keyboard system or an electronic keyer with memory. W1ZR writes: "Both are real options until your proficiency increases, and I expect it will once you realise how much fun it can be!" I would add a plea to experienced operators to show the utmost tolerance to novice operators and to slow down – the appropriate but little used signal is QRS!

**UNLICENSED TRANSMITTERS & OFCOM.** By 'unlicensed transmitters' I am not referring this time to the many 'pirate broadcasting stations' that have proliferated over many years on the VHF/FM band, particularly near my QTH in South London and which have largely defeated the efforts of successive regulators – GPO, Home Office, OFTEL and OFCOM – to eliminate them, despite the interference they often cause to the authorised BBC and commercial stations.

Instead, I am referring to the increasing number of low-power licence-exempt devices, with the current and future pace of growth being actively encouraged by OFCOM. This encouragement is made clear in another article in *IET Communications Engineer* (April/May 2007, pp22 – 27) by Professor William Webb, Head of R&D at Ofcom, this time in conjunction with Dr Reza Karimi, Spectrum Policy Advisor at Ofcom. To quote the introduction: "In our everyday life we use a variety of wireless devices that do not require a licence to transmit radio waves. These licence-exempt devices range from cordless (DECT) phones and car key fobs to baby monitors, garage-door openers, wireless headsets and WLANs in the home and office. Licence-exempt devices are also used widely by businesses for applications such as wireless alarms, anti-theft systems, radio-frequency identification (RFID) chips for activating doors and ticket barriers or used for stock management in retail, radar level gauges, and even data links to remote base stations in cellular networks. The pace of growth in this area

# TECHNICAL TOPICS

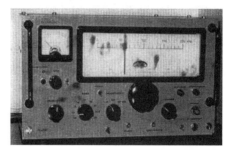

The RX-9B-DC receiver, pictured on the QSL card of Roelof Bakker, PA0RDT. This is a unique homebuilt direct conversion receiver. PA0RDT designed and built it based on the now discontinued Plessey active mixer type SL6440 high-level cross-coupled balanced-mixer chip. Using nine relay-switched oscillators with equal bandspread with a single calibrated dial covering 200kHz segments of the 1.8, 3.5, 7. 10, 14, 18. 21, 24 and 28MHz bands. The oscillator is stabilised using PA0KSB's classic 'huff and puff' stabiliser (often featured in 'TT') with a 25Hz grid. Passive audio filters are used throughout. MDS –126dBm, dynamic range 87/96dB, IP3 =7/-17dBm. BW 4000Hz, selectivity 4, 2, 0.3 and 0.1kHz. Receiver gain 120dB. S-meter S0 to S8 +30dB +/- 1dB.

looks set to continue with the emergence of new technologies for applications such as ultra-high-speed personal area networks, home automation and short-range anti-collision radar."

The Ofcom authors note that from the point of view of a spectrum regulator there are broadly two reasons for licence-exemption. (a) First, if the economic benefits of the exempted use are greater than those of alternative, licensed use. (b) If the demand for spectrum in a given frequency band is less than the supply.

Generally, the licensed radio amateur suffers little interference from the exempt devices which transmit only intermittently; rather a problem can arise where the amateur transmissions interfere with the exempt device. Remember the problems with car fobs which led, I seem to recall, to the loss to amateurs of part of the 70cm band? And the warnings, particularly for early models but still sometimes relevant, against using implanted pacemakers in close proximity to high power transmitters. I cannot recall reading of any problems arising from interference to or from 2.4GHz Bluetooth and Wi-Fi devices, but it seems possible for amateurs using the 2.3GHz band. Admittedly, users of licence-exempt devices have no legal guarantee that the spectrum will be free from interference and are not protected from interference caused by other regulation-compliant users. But social consequences and a wish to live in peace with neighbours can exert powerful pressures.

What seems potentially worrying is that in April this year Ofcom published "The Licence Exempt Framework Review (LEFR)"

consultation which explores ways in which licence-exempt devices might be managed in the future and whether there would be merit in dedicating more spectrum to them. Indeed, the Ofcom authors in their conclusions even suggest that very low power devices should be exempt from licensing, regardless of the band in which the operate, on the basis that they are highly unlikely to generate interference.

Peter Chadwick, G3RZP, is professionally concerned, as Chairman of the ETSI ERM TG30 committee on Wireless Medical Applications, in producing standards for medical devices using wireless, including radio in implanted devices such as pacemakers and the like. He writes: "All very interesting although a world in which a QRO transmitter produces about 1mW of RF and a 'high gain' antenna is –10dBi.".

He was interested to read the comments in the April 'TT' pp 82-83 on Ofcom's interest in flexible bands ("A licence to do (almost) anything you want") and writes: "It's an ongoing saga with which I've been somewhat involved professionally. Paradoxically (and this doesn't seem to have registered with administrations) the lighter the licensing regime, the greater the regulation required if there isn't to be chaos! That regulation can be voluntary, as in the amateur bands, or lightly enforced, as for short range devices in the 868MHz region.

" Incidentally, when CEPT PT SE42 started looking at flexible bands, one of the first points made was that the only truly flexible bands are the amateur bands and they need regulation [band planning] albeit on a voluntary basis [in most IARU countries and by licence terms in North America]. I know that the RSGB and then IARU band plans are post-war but I recall reading that even pre-WW2, CW was at the top end of 14MHz and 'phone at the bottom end, so I suppose there was informal band-planning."

In fact, pre-WW2 on the then wider 20m band (14,000 to 14,400kHz) most CW operation was at the top and bottom ends with most 'phone in the middle. The RSGB Council endorsed (with changes) a formal band-plan drawn up by an *Ad Hoc* sub-committee, chaired by Council Member Ian Aucterlonie, G6OM. It was announced in the *RSGB Bulletin*, April. 1949: "Experience had shown beyond any doubt that the average telegraphy signal stands little chance of survival at a distance if it has to compete with the much broader signals emitted by telephony stations. ... To be effective the plan requires the active co-operation of every amateur. ... With goodwill all round, the RSGB Band Plan will encourage the newcomer, safeguard the low-power operator, permit full experimental work, and above all, help YOU to enjoy Amateur Radio at its best!"

The RSGB Plan later formed the basis of the Region 1 IARU plan which, with subsequent modifications, should be observed by all amateurs. In connection with my comments on the earlier article by William Webb ('TT' April), G3RZP explains some of the terminology used in this field by the ITU committees. He writes: 'Out of band emissions' has a particular and defined meaning; they are emissions in the domain between the necessary bandwidth of an emission and the onset of the spurious emission domain. The spurious emission domain is the frequency range separated by more than 250% of the necessary bandwidth from the emission. So, for a GSM signal, which has a necessary bandwidth of about 180kHz, the 'out of band' domain is from 90 to 225kHz away from the carrier; anything further away is a spurious emission. ..."

COAX IMPEDANCE TRANSFORMERS AND BALUNS. The use of quarter-wave (electrical) sections of coax to form monoband impedance-matching transformers is revisited by Joel R Hallas, W1ZR in *QST* March 2007, p57. (An electrical length is the physical length multiplied by the velocity factor of the cable; **Figure 1** shows one method of checking resonance.) For example, to provide a match between a 50-ohm feeder and an element feed point impedance of about 100 ohms (close to the feedpoint impedance of a quad loop) an electrical quarter wave length of 75-ohm cable can be used: **Figures 2 and 3(b)**. Two parallel sections of the same type connected as in **Figure 3(b)** have a resultant $Z_o$ equal to half the $Z_o$ of the cable. When connected in series, as in **(c)** or **Figure 2** it doubles

Figure 1: Using an antenna analyzer to trim a quarter-wave section. W1ZR points out that the most accurate results are usually obtained by leaving the far end open and tuning for near 0-ohm resistive and 0-ohm reactance at the desired frequency. (Source QST)

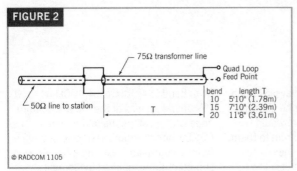

**Figure 2**: Use of a quarter-wave of 75-ohm coax cable as an impedance transformer to provide a 100-ohm match to a single quad element with typical dimensions.

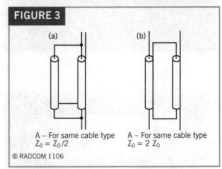

**Figure 3**: (a) Two parallel line sections (any length) of the same type of coaxial cable have a resultant impedance half that of the cable. (b) Connection of two quarter-wave cables in series to double the resulting impedance. (Source QST)

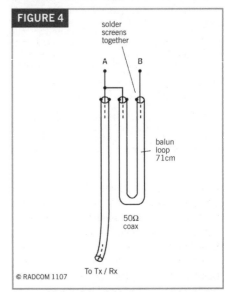

**Figure 4**: A 4:1 half-wave coax balun for 144MHz. Source: G1EXG on www.creative-science.org.uk/balun.html

the resulting $Z_0$.

Johnathan Hare, G1EXG, who has a "Creative Science Centre" web site aimed at helping people explore science through experiment draws attention to the 4:1 half-wave coax balun which can transform an unbalanced 50-ohm coax feed to match a 200-300 ohm balanced feed-point. **Figure 4** shows his suggested arrangement for roughly matching the element of a 144MHz folded-dipole to a 50-ohm coax feeder. Actual length of the balun loop section is again governed by the velocity factor of the cable.

**COMMUNICATIONS: INVENTIONS & INVENTORS.** An Argentine technical journalist – Juan Pablo Conti – recently contributed to *IET Communications Engineer* (February/March, 2007) an article in which he listed what he believes to have been "The 10 greatest communications inventions" and some of those he considered responsible. His list comprised:

**(1)** *Cable communications.* First commercial electrical telegraph built by Sir Fothergill Cooke, Paddington to West Drayton, in 1839.
**(2)** *The telephone* (usually credited to Alexander Graham Bell who was granted a US patent in March 1876, although he points out that the concept and in some cases prototypes already existed) – Meucci, Bourseul, Reis. Gray and Edison who "crucially developed the carbon grain microphone which remained a key component until the 1980s".
**(3)** *Radio.* When Heinrich Hertz was asked in the late 1890s what the ramifications of his discovery [of electromagnetic waves] were, he shrugged and said: "Nothing, I guess. It's no use whatsoever this is just an experiment that proves maestro Maxwell was right". Conti claims "Wireless telegraphy was officially born in 1897 when Guglielmo Marconi established the world's first radio station on the Isle of Wight.... it must be noted that Marconi actually formed part of a group of early experimenters...." He named only Alexander Popov and Nikola Tesla.

In brief: **(4)** *Television.* John Logie Baird's 1925 public demonstration of moving silhouette images based on the earlier discovery of selenium by Willoughby Smith (1873) and the scanning disk by Paul Nipkow (1890). Farnsworth's all-electronic system shown in Philadelphia in 1936. **(5)** *The Satellite.* Credited to Arthur C Clarke and his 1945 *Wireless World* article. **(6)** *Fibre Optics.* Victorian idea developed by Charles Kao and George Beckham. **(7)** *The Mobile Phone.* Credited to Martin Cooper of Motorola. **(8)** *The Internet.* **(9)** *The Wireless Internet.* Finally, **(10)** *Convergence* – the coming together of the earlier systems as currently underway.

I cannot help feeling that Conti's article, while cleverly picking out the major ten advances in communications systems, underlines the problem of singling out individuals as 'the inventor' or inventors of a system. Is it the person who first publicly demonstrates an early prototype? Or describes the concept? Or takes out a Patent (which does not guarantee that the idea will work in practice)? Or sets up the first commercial application?

Conti's credits seem to embrace all of these criteria, accepting uncritically the populist view of the 'inventor' rather than attempting historically consistent evaluations. The result is the omission of names that must surely deserve to be recognised as having made absolutely key contributions to radio telecommunications and television systems. One thinks off-hand of Samuel Morse/Vail, Lord Kelvin, Sir Oliver Lodge, Prof. Reginald Fessenden (see below), Lee de Forest, Campbell-Swinton, Howard Armstrong, H J Round, Zworykin, Blumlein, Alec Reeves, Shannon and others at Bell Telephone Laboratories. And then what about antennas (Yagi, Brown etc) and the mechanical and later electronic keyboard systems or the development of the gamut of semiconductor and piezo-electric devices?

**MARCONI AND FESSENDEN.** The April 'TT' included a preliminary report from G3AGA on the experimental 1900kHz beacon transmissions from GB3SS at Poldhu during December/January. The idea was to check whether a radio path could possibly have existed in daylight between Cornwall and Newfoundland in December 1901 when Marconi claimed to have received the famous three dots – a claim that gave rise to a controversy that still exists a century later. The GB3SS tests showed clearly that such paths often exist on 1.9MHz across the North Atlantic even under all-daylight conditions. This does not prove that Marconi actually received the Poldhu 400/500-metre transmissions. But it does lend some support, as noted in the April issue, to the view, expressed many years ago by Gerald Garrett, G5CS and J A Ratcliffe, a leading propagation scientist, that the signals received on Marconi's untuned receiver with a makeshift kite antenna could have been harmonics rather than the fundamental transmissions.

I noted also that Dr John Belrose, VE2CV has for some years been sceptical of the Marconi claims and has done much to promote the pioneering work of Reginald Fessenden, The April 'TT' also contained a short report of an American article which challenged the usual accounts of the 'world's first radio broadcast' having been made by Fessenden from Brant Rock at Christmas 1906.

These items have had the result that VE2CV has sent me a large batch of his recent papers expressing doubts about Marconi's claims, together with some emphasising his support for Fessenden's achievements, particularly those achieved using the high speed alternator that generated continuous wave AC at low radio frequencies.

It would take a whole issue of *RadCom* to explore fully the results of his detailed

## TECHNICAL TOPICS

Marconi at Signal Hill, Newfoundland, December 1901

research papers which have been backed up by model experiments and computer simulations and analyses.

My April 'TT' item, in common with those of other writers, commented: 'Because of legal action Marconi was prevented from repeating the experiment in Newfoundland ....' VE2CV writes: Marconi was not prevented from carrying out follow-on experiments by the cable company, in fact it is my understanding that they offered to assist and observe – he was prevented from establishing a [commercial] service. ....Regarding [the belief in] Marconi's success in receiving signals *and messages* during his transatlantic crossing [on the *Philidelphia*] in February 1902 read the addendum to my original on-line article "Sounds of a Spark transmitter" – this shows that with the low spark rate he was using [at Poldhu] at that time – *impossible*. ... A lot has been written about spark transmitters by authors who never ever heard a spark transmitter [I still have his audio tape 'Sounds of Spark – G3VA]. I try to put this into words which non-technical people can understand on page 2 (bottom half) in my paper 'Fessenden on Roanoke Island and the Outer Banks, NC, 1901-1902'. I wonder if readers heard the BBC broadcast on Fessenden's Christmas Eve broadcast – if not find someone with a Computer and Web access – it can be heard."

VE2CV dismisses the hypothesis that Marconi at Signal Hill may have received harmonic signals from Poldhu, mainly it seems because he is convinced that the substitute fan-type antenna at Poldhu (in conjunction with its matching to the high-power transmitter) would have been a poor radiator at harmonic frequencies. He also believes that the company photographs showing this antenna – following the blowing down of the intended circular ring of masts – are in effect heavily retouched photographs based on those of the original antenna.

Personally, I would not dismiss the 'harmonics' hypothesis lightly. It would then leave us with only three possibilities: **(1)** Marconi did receive the Poldhu 400-500 metre signals as claimed, no matter how unlikely that has seemed to so many scientists. **(2)** The clicks he and Kemp heard were atmospherics or local electrical interference genuinely mistaken for the Poldhu signals. Or **(3)** under intense commercial pressure, they made claims that were knowingly false. I doubt whether we will ever know for sure.

There seems little doubt that Marconi and the company remained sensitive to the accusations that their 1901 claims were disputed. The written evidence in the form of the scribbled note in Marconi's diary: "Sigs. at 12.30, 1.10 and 2.20" are far from constituting a running log of an historic experiment otherwise surrounded with so much photographic publicity, etc. Key features of the equipment still remain uncertain. And, as W J Baker in his "History of the Marconi Company" puts it, "Kemp, who throughout his working life meticulously kept a detailed daily journal, records the victory as if it were scarcely of more account than the putting on of his boots."

Geoff Voller, G3JUL, writes: "You may recall that I was assistant to Gerald Garratt, G5CS when he was the Keeper of the Communications Section of the Science Museum. Both being Radio Amateurs, we often discussed propagation and experimented with various wire aerials for harmonic use before settling for a large 'doublet' as the first aerial for GB2SM.

"Prior to the special exhibition at the Museum held to celebrate the 60th anniversary of Marconi's achievement, G5CS tried to persuade the then Archivist at Chelmsford [W J (Bill) Baker] that the [harmonic] hypothesis should be mentioned in the display to provide balance. But the Marconi Company did not want to consider this theory and Gerald was persuaded not to pursue the matter in any publication.

"Incidentally, a GB station was set up at the premises where Marconi's widow was staying and it was agreed that she should speak over it to GB2SM at the Museum prior to opening the Exhibition. However, although we had a QSO with the special GB station, she did not take part.

"The Science Museum has lost the GB2SM logs and I seem not to have a copy of the brochure that was published in connection with the Exhibition."

I would add that I still think it was a great pity (not only for Amateur Radio) that the Science Museum years later (long after G5CS retired) closed down GB2SM. Indeed they also downgraded the Museum's radio collection of historic exhibits on the third floor, moving just some of the exhibits to the Telecommunications section and putting many into store. *Sic transit gloria mundi!*

**FESSENDEN'S FIRSTS.** Dr John Belrose, VE2CV, has long been active in seeking to widen public appreciation of the historic importance of Canadian-born Reginald Aubrey Fessenden (1866-1932). In "Remembering the 100th Anniversary of the First Radio Broadcasting" addressed to Friends of the CRCC [Communications Research Centre Canada], he stresses that Marconi, until 1912, concentrated on spark-gap transmitters and a curious magnetic detector, and was not "a pioneer of radio as we know it today" – a rather harsh judgement though similar to that of many historians in regard to Baird and his mechanical television.

Instead he compares Fessenden's pioneering work with that of Alexandra Graham Bell: Bell's aim was 'words over wires'; Fessenden's was 'words without wires'. VE2CV considers that "Fessenden is 'The principal pioneer of radio as we know it today' and lists his impressive pioneering achievements as follows:

*The first to use the word and method of continuous waves (circa 1897). The first to transmit voice over radio (December 1900). He devised a detector for continuous waves (circa 1902). Transmitted tones for telegraphy over a CW-like wireless link (April 1902). First to use the word and method 'heterodyne' (circa 1902). The first to send two-way trans-Atlantic wireless telegraphy messages (circa January 1906) and the first to record the night-to-night variability of wireless transmission (propagation studies). The first to discover evidence for long path as well as short path signals (Autumn 1906). The first to send wireless telephony (speech) across the Atlantic (circa November 1906). The first to demonstrate wireless telephony in practical use, ship-to-shore and shore-to-ship, between Brant Rock and a small fishing vessel 20km out in Massachusetts Bay (circa 3 November 1906). The first to demonstrate wireless transmission in conjunction with wire lines (telephone-to-telephone via radio), 21 December 1906. The first to make wireless broadcasts, speech and music, 24 and 31 December, 1906.*

Fessenden developed the principle of amplitude modulation (AM) as an example of heterodyning. The spate of firsts in 1906 resulted from his pioneering use of a high-frequency alternator developed for him by Swedish-born E F Alexanderson (1878-1975) of General-Electric which eventually could generate alternating current with a frequency as high as 80kHz, providing Fessenden with a source of continuous-wave energy.

# TECHNICAL TOPICS

PAT HAWKER, MBE
37 DOVERCOURT ROAD,
DULWICH, LONDON SE22 8SS

G3VA

# Technical topics

*50 years of technical progress*

**WHAT HAVE WE LEARNED?** With 'TT' now well into its 50th year, it is surely time to consider how the technology has advanced and how this has affected the hobby. In April 1958, Amateur Radio was long matured; its operational customs, procedures, awards and contests etc well established. SSB was gradually becoming popular, although AM was still the major HF phone mode, and NBFM was by then in widespread use on VHF. Electromechanical RTTY using FSK had proved its value; slow-scan and real-time ATV had an enthusiastic, if minority, following. CW was still the dominant mode for HF DX. Perhaps the main technical difference from today was that amateurs were still using separate transmitters and receivers, both on HF and VHF. The four or six-foot rack assembly was being progressively replaced by table-top transmitters, well-screened to combat TVI. Many transmitters were home-built or modified wartime surplus.

Surplus wartime communications receivers, such as the AR88, HRO, BC342, R1155 etc, were widely used. Most of the early post-war designs followed basically similar circuitry although there was increasing use of double- or even triple-conversion superhets. The long-established standard IF of 455kHz remained popular, with a few amateur models using higher IFs to reduce 'image'. The 'straight' regenerative home-built receivers that had dominated the 'thirties had largely vanished and did not re-appear until the 1970s in the form of the simple homodyne solid-state direct-conversion HF receiver.

But change was on the horizon. 1957 saw the appearance on the US market of the first SSB/CW factory-built compact transceiver with 455kHz mechanical filter and intended for fixed or mobile use. This Collins KWM-1 model soon influenced other firms including KW Electronics in the UK with their KW2000-series and the early Japanese black boxes, originally marketed in Europe under such brand names as Trio and Sommerkamp, and kit transceivers by Hallicrafters. All these models were based on thermionic valves, although by 1958 a few transistors were finding use in auxiliary and test equipment, with the silicon diode rapidly replacing the rectifier valve.

Let there be no mistake. Many amateur stations in the 1950s were capable of highly effective and efficient operation. It was the combination of double-conversion, SSB and the very strong signals from HF broadcast stations or nearby amateurs that began to cause disaffection with classic receivers. The long warm-up drift and vulnerability to very strong signals, the spurii with multiple conversion were becoming irksome.

An iconic article by Byron Goodman, W1DX "What's wrong with our present receivers?" appeared in *QST*, January 1957 drawing attention to the importance of gain-distribution, and the limitations of multi-electrode valve-mixers with their extremely high equivalent noise resistances. Such mixers as the 6K8 (ENR 290,000 ohms) required high-gain RF stage(s) in front of them to achieve good sensitivity. This limited the dynamic range, even with several tuned circuits ahead of the mixer. Double-and triple-conversion increased the number of spurious responses etc.

High-gain pentode valves used as tuned RF amplifiers combined with multi-electrode valves do not cope well with strong signals, Collins in 1957 introduced the 75A4 model using a variable-mu 6DC6 RF stage in place of the 6AK5 in the earlier 75A models.

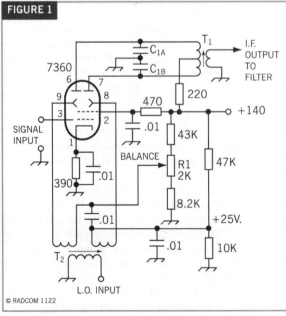

**Figure 1**: Circuit diagram of a 7360 double-balanced low-level mixer

**DOUBLE-BALANCED BEAM-SWITCHING MIXERS.** In the UK, almost 1000 amateurs tackled the construction of a high-performance receiver described by Dick Thornley, G2DAF which followed many of the design features of the 75A3. Therein lies a story.

By 1957, RCA had introduced beam-switching valves for TV receivers, adding, in 1960, the 7360, intended as a high-level modulator for SSB transmitters. Beam-switching valves were soon being used by amateurs as product detectors. But there remained a need in the receiver chain for greater front-end linearity. The weakest spot was the use of noisy mixers preceded by high-gain RF stage(s), What was required was a low-noise linear mixer that could cope with very weak signals (less than 1µV across 50 ohms), so requiring no pre-amplification, yet coping – without blocking or intermodulation – with very strong signals, up to about 1V, a dynamic range of over 120dB.

The valve mixer that can come closest to this is an active double-balanced commutating (switched) mixer, driven by an accurate square waveform, and using a beam-deflection valve such as the 7360. But in 1957 this still needed to be generally recognised.

In 1957-59, Brian Mitchell, G3HJK, built an experimental modular front-end using a beam-switching valve as a mixer without the traditional RF amplifier stage(s). He found this had a noise factor of about 10-11dB (a figure low enough to provide more than adequate sensitivity up to at least 25-30MHz. He tried to persuade G2DAF to develop a new front-end for the G2DAF receiver using this approach. This never happened, and G3HJK's pioneering work has gone unreported until now.

It was several years later that an article "A New Approach to Receiver Front-End Design" by William E Squires, W2PUL, appeared in *QST* (September 1963, pp31-34). This reiterated W1DX's general principles: (1) as little gain as possible before taking full selectivity (2) superb linearity in any stage preceding [final] selectivity [whether this is achieved at IF or AF].

W2PUL stressed that the ideal receiver should have *no RF stage* and have *as few frequency conversions as possible*. But the major point made in his article was to advocate the use of an RCA 7360 valve as a front-end mixer: **Figure 1**. Within a few months

# TECHNICAL TOPICS

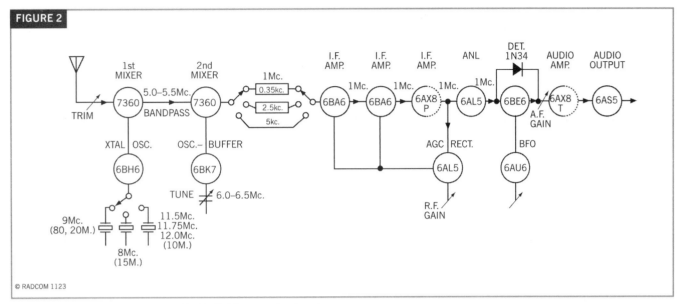

Figure 2: Simplified block diagram of the 1964 SS-1R HF amateur bands receiver

this article was shown to have provided a preview of the Squires-Sanders SS-1R receiver, reviewed in *QST*, May 1964, pp54-56: **Figure 2**.

This review was well before the time that *QST* routinely provides detailed performance measurements but it is clear that the SS-1R was "some receiver" on the then existing HF ham-bands, provided it was connected to a well-matched 50-ohm feeder. With a price tag in the order of $900 in those pre-inflation days, it was not a cheap receiver. Relatively few, reputedly under 100, were sold (I have never heard of any in the UK) before the tragic death of both Squires and Sanders in an aircraft crash.

But the 1963 *QST* description by W2PUL of the advantages of a 7360 front-end mixer influenced other amateurs, and a number of designs soon appeared. I recall the ingenious "Miser's Dream" design by W1DX (*QST*, May, 1865) using a signal frequency Q-multiplier to sharpen the RF response before the 7360 mixer: **Figure 3**.

By the late 1960s, small-signal valve manufacture was in terminal decline; attention was focussed on solid-state receivers, despite their initial severely limited dynamic range. A final effort to combine the 7360 and other front-end valves with transistors in a high-performance HF receiver was described by Peter Martin, G3PDM, appearing in a series of articles in *Radio Communication* and later reprinted in a booklet "Plagiarise and Hybridise". This featured a push-pull transistor oscillator with low phase noise that by then was becoming recognised as a limitation on the close-in performance of receivers offering high dynamic range.

Looking back, there can be little doubt that the 7360 was the best-ever mixer based on valves and barely, even now, on solid-state devices. One snag was that it needed to be protected from varying external magnetic fields.

## IMPROVING THE SOLID-STATE FRONT-END.

The classic HF communications receivers based on valve technology and crystal filters were reasonably well suited to narrow-band reception of amateur signals. By 1965, as noted above, further improvement in dynamic range was becoming possible. But by then the 'silicon revolution' was in full-swing. One result was to set back the performance of communications receivers. As late as 1981, VK5AR, wrote: "Solid-state technology affords commercial manufacturers cheap, large-scale production and is ideally suited to logic and non-linear applications, But for [amateur radio] receivers, transceivers, transverters and transmitters of practical simplicity, valves remain incomparably superior for one-off, home-built projects."

But there could be no going back. In the 1960s, Racal marketed the RA217, a solid-state version of its classic RA17 with its triple-mix Wadley loop. It proved a disaster. One outcome was that Racal sponsored Dr J G Gardiner to undertake an investigation into solid-state mixers at Bradford University. His team soon endorsed the cross-coupled double-balanced mixer configuration.

With D C Surana, he wrote (*Proc IEE*, November, 1970) "In many applications, the single-transistor mixer fails to provide certain features of performance which are

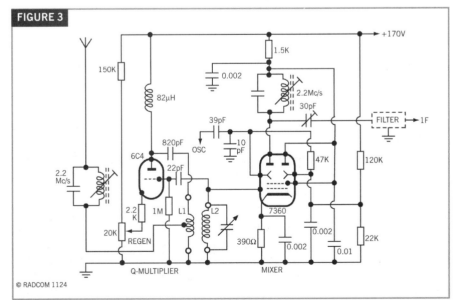

Figure 3: Front-end of W1DX's 3.5MHz Miser's Dream receiver with Q-multiplier to sharpen the pre-mixer selectivity.

of great value in the design of communications equipment, especially receivers. Of particular importance is the property of double balance, ie high suppression of both local oscillator and input signals at the mixer output, which is readily available in the ring-diode mixer for instance, and which prevents high-level input signals and the local oscillator from overloading subsequent stages of the receiver. The ring modulator has the disadvantage that it introduces significant conversion loss, typically 6dB (with hot-carrier diodes) but the properties of double balance and useful conversion gain can be obtained simultaneously from the two-transistor 'cross-coupled' mixer of **Figure 4(a)**."

The cross-coupled mixer is well suited to integration. In the late 1970s Plessey introduced a number of linear ICs specifically for communications applications (SL600 and SL1600 series). Signetics/Philips marketed the NE602, widely adopted for battery-powered HF receivers. Then, in 'TT', June/July, 1980, Peter Chadwick, G3RZP, as a senior applications engineer with Plessey Semiconductors, drew attention to a new high-level SL6440 mixer IC based on the 'tree' form of cross-coupled transistor approach: **Figure 4(b)**. This device provided a convenient high-performance linear mixer (30dBm intercept point, reasonably low noise, +15dBm compression point (1dB) with conversion 'gain' of –1dB).

In *QST*, (April. 1981) with Doug DeMaw, W1FB, G3RZP outlined the design of a simple high-performance 3.5MHz receiver using the SL6440 without an RF amplifier stage. It had a 455kHz IF with pre-mixer selectivity using gang-capacitor tuning of two resonant circuits. Although the noise figure was put at 20dB this was adequate for 3.5MHz. The SL6440 remained a good choice for amateur receivers until the late 1990s when Plessey Semiconductors was taken over by Siemens and production of the device ceased.

But meanwhile other approaches to high-performance double-balanced mixers had continued. Ulrich Rohde, while DJ2LR, did a lot of work on double-balanced bipolar junction transistors, see, for example, "Performance capability of active mixers" (*Ham Radio*, March 1982). Since solid-state devices act well as switches, it is not surprising that the main efforts have been concentrated on switching-mixers, both balanced and double-balanced. Packaged diode-quad packages using hot-carrier diodes, such as those marketed by Mini-Circuits, have on the plus side good performance but have intrinsic conversion loss and require significant drive power. Ed Oxner, KB6QJ, a senior applications engineer with Siliconix, concentrated from 1973 on the use of power JFETs in double-balanced mixers. This led, in 1986, to his "Super-High Dynamic Range Double-Balanced Mixer" ('TT', March 1986, or *TTS 1985-1989*, pp222-3): **Figure 5**.

It had outstanding intermodulation performance and overload characteristics and was based on a monolithic quad small-signal DMOS FET (Si8901). This configuration was capable of a two-tone IMD exceeding +36 dBm with less than 50mW of local oscillator drive and a noise figure less than 8dB. Jacob Nicholson, N6NWP, presented a high-performance front-end using the Oxner Si8901 mixer. Unfortunately, changes at Siliconix meant that the Si8901 soon went out of production although a rather similar device, the SD5000, became available from Signetics/Philips.

Move forward to 1993 and enter Colin Horrabin, G3SBI. As Wes Hayward, W7ZOI, puts it in *Experimental Methods in RF Design* (ARRL, 2003) in a section on high-level FET mixers: "Perhaps the most exciting work published in the past decade was a note in 'TT' (Sept/Oct 1993) where Pat Hawker, G3VA, presented previously unreported work on a new ['H-mode'] mixer topology by G3SBI. This four-FET mixer, **Figure 6**, differed from earlier circuits. Oxner's design used FETs as series switches, while Horrabin used the FETs as grounded switches. This is still a commutating mixer, but transformer action now generates the needed signals. His circuit used a monolithic quad of MOSFETs, the Philips SD5000, which is

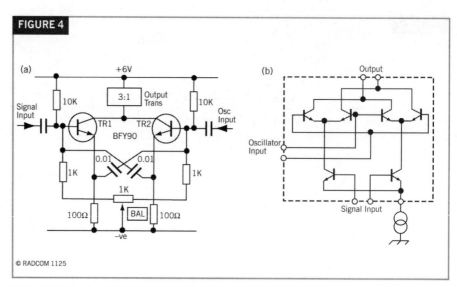

**Figure 4**: (a) Cross-coupled mixer using two bipolar transistors. (b) Tree form of cross-coupled mixer as used in the integrated SL6440 high-level mixer.

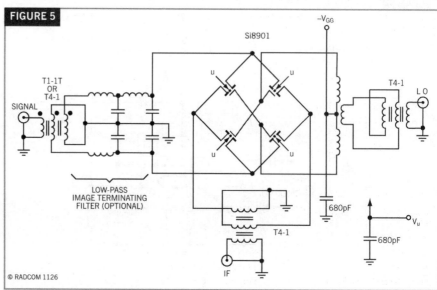

**Figure 5**: KB8QI's super high dynamic mixer as presented in 1986

# TECHNICAL TOPICS

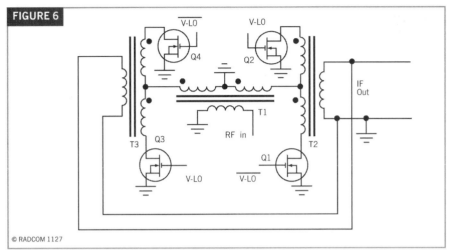

Figure 6: The basic H-mode mixer as developed by G3SBI in 1995

essentially the same MOSFET as used in Oxner's Si8901."

W7ZOI notes that this configuration has produced third order inputs as high as +55 dBm. In practice, such performance is limited by other components in the receiver chain such as the RF transformers, RF filters, crystal filters etc. The close-in performance is governed by the local oscillator phase noise (leading G3SBI to develop a novel two-resonator, grounded-grid, low-phase-noise oscillator first presented in 'TT', July 1994). Currently G3SBI is seeking to improve the performance of DDS synthesisers without using a PLL loop. He writes "Jeff Keip, who is head of the Analog Devices DDS design team, said the spurious performance of the 1GHz AD9912 has been shown in laboratory tests to be superior to the 1GHz AD9910, although the reason has not yet been established. Clearly, for amateur radio applications, the AD9912 should be used."

Gian Moda, I7SWX takes up the story: "Then, at the beginning of 1998, my idea of using computer Fast Bus Switches (FST3125 etc) simplified the G3SBI H-Mode mixer, removing any need for bias adjustment controls, lowering the conversion losses and still keeping high IP3 performance; a very hot front-end. Two high-performance HF transceiver projects have been published in RadCom using the H-Mode mixer: the CDG2000 (G3SBI, G8KBB, G3OGQ) and the Pic-A-Star (G3XJP).

"Later a simplified version [by I7SWX] was the two-transformer H-Mode mixer ('TT' April 2003) permitting home-brewers to save money winding their own transformers and still keeping high performance [though slightly down on the three-transformer design – G3VA]. It is well known that the H-Mode Mixer has been implemented in many ham projects and for improving factory-built equipment, particularly in Europe."

Both G3SBI and I7SWX (and many others world-wide) pay tribute to the current detailed work by Martein Bakker, PA3AKE, in measuring and comparing the ICs, ferrite cores and crystals etc used in the three- and two-transformer H-mode mixers and synthesisers. It is hoped to refer further to the latest developments arising from this work, including a two-transformer version using the FSA3157. Meanwhile readers are referred to PA3AKE's website given in the May 'TT' p80.

IMD IN DIGITAL RECEIVERS. With the H-mode mixer now capable of an IMD performance comparable with the 7360 valve mixer/product detector, it may appear that our front-end problems are nearly solved, even though G3SBI, I7SWX, PA3AKE are still bent on further improvements. But along comes the DSR and DDR digital revolution with its new set of front-end problems and limitations set by currently available high-frequency analogue-to-digital converters. The DSR receiver with A/D conversion at IF or AF can use the FST3225 etc as a switching image-rejection (Tayloe) I/Q mixer/detector before A/D conversion but the direct digital receiver (DDR) eliminates the mixer altogether with its performance affected in a different manner than with a conventional mixer. Performance of a DDR cannot be specified by means of a straightforward two-tone test.

"IMD in Digital Receivers" by Leif Åshrink, SM5BSZ (QEX, November/December 2006, pp18-22) is sub-titled "Performance limitations of receivers with the A/D converter at the antenna – how to measure and work around them".

This is introduced as follows: "Today, most receivers have an A/D converter at some point in the signal path because digital technology can provide better filters at lower cost compared to analogue technologies. The A/D converters gradually move nearer the antenna and digital technology takes over a larger fraction of all the filtering work. Someday receivers will be completely digital and sample directly at the signal frequency – at least for HF bands. Sampling directly at the signal frequency is already possible today with amateur equipment. The first commercially available [amateur] equipment is the SDR-14 from RFSPACE. This article highlights the problems of characterising this new class of radio receivers."

SM5BSZ continues "Standard methods of characterising performance in terms of third-order intercept and intermodulation-free dynamic range fail badly and cannot be used at all to give an adequate representation of how all these new receivers perform as compared to conventional radios when used on the same antenna with real signals. New ways of doing measurements are called for, and this article is intended to shed some light on the problems of measurement and on the methods we can use to improve performance by adding preselectors and reducing out-of-band signals. Digital technology is developing rapidly, better A/D converters will become available and performance will improve drastically in the future but still the problems of properly characterising a radio receiver will remain."

For a description of SM5BSZ's methods and findings readers are referred to the five-page QEX article. The point made here is that SM5BSZ shows that the production of "intermodulation-type" false signals (spaced every 5kHz) when an SDR-14 receiver is tuned to 7MHz is much *reduced* (virtually eliminated) when fed with strong signals rather than with weak signals. SM5BSZ states "The SDR-14 receiver is an excellent receiver provided it is preceded by a suitable pre-selector that removes most of the HF signals outside the band of interest and amplifies what passes the filter to a level close to the maximum level the unit can handle. There is an LED on the SDR-14 showing if too much gain is applied. More colour images and detailed information about this real life test can be found on Internet (www.sm5bsz.com/digdynam/practical.htm).

As noted in the May 2007 'TT', the professional R/S EM510 direct digital receiver, has the antenna signals feed the ADC via a bank of switched RF high-pass and low-pass filters and a very linear RF pre-amplifier.

# Technical topics

*Mixers, Morse and methods of generating AM are some of the topics this month.*

**I7SWX's FSA3157 2T H-MODE MIXER.** The August 'TT' included an account of the development of high-performance mixers, including the doubly-balanced 7360 beam-tube mixer; KB6QJ's quad-FET mixer; the Plessey 6440 tree-mixer; G3SBI's ground-breaking H-Mode mixer and its later implementation using fast-bus switches such as the FST3125; and the simplified two-transformer version devised by I7SWX. Earlier, the May 'TT' outlined some of the recent detailed work by PA3AKE in measuring the performances of the latest generation of IC switches, underlining the value of the FSA3157. It is clear that the mixer device need no longer represent a practical limit to strong-signal handling of receivers. Performance is now more likely to be limited by the phase noise of the synthesiser or local oscillator, the ferrite materials used in the transformers and/or pre-mixer selectivity filters; the crystal roofing filters; or the amount of pre-mixer gain required for maximum usable sensitivity. For software defined radio, there is still much room for the improvement in the analogue-to-digital converters (ADC). Until all of these factors have been sorted out, there is still a role for home-constructed mixers being chosen not only for performance but also on grounds of simplicity and cost.

Giancarlo Moda, I7SWX writes *inter alia*: "After the publication in the May 2007 'TT' of the measurements being made by Martein Bakker, PA3AKE, I could not resist asking him if he would test the two-transformer version of the H-mode mixer using a FSA3157. I sent him the two home-brewed transformers and also a 2T H-mode mixer using the Pericom PI5C3125 with a 74AC86 squarer; the PI5C3125 is probably the best 3125 switch in production, having well-balanced minimum 'on' and 'off' values of 0.5nS; the FST3125 of other manufacturers are not so good, with values similar to the FSA3157.

"PA3AKE was happy to make the measurements and has added the results to his web page www.xs4all.nl/~martein/pa3ake /hmode. Paragraph 5 Homebrew transformers ° BN43-2402 2-Transformer H-Mode Mixer.

"The results are very interesting with both the FSA3157 and the PI5C3125, using 1:1 and 1:4 transformers and also the comparisons with the 3T configurations. To quote selectively from the PA3AKE site:

*IP3 performance is comparable with the results obtained with the Minicircuits TT4-1a, except for 160m and 80m. Especially on 160m the IP3 is less good and the transformers seem to be the limitation. IP3 is peaking on 40m. Performance is still good on 6m although spur levels are limiting the MMS. With regard to the observed average spur levels on 15m, Gian's transformers perform really very well. At –7dB average, they are second only to ADTT1-1 (-9.3dB). Perhaps having the two secondary windings on one core gives a natural symmetry as it eliminates any core differences.*

*RF-IF isolation is good. With band-specific adjustments it is possible to reach 50dB on most bands.*

*The 2-transformer H-Mode mixer is clearly a very good alternative to the more expensive Minicircuits transformers. If IP3 performance is not the first priority and good spur rejection is important and if one is prepared carefully to duplicate Gian's transformers then this is a good alternative! A drawback could be that an adapted PCB layout for best performance is needed to keep good symmetry and short connections between the FSA3157 switches and the two transformers, as is possible for stock transformers.*

I7SWX continues: "A sketch showing how to wind the home-brewed transformers, with pictures, is given in PA3AKE's reports. Also of interest are PA3AKE's measurements in Paragraph 9 –Appendix: AD9951 DDS spurs on 15m FSA3157 mixer schematic and PCB layout.

"I believe these are the first ever measurements made on how a mixer responds to DDS spurs. Martein has also added the FSA3157 diagram to help his readers understand how his H-mode mixer is configured. This shows that PA3AKE has not only done measurements but also implemented an intelligent mixer assembly that uses a 12-bit DAC to achieve the optimum 'on/off' ratio of the frequency squarer on each band, which can be managed by the receiver controller. That is the best IP3 on/off for each band.

"I suggest that 'TT' readers visit Martein's web site from time to time as I believe he is still adding to his information, including a complete high-performance front-end. In the meantime, I have revisited my 2T H-Mode mixer using the new FSA3157, at the same time simplifying the LO squarer using a SN74LCV2G86 double XOR device: **Figure 1**.

"While readers may feel that SDR and then DDR systems will, before long, eliminate the need for mixers and analogue stages, I am sure there will still be interesting and important projects using old and new analogue mixers; possibly the H-Mode Mixer as a first conversion mixer front-end with an SDR IF".

**MORE P-E GENERATOR LORE.** A 'TT' item "Petrol-Electric Generator Lore" (June 2006, pp72-73) provided a number of suggestions on achieving reliable and safe running of petrol-electric generators. As these devices remain the prime source of power for field-days and emergency operations, it seems worth while, despite some repetition, quoting selectively from "More Power to You" by H Ward Silver, NOAX (*QST*, June 2007).

"All generators are not created equal. Along with the power rating, consider waveform quality and regulation. 'Contractor grade' generators for powering tools have poor regulation and distorted waveforms, particularly near full rating. Keying a radio can cause large voltage swings, risking damage to a power supply or improper operation. If you can, use a generator intended to power electronic equipment. Poor regulation can be helped by loading a 2kW generator with a pair of 100W light bulbs at all times.

"Test generators well before field use unless you want to learn field generator repair! A generator with old fuel in the tank and carburettor is likely to run poorly, if at all. Generators should be stored with the fuel line and carburettor dry and a stabilising agent added to the stored petrol. Replace black and dirty oil. Inspect the air filter and silencer for clogging by dust or debris. Some sites may require a spark suppressor, so be sure your generator meets the [site] rules.

"Monitor fuel consumption, devising a fuelling schedule so the lights don't go out unexpectedly. Your 'Generator Czar' should provide additional oil of the proper viscosity, fuel in safety containers away from the exhaust, a fuelling funnel, mopping rags, and a flashlight for night-time maintenance. Review the instructions for each generator, such as those found on the web at

# TECHNICAL TOPICS

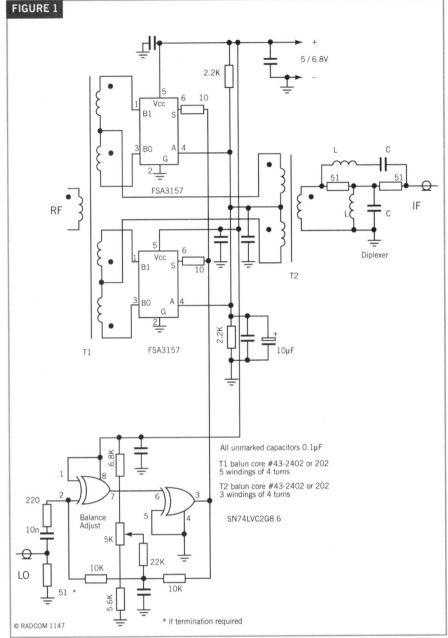

Figure 1: I7SWX's improved two-transformer H-Mode mixer using two FSA3157 fast switches.

mayberrys.com/honda/generator/html/operation.htm.

"Take generator safety seriously. Never run a generator in an enclosed space – be sure there's plenty of ventilation. Keep flammable materials such as dry grass or cloth clear of the exhaust. Keep a fire extinguisher at each generator. Never fuel a running generator – insist that it be stopped first and try to have two-person crews do the job.

If you operate near homes, consider the neighbours! If your generator is noisy, use plywood sheets to make a sound baffle. Try to direct or deflect noise up and away from people trying to sleep."

NOAX also provides information on the AC wiring, battery operation and alternative power sources such as solar power, bicycle- and pedal-operated generation. *Inter alia* he suggests that "Of the different battery types, deep-cycle marine or RV batteries and gel-cells are the best choice for Field Days. ... Many radios do not operate properly at less than 12V. That means the batteries may need to be charged frequently or continuously, or you may have to use a more tolerant radio!"

Personally, since 1944, I have always viewed P-E generators with some misgivings. First, due to a stupid mistake, I managed, while at Nymegan with IS9 (MI9) to write off completely my SCU9 Onan 150W P-E generator. Later, I was constantly frustrated by a repeatedly moist sparking plug in a replacement generator used close to the Rhine, often frantically pulling the starting rope repeatedly in an effort to meet transmission schedules.

In one of my first post-war NFDs, the generator provided an output dropping to about 20Hz, burning out the receiver PSU transformer. It pays to follow the advice given above by N0AX and the additional hints given in the June, 2006 'TT'!

Surprisingly, Onan 150W P-E generators were occasionally used, despite the noise of their two-stroke engines, by clandestine stations in Occupied Territory, including Norway. Dave Williams, G3CCO, has told me that he advised agents going into Malayan jungles always to dig pits for their generators in order to deaden the noise.

MORE ON ERGONOMIC CONTROLS. The June 'TT' item on the placing of receiver controls so as to leave the right hand free to write down incoming CW messages and to use the key while making operational adjustments with the left hand, resulted in several comments that for phone operation it is preferable to have the tuning control on the left-hand side as is usual practice for SSB/CW transceivers. This allows holding a fist microphone in the left hand while using the right hand for adjusting the controls, or entering information in the log etc. I suppose the choice depends on which mode you use most.

Ian Brothwell, G4EAN, comments: "When I first learned Morse Code I realised that using the same hand for both the Morse key and a pen would be rather awkward. I had a simple solution: I am right-handed so I learned to use the Morse key with my left hand. No need to put down my pen in order to use the key.

"This turned out to have benefits outside amateur radio. For example, I use a left-handed mouse with my computer and this evens out the workload between my hands. Interestingly, although it is easy to set up a mouse for left-handed use, my left-handed friends all use a right-handed mouse."

Personally, I have known at least one other instance of s left-handed amateur training himself to operate a Morse key with the right hand. This had the additional advantage that most semi-automatic ("bug") and electronic keys are intended for right-handed use. Apparently, it is easier to acquire the ability to use your 'minor' hand to use a key than it would be to learn to write with it.

CONSTANT-MODULATION CONTROLLED-CARRIER AM. A series of 'TT' items during 2006 and February 2007 discussed various classic and more modern forms of amplitude modulation, including pulse width modulation (PWM/PDM), Doherty-. Terman-Woodwood-, Series–, Constant- and Variable Efficiency modulation.

John Pegler, G3ENI, brings to attention a further relatively little known AM system. In "Constant Modulation, Controlled Carrier Working" (*Short Wave Magazine*, August 1951, pp336-339) he explained the advantages and disadvantages of controlled carrier operation and presented a practical system for amateur working. His introduction stated:

"The constant modulation controlled carrier system has not been very popular in amateur radio circles in spite of the fact that it has a lot to recommend it and is a well established system of speech communication. The basic principle is that the RF drive to, or amplification of, a modulated stage is controlled in such a manner that the RF output is proportional to the average strength level of the AF source to be transmitted: see **Figure 2**. If as a result of this control, the RF output voltage is swung between zero and double its average value, then 100% modulation will result at all output levels, eg a whisper will fully modulate and a shout will not overmodulate.

"Its chief advantages may be listed as follows:
**(a)** A predetermined level of modulation can be maintained over wide ranges of audio power.
**(b)** Only sufficient RF energy is generated to accommodate the AF component at the given level of modulation.
**(c)** Greater intelligibility at weak signal strengths, and under conditions of high background noise.
**(d)** Reduction of heterodyne interference between adjacent stations.
**(e)** Efficiency systems of modulation can be used, thus economising in audio equipment.
**(f)** Such efficiency systems can be worked at their maximum efficiency at all times.

"The disadvantages are mainly of a design nature:
**(a)** Good regulation of power supply and bias supply is required.
**(b)** Varying load on RF driver stages and modulator.
**(c)** Weak [controlled carrier] stations may be difficult to locate owing to absence of steady carrier.

"The following methods of obtaining constant modulation may be employed:
**(a)** Anode modulation plus control of HT either by a series valve or by a saturable reactor in the power supply.
**(b)** Anode modulation plus control of grid, screen-grid or suppressor-grid voltages.

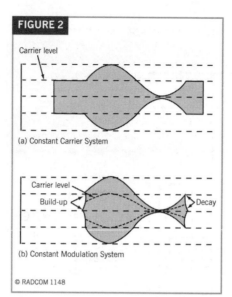

**Figure 2**: Envelope shapes illustrating the difference between (a) a conventional constant carrier amplitude modulated system, and (b) controlled carrier constant modulation as described by G3ENI in 1951.

**(c)** Control-grid, screen-grid or suppressor-grid modulation with control of one or more electrodes.

"Of the above methods, (c) deserves the greatest popularity and is the simplest and most economical. Control and modulation of the screen grid is the system that will be described. With efficiency modulation systems ... the anode current and efficiency both double on modulation peaks. Thus the 35% efficiency normally obtained in the unmodulated condition rises to 70%. Using controlled carrier, with constant modulation near to 100%, enables the PA anode efficiency to be kept in the region of 60 to 70% at all times."

G3ENI's SWM 1951 article describes the implementation and setting-up procedure for controlled carrier applied to a 144MHz transmitter (**Figure 3**), although equally applicable to HF transmitters. It would seem that the system provides some of the benefits of the double-sideband suppressed-carrier (DSBSC) mode but without the need for a phase-accurate inserted carrier at the receiver and a linear PA in the transmitter. Receiver AGC might require some consideration. For a given valve PA limited by anode dissipation, controlled carrier would permit the use of higher than its specified HT (as with but not to the same extent as SSB); this would not apply to a voltage rather than a current limited solid-state PA. I am not sure whether the system has ever been adapted for use with solid-state PAs but to do so might prove an interesting project.

In the now distant past, one of the significant benefits of AM, since it was receivable on domestic "all-wave" receivers, was that the amateur bands acted as a powerful recruiting aid for amateur radio. With talk of all-digital radio broadcasting. it may seem the days of AM (including SSB) are limited, though many of us suspect it will not be in our time.

**MORE ON MORSE.** The July 'TT' quoted from *QST* some of the technical reasons why those amateurs who have gained access to the HF bands without having to pass a Morse test should nevertheless be encouraged voluntarily to become proficient CW operators. As someone who some 70 years ago struggled to achieve the then required 12 words per minute, I have never regretted the time taken from my schoolwork. Yet undoubtedly there has always existed an "anti-Morse" brigade, now being reinforced by those who claim that it is a waste of time to learn an "obsolete" technique. By comparison, the pro-Morse and mixed-mode amateurs

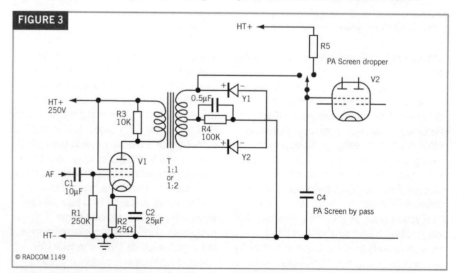

**Figure 3**: Controlled Carrier screen modulated system as applied by G3ENI to a 144MHz transmitter using a 832 (V2) dual-tetrode power amplifier, but applicable to any tetrode PA on any band, V1 6V6. Y1 andY2 were described as low-current rectifiers (in an era before general use of silicon diodes).

# TECHNICAL TOPICS

seldom, if ever, attack the use of voice or the various digital modes!

Recently, at a street party to mark the 100th anniversary of Dovercourt Road, I set up a Morse demonstration stand. This was intended primarily for the children, and included four of my ancient and modern "straight" keys with a crude tone generator, a tape recorder on which I had recorded professional and amateur use of Morse, plus some photographs and copies of the code, etc. It was interesting to find the considerable interest shown by young children, less so by teenagers, in using the keys if only to laboriously spell out their names.

A number of adults also showed interest, although I often found it necessary to prove to them that they were wrong in their belief that the code is obsolete. This I did by playing out my tape of enciphered military automatic Morse, recorded a few days earlier. These, probably French, Morse transmissions at about 18 groups per minute are transmitted over long periods on various frequencies (including within the shared 3.5MHz band and around 5.2MHz). The publicity that was given a few years ago to the closing of the coast stations etc has mistakenly led many people to believe that Morse is dead, or confined just to amateur radio.

QST (June, 2007) carries a report that the Dutch Military Aviation Museum at Soesterberg has a vintage radio shack in a mock-up of a B-25 Mitchell WW2 bomber, operated by PA0AAJ under the call PI9MLM, mainly on CW on 3.5, 7 and 14MHz. The museum provides 'workshops' for children aged 8 to 12 years on jet engines, flying and Morse code. The Morse workshop takes about 90 minutes at a cost of about €5 per child. It is proving very successful – "the kids like the 'secret code' and they also start writing in dots and dashes". They are given an explanation of Morse code history and a demonstration of PI9MLM; shown how to assemble a 'sounder kit' which consists of a pre-drilled wooden base, screws, buzzer, battery holder, plus batteries and keyer, made from a strip of PCB and a doorknob. The sounder can be taken home together with an explanatory booklet about Morse.

QST (July 2007) includes an item on how XYL WB9ZHC recently taught a young girl, now licensed as KC9KEW, to learn Morse with the aid of a gaily decorated tone generator and key base and then presented her with the equipment. There can be little doubt that the younger you are, the easier it is to learn Morse.

In 1994, an Ad Hoc committee of the IARU set out a reply to the question: "What progress in amateur digital communications and amateur voice communications are likely to affect the future use of Morse by amateurs?"

The report, I believe, was never published in RadCom but still seems relevant. To quote briefly from a few of its findings, although the committee clearly did not foresee the profound changes in licensing regulations:

"There are three factors that influence a radio amateur's choice of operating mode: (1) Suitability for the intended communications task. (2) Cost and availability of equipment. (3) Personal preference. A consideration of these factors suggest that progress in digital and voice communications is likely to have only a very limited effect on the future use of Morse code by amateurs.

"Today [1994] radio amateurs use Morse code in preference to some other mode for one or more for one or more of a number of reasons:
- The equipment is simpler and therefore less expensive.
- The equipment can be home constructed more easily.
- Communication can be achieved with relatively little effective radiated power (a characteristic that some digital systems also possess)
- Under certain conditions, such as weak signals or heavy interference, other modes are not feasible or may not be available to all of the operators.
- Some propagation media, such as auroral reflection, distort the signal too much for any other mode to be used [Consider also moonbounce, meteor scatter and the LF assignments – G3VA].
- The operator gains personal satisfaction from using this personal skill.
- Language barriers can be more easily overcome.

Among its enthusiasts, Morse code operation is perceived as more pleasurable, for a wide variety of reasons that are quite subjective, but are nonetheless perfectly valid to the individual."

**MARCONI & MARINE SPARK.** John B Tuke, G3BST, writes with reference to the recent discussion whether the dits 'received' by Marconi and Kemp in December 1901 might have been harmonics:

"I was a marine radio officer from March 1937 to June 1939. At first, for six months, as third operator on large ocean liners taking the 0000-0400 'graveyard' watch; then as sole RO on tramp steamers. On the big ships, we had the Marconi rotary spark transmitter, I think it was rated 5kW on MF. On the tramps, I had the Marconi quarter-kW quench-gap transmitter. In both cases, the frequency 'spread' was very large, and would easily have gone into the second harmonic. To give an idea of the frequency spread: if we were within 30 miles of a coast station and requested to change from the 600m (500kHz) calling frequency to the 705m (425kHz) traffic frequency, we did not bother to retune the transmitter which involved altering the ATI taps, etc. Naughty of course but all a long time ago!

"Even if the [Poldhu] aerials were roughly resonant at the fundamental (450m ?) wavelength I would expect there to be a large degree of capacity coupling between the transmitter and antenna – the right impedance to couple to a second harmonic."

**CORRECTION – COAX TRANSFORMERS.** My apologies for the minor drafting errors that crept into Figures 3 and 4 of "Coax impedance transformers and baluns" ('TT' July 2007, p79). Amended diagrams are here as **Figures 4** and **5**.

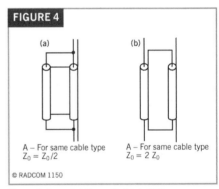

(a) A – For same cable type $Z_0 = Z_0/2$

(b) A – For same cable type $Z_0 = 2 Z_0$

Figure 4: Amended Figure 3 of July 'TT'

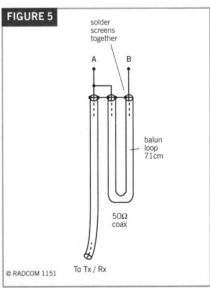

solder screens together

balun loop 71cm

$50\Omega$ coax

To Tx / Rx

Figure 5: Amended Figure 4 of July 'TT'

# Technical topics

*Pat Hawker continues his look at the diverse world of radio engineering*

**500kHz OFFERS EXPERIMENTAL OPPORTUNITIES.** I have to confess that I have not yet applied for a 500kHz NOV and have no practical experience of transmitting on this frequency. For most of the 20th Century, 500kHz (600 metres) has been a maritime calling and distress frequency and has been the means of saving literally thousands of lives.

This does not imply that I have never listened on 600m, both in its heyday and more recently. But so far I have not heard an amateur signal. Perhaps activity is too low; perhaps the noise from television and other switched-mode power supplies too high: perhaps I have done too little to make my receiving antenna less immune to electrical interference; perhaps my general coverage receiver loses sensitivity at low frequencies. I really must make more effort.

It was different in the 1930s when, growing up in a Bristol Channel coastal resort, I soon became interested in listening to the many Coast Stations and ships using 600 metres. And also the large liners using the 143 kHz (2100 metres) calling frequency with their Radio Officers rattling off stacks of passengers' telegrams – a lesson in good CW operating. There was also of course the R/T trawlers, lighthouses and coast stations sharing the 1.7MHz band (I was once politely asked to stop transmitting for a time by Burnham Radio as I was causing interference). Pre-war the calling and distress frequency was 1650kHz, later changed to 2182kHz, within the band 1.6 to 3.7MHz. The trawlers often provided a lesson in strong language!

For the 600 metre band (405-525kHz) I built a simple battery operated two-valve receiver with two switched variable tuning capacitors so that I could listen to both transmissions when ships and coast stations changed to their traffic frequencies. Spark was still being used on 600 metres by a few ships, and I still recall listening to an Egyptian-registered vessel, which had struck a mine, sending SOS on a rasping spark transmitter. (It had been intended that maritime spark should be prohibited by January 1940 but, due to WW2, this was postponed).

In daytime one could hear coast stations all over the UK as well as those in France, Belgium and inevitably the famous Dutch PCH coast station on LF/MF/HF, often sending its callsign as P and then four dashes (CH).

All this is now history, with 143 and 500kHz silent – and relatively little use of the HF maritime bands – HF, VHF, and satellite – with GMDSS (Global Maritime Distress Signalling System) replacing SOS. But why should we, as amateurs, seek to explore anew the former MF and LF maritime frequencies? A convincing answer can be found in an article "The ARRL 500kHz Experiment: WD2XSH" by Frederick (Fritz) H Raab, W1FR (*QEX*, July/August 2007, pp3-11). This long article is sub-titled "Twenty-one radio amateurs begin exploration of a historic part of the radio spectrum".

Of the 600m band, W1FR writes: "This band is of interest to radio amateurs for quite a number of reasons:

- Ultra-reliable emergency communications via ground wave.
- Unique propagation and noise environment, and
- Experimental work with antennas, modulation, and signal processing

The WD2XSH experimental licence allows a group of 21 American amateurs to explore this unique part of the spectrum, hopefully paving the way for a future amateur band. The two key objectives of the licence are:

- Demonstration of non-interference with other services, and
- Experimentation with regional ground-wave communication.

"Naturally, the participants also want to determine what kind of DX can be achieved using both normal CW and QRSS (very slow speed CW designed to be copied using a computer program) and this will add further to our understanding of the capabilities of this band."

The WD2XSH project, with the 21 stations (20W ERP, 505 to 510kHz, CW including QRSS) spread over the continental USA, and licensed until September 2008, is not the only current 600m experimental project. It is planned to make CW contacts, CW beacon transmissions and QRSS beacon transmissions. All the 21 stations use the call WD2XSH with an individual suffix number, eg WD2XDH/14. CW beacons use the sub-band 505.3 to 506.3kHz.

According to W4FR other special permits have included a number of US stations in the band 440 to 495kHz, although this facility was withdrawn at the request of the US Coast Guard. However, WA2XRM (Colorado) operated by W0RW has been operating on 480kHz with an ERP of up to 100 watts since 2004, and this has been renewed through 2009.

SM6BHZ is permitted to transmit with an EMRP (effective monopole radiated power) of 20W between 502.5 and 505.0kHz until 20th November 2007. German amateurs DJ2LFG and DK8KW have experimental licences as DI2AG and DI2BO between 505.0 and 505.2kHz.

Since March 1st, 2007, UK amateurs have been able to apply for an NOV to use 501 to 504kHz but with an ERP of only 0.1W. At the 2000 IARU Region 1 Conference, following an RSGB presentation, a working group was formed "to investigate the possibility of a frequency allocation of approximately 10kHz between 470 to 490kHz to investigate propagation and the use of new communication technologies." IARU Region 1 are continuing to co-ordinate international efforts in this area.

Although the efficiency of an average amateur antenna is low at these

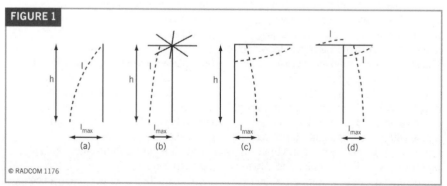

**Figure 1**: Voltage and current distribution along short resonant quarter-wave antennas fed against earth: (a) vertical monopole (if necessary inductively loaded at base although preferably loading should be two-thirds high); (b) umbrella; (c) inverted-L (d) T-antenna (top section may be two or more spaced wires). Adapted from "Marine Radio Manual"

# TECHNICAL TOPICS

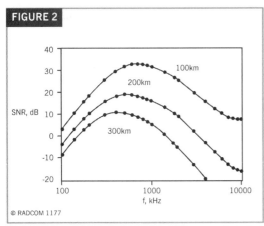

**Figure 2**: Ground-wave SNR as a function of frequency for typical amateur use. For communication with a vertical antenna of 40 to 50ft height, the best signal-to-noise ratio per watt of transmitter power output occurs in the range 400 to 600kHz. The SNR of a ground-wave signal depends upon a combination of (a) antenna gain; (b) surface-wave attenuation; and (c) atmospheric noise level. Predictions in the diagram are based on 15m (50ft) monopole with 16 30m radials; ground 0.01 S/m and ε = 10.1 watt delivered to the antenna; 1 Hz bandwidth. Median atmospheric-noise factor for US spring and autumn (70Db)/ median atmospheric-noise level (50%). (Source W1FR, QEX).

frequencies, the restriction to 0.1W ERP seems unduly low and may account for the relatively slow take up – but it is at least a beginning and a challenge.

A realistic 600m amateur band would offer unique opportunities for experiments with electrically short antennas (**Figure 1**); propagation of ground (surface) waves as well as night-time ionospheric paths; countering atmospheric and man-made noise; modulation and signal processing. W4FR shows that for practical vertical antennas (eg 40 to 50ft aluminium tubing) and average ground, 500kHz offers near optimum ground wave propagation in terms of SNR/watt over distances of 100 to 300km (60 to 200 miles). **Figure 2** shows that the frequency range 400-600kHz can provide optimum ground wave transmission for amateurs.

Ship stations had the advantage of near perfect 'ground', denied to land-based stations. There will clearly be an advantage in having a rural site of good ground conductivity. W4FR suggests that the limited real estate available to the average amateur precludes the installation of ideal grounding systems: "In poor soil, the common ground to the power grid and water system are likely to be more effective than ground rods and radial systems of limited size".

He notes that short top-loaded verticals appear to suffer significant losses from nearby trees and that this problem has not been well explored. In contrast, magnetic loop antennas can be placed in forested areas without significant losses, although magnetic fields flowing in the conducting ground cause losses much as do the electric fields surrounding a monopole antenna. As with 137kHz antennas, inductive loading and matching coils should have the highest possible Q. In general, 500kHz antennas will have higher radiation efficiency than can be achieved at 137kHz.

During much of daylight, propagation is by ground-wave giving a reliable coverage area, with skywaves blocked by D-layer attenuation. The D-layer vanishes from about dusk to dawn and the coverage area expands, but often with severe fading and phase distortion in that part of the coverage area where both ground and skywaves are received at roughly the same strength, an effect evident on medium-wave broadcast stations.

On modulation and signal processing, W1FR writes: "The limited bandwidth available at 500kHz makes narrow-band digital modes of great interest. BPSK and QPSK provide the lowest bit-error rates for a given amount of signal power. PSK-31 is therefore a natural candidate for this application.

"Improvements may be possible. PSK-31 uses sinewave shaping of its data pulses to provide synchronisation as well as to keep the signal in a very narrow bandwidth. This AM necessitates a linear RF-PA and results in an average transmitted power that is only half the peak-power capability of the PA.

"Minimum-shift keying (MSK) is a form of QPSK that (like PSK-31) employs sinusoidally shaped data pulses to constrain bandwidth. Delaying the modulation on the quadrature carrier by half a bit results in a constant-amplitude composite signal. The average power is the same as the peak power, and the signal can be amplified by a nonlinear PA. Establishing synchronisation is, however, more difficult. The Spectrum software contains an MSK-31 mode that otherwise follows PSK-31 protocols. MSK has been little used in amateur applications, so evaluation of its capabilities is certainly an area for investigation.

"Development of a synchronisation scheme suited to short amateur-type transmissions will probably be needed. This might be embodied in software tailored to this frequency range, much as the WSJT software is tailored to meteor-burst and EME communication.

"At 500kHz, signals may travel significant distances by both ground-wave and sky-wave propagation. In some cases, the resultant 'multipath' signal reception may make it difficult or impossible to receive phase-modulated signals such as PSK-31 ... this phenomenon has been observed in reception of PSK-31 signals from experimental 137kHz station WC2XSR/13 after sunset. If this proves the case for 500kHz, it may be necessary to develop more sophisticated signal processing or to revert to something like FSK-31 that does not use phase information.

"The more impulsive character of the atmospheric noise at 500kHz means that there is more opportunity to reduce its effective level through non-linear signal processing such as clipping and blanking. The presence of higher levels of man-made noise makes techniques such as noise cancellation of great interest."

## VARIABLE DIELECRIC CAPACITORS.

Brian Austin, G0GSF draws attention to an article "Variable Dielectric Capacitors" by Harry Brash, GM3RVL (*Sprat*, Issue Nr 131, Summer, 2007, pp10 to 12). This describes the construction of variable capacitors in which the rotary plate comprises a slice of dielectric material that rotates between fixed copper plates. This provides a variation of capacitance that is proportional to the dielectric constant of the rotary vane, although there will be an appreciably high minimum capacitance. Consequently, there is a significantly lower maximum/minimum ratio than with a conventional variable capacitor in which a metallic rotary vane rotates within fixed metal plates.

However, GM3RVL believes there are 'lots' of applications where this is not important, and that there are some inherent advantages. He writes: "Since it is the proportion of the gap which it is filled by the moving dielectric which matters, lateral movement of the moving vane along the shaft axis is relatively unimportant. Whereas, for the conventional variable capacitor, it is critical, due to the inverse relationship with the plate separation. This should make the mechanical construction less demanding. Another important advantage is that there are no moving electrical contacts to generate noise or to introduce losses in an ATU."

**Figure 3(a)** and **(b)** shows the basic form of the dielectric variable capacitor, GM3RVL also describes in detail the practical construction of an improved design (**Figure 3(c)**) which incorporates a number of refinements. He writes: "The project was aimed at constructing a variable capacitor suitable for a VFO. The capacitor plates were formed from pieces of double-sided glass fibre PCB with a single rotating vane of suitable high-dielectric material such as plastic moving between the PCB plates to vary the capacitance. Initial attempts were very crude. The outer copper surfaces were isolated close to the base by saw cuts leaving just enough copper for soldering to the baseboard PCB. The inner surfaces were similarly isolated by saw cuts unless one plate is to be grounded. The rotating dielectric vane was cut from an available piece of plastic sheet (2mm Darvic) and drilled to give a tight fit onto a Paxolin 1/4-in rod as the drive shaft. The two fixed plates were clamped together and drilled with clearance holes for the drive shaft. The capacitor was then assembled and lightly clamped with paper spacers between the two fixed plates and the dielectric vane. The whole assembly was then soldered along the isolated base strips to the PCB baseboard. The two paper spacers were removed and the variable capacitor was operational."

In a more finished design (Figure 3(c)) using 1.5mm thick Plasticard for the dielectric vane, GM3RVL achieved a capacitance variation of 84 to 94pF. Subsequently, using a moving vane made from glass fibre PCB with all the copper etched off, he obtained a capacitance variation of about 86 to 106pF, He considers this would often be suitable for tuning a VFO. He concludes: "It is surprising that this type of capacitor is not used more widely. Mechanically, they are quite simple to construct and, provided that the dielectric material has passed the 'microwave oven' test, it should be relatively loss free, despite the high voltages that can occur in ATUs – even at QRP. You have the added benefit of being able to design the capacitor to fit the task, rather than the usual requirement of designing the circuit to visit the available variable capacitor."

G0GSF comments: "Before reading GM3RVL's article I had never thought of a variable capacitor using the dielectric as the rotor. I was intrigued enough to do some calculations to satisfy myself as to the possibilities. I can see an immediate application for the technique as the tuning capacitor in an electrically small loop antenna – the so-called 'magloop'.

"As you may remember, in 'TT' (February. 2006. p73) Jack Belrose, VE2CV put his finger on the contact resistance of the rotor coupling mechanism as the reason why no one has ever managed to achieve, in practice, anything like the theoretical Q with these antennas, despite the claims made in some quarters. In theory at least, it should be possible to achieve a much higher Q from a variable capacitor if all friction- type electrical contacts are replaced by well-soldered joints to two sets of stationary plates. The insulator rotor of appropriate material, with very low loss and as high a relative permittivity (dielectric constant) as possible is then rotated between the plates to vary the capacitance. As GM3RVL makes clear high permittivity material is attractive. Of more importance though is the loss tangent, which should be as low as possible. This certainly sounds like an idea worth exploring. "There is, I would suggest, a basic problem in making variable dielectric capacitors for use in loops required to cover more than one amateur band, on account of the high minimum capacitance and restricted capacitance variation. GM3RVL developed his design primarily for use in a VFO or QRP ATU. I suspect his designs might require modification for use even in a single-band loop antenna to be used with, say, a 100W transmitter. VE2CV has made it clear that in such circumstances the tuning capacitor has to withstand very high voltages and very high RF currents. Nevertheless, I would agree with G0GSF that the variable dielectric capacitor has interesting possibilities.

SSB TRANSCEIVER BASED ON THE 7360. The August 'TT' described how the RCA7360 beam-switching tube (and RCA's earlier 6AK8) was introduced in 1960 as a high-level balanced for SSB transmitters, but also offered the possibility of a major advance in receiver design when used as a front-end mixer. It was noted that the 1963 QST article by W2PUL, in effect pre-viewing the Squires-Sanders SS-1R receiver, influenced other designers, but came too late to stem the terminal decline of valves in favour of solid state circuitry although this provided inferior performance for many years.

I did not know, when I wrote that 'TT', that Adam Parson (as ZS6XT but now VA7OJ/AB4OJ), had published two articles in Radio ZS, October 1954: "The RCA-7360" p11 and "An SSB Transceiver" pp12-18. The second article described in detail the construction of a 14MHz SSB transceiver designed and built around a 7360. As he puts it in a recent letter: "This rig expanded on the Squires-Sanders concept in that a single 7360 had a dual role as receiver front-end mixer and as transmitter balanced-modulator. The transceiver's performance was surprisingly good for its era."

In Radio ZS he wrote: "This article will describe a 14MHz SSB transceiver which was designed and built by the author in order to illustrate the superior performance of the RCA-7360, both as a sideband generator and as a receiver front end…"

Figure 4 shows the block diagram of the SSB/CW transceiver (PSU not included). At 1964 prices it would have cost about 75 Rands, of which about half (R36) represented the McCoy 48B1 9MHz crystal filter. Maximum PEP output was 65W for 100W PEP input. Power requirements: 12V AC for all heaters except those of V1 and V6; +225V DC for low-level stages; +600V for the QQE05/40 (US 5894) PA; -50V DC for bias; -24V for relays; and –12V DC for transistor circuits and for the heaters of V1 and V6. All were provided from a single unit on a separate chassis.

The performance figures for this 1964 home-built model remain impressive: Receiver sensitivity 0.6µV for 20dB (S+N)/N at 14.2 MHz (with no RF amplifier); adjacent channel selectivity –6dB at 1.35kHz and –55dB at 1.95kHz from centre of filter passband. Image rejection better than –50dB, IF rejection

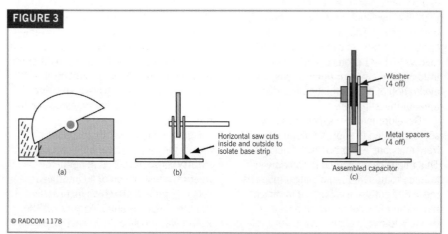

**Figure 3**: GM3RVL's variable dielectric capacitors: (a) and (b) prototype; (c) working design. For full details including design formulae see Sprat, (Journal of the G-QRP-Club) Issue 131, Summer 2007.

# TECHNICAL TOPICS

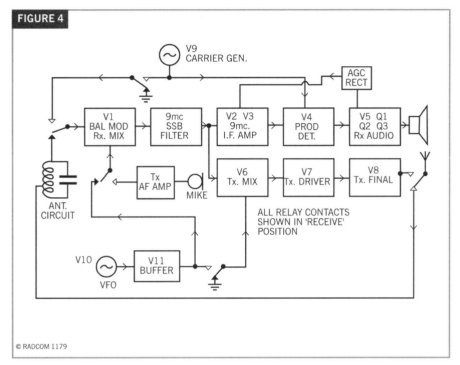

**Figure 4**: Simplified block diagram of the 14MHz SSB/CW transmitter using a RCA-7360 as both receiver 1st mixer and transmitter modulator as built and described by Adam Parson, now VA7OJ/AB4OJ in 1964 while ZS6XT. Receiver: V1 7360 (mixer); V2 EF183 (1st IF); V3 EF89 (2nd IF); V4 ECC82 (product detector); V5 ECC81 (AF amplifier); TR1, 2, 3 transistor AF amplifier and output. Transmitter: V1 as modulator; (V6) ECC81 (mixer); V7 EL83 (driver); V8 QQE )06/40 (power amplifier). Transistor speech amplifier. Joint 5-5.5MHz VFO: V10 EF91 (oscillator); V11 ECC82 (buffer). Break-in CW facilities etc.

−80dB. Cross modulation: a 100mV signal at the input causes just perceptible interference with a 2µV signal about 23kHz away. *Transmitter* – power output 65W into 50 ohms; carrier suppression −60dB; unwanted sideband suppression −55dB; spurious signals −50dB or better.

As VA7OJ puts it: "The circuit topology drastically reduced the valve count. The transceiver's performance was surprisingly good for its era."

## MIXERS – HOW GOOD IS GOOD ENOUGH?

In connection with the August 'TT' notes on the progress of mixers, Peter Chadwick. G3RZP, has commented on several points. First he corrects my mistake in saying that Plessey Semiconductors (who made the SL6440 high dynamic range mixer chip), was taken over by Siemens. He writes: "Plessey Semiconductors did *not* become part of Siemens. When GEC and Siemens bought Plessey, Plessey Semiconductors fell firmly into the GEC part, combined with Marconi Electronic Devices Ltd (MEDL) as GEC-Plessey Semiconductors (GPS) ... becoming Zetex, and later Zarlink Semiconductors after being spun off from GPS.

"Incidentally, July saw the 50th anniversary of the Cheney Manor plant in Swindon, which was Europe's first purpose-built semiconductor facility. It was originally a joint effort with Philco for making surface-barrier transistors [the first type of transistor capable of working up to VHF–G3VA]. It is Zarlink's only fabrication plant, It works on advanced bipolar processes; these days practically no standard products, all custom designs.

"In regard to mixers, I think we have got to the stage where mixer performance is not much of a limitation any more; it is now oscillator phase noise that is limiting performance. As noted previously in 'TT', I am not convinced that, for most amateurs, there is actually a need for dynamic ranges in excess of 100dB. One arrangement I intend to play with sometime is a mixer with a twin-triode acting as a switch, with the cathodes driven in parallel, cascode style, from a high gm triode with push-pull local oscillator drive to the grids, and push-pull output. My intention is to drive it hard to discover its IMD performance. Obviously of no real practical use but an interesting project!

"Personally, I can't summon up much enthusiasm for digital SDR receivers, maybe because that is too close to work, although the difficulty with SDR seems to appear when extremely low power consumption is required. Such as a fully synthesised 800MHz transceiver operating down to 1.1V and drawing under 1.5mA, for hearing-aid applications. With the increasing degree of specialisation in analogue ICs these days, I can see it getting harder and harder to use ICs in homebrew projects, so we may well be forced into doing things digitally. Alternatively, of course, it's back to discrete transistor design."

## SUPER-LINEAR CRYSTAL FILTERS.

The June 'TT' included a detailed discussion on the dynamic range of crystal filters with non-linearity evident on very weak and very strong signals. This included a contribution by G3UUR who showed that surface contamination of the crystal plate during manufacture was the prime problem. He added: "I am surprised that anomalous non-linearity is still a big problem in modern crystal filters. Manufacturers have the knowledge and means to keep the surfaces of their resonators clean and damage free these days, but the cost of doing so is not economical for them."

The validity of G3UUR's comments is shown by the measurements made by Martein Bakker, PA3AKE on his latest 9MHz roofing filter using crystals made for him by the specialist German firm QT GmbH, at the suggestion of Helmut, EA5GNA, via a friend of his who works for QT.

Colin Horrabin, G3SBI writes: "I asked PA3AKE to make IP3 measurements at different two-tone spacings going out from 500Hz. These crystals are super-linear in a four-pole filter and have perfect IP3 behaviour. He reached 46dBm at 2kHz spacing which, with the 5dB loss of the mixer, would give over 50dBm IP3 at the antenna. The spectrum analyser plot shows IP3 "at a 500Hz tone spacing" of 39dBm. This is exceptional performance; you could not follow it with an amplifier and CW filter without degrading the 'in band' performance. However, as the QT crystals have such a low resistance, a 500Hz roofing filter could be used on CW immediately after the mixer, with very low insertion loss before any amplification to give exceptional results."

For the two-tone 500Hz test the input level is −0.5dBm and IP3 is +38.9dBm. Other measurements by PA3AKE include out-of band frequency with the first figure that measured at the lower side of the filter and the second at the upper side: 1kHz 42.5 – 38.6; 2kHz 46.0 – 46.5; 5kHz 49.5 – 48.5; 10kHz 53.5 – 50.5; 20kHz 53.5 – 52.5; 50kHz – 53.0. PA3AKE writes; "Interestingly the upper side, that happens to be the sharp edge of the filter, has a bit more difficulty in reaching the really high values. From 2kHz and onwards this filter is a killer!"

# TECHNICAL TOPICS

PAT HAWKER, MBE
37 DOVERCOURT ROAD,
DULWICH, LONDON SE22 8SS

G3VA

# Technical topics

*More on oscillators, mixers, RFI and a novel way of distributing audio on a train*

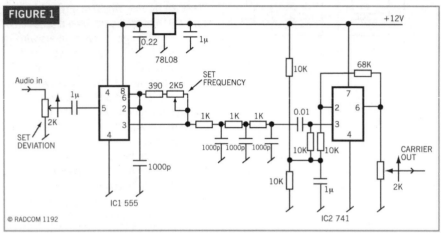

**Figure 1**: G3KKD's 250kHz FM generator developed as part of a passenger train public address system distributing speech and music over AC power leads to loudspeaker amplifiers in the carriages.

**250kHz FM GENERATOR.** Ian Waters, G3KKD, recently needed a source of frequency modulated carrier on 250kHz as part of a system for transmitting public address and background music along the 240V 50Hz mains cable that runs the length of trains on the Ffestiniog and Welsh Highland Railways (but could find other applications).

G3KKD writes: "The FR is another of my interests, and I have been a volunteer worker on this railway for 40 years. I originally built a conventional LC oscillator modulated by either a variable reactance modulator or varicap diodes but then began to 'think outside the box' and came up with the circuit shown in **Figure 1,** which may well be an 'original'.

"A 555 timer IC is configured to run at 250kHz and provides a square wave output with an equal mark/space ratio. This is then integrated into a near sine wave and amplified. The 741 op-amp just about works at this frequency but a more modern IC op-amp would be better.

"This circuit works well with adequate modulation linearity and audio frequency response. The frequency stability is not good but in this application this does not matter as the PLL demodulators in the receivers used will track the carrier over quite wide limits.

"There is nothing unusual about the rest of the system, which superimposes the carrier on the mains, picks it off in each carriage and demodulates it to feed loudspeaker amplifiers."

**RFI TO VEHICLE ELECTRONICS.** The effect of strong RF fields on the many microprocessor systems, as fitted in many modern cars, was noted last year ('TT' January and May 2006) in connection with the use of high-power amateur mobile transmitters, or when passing radio broadcast stations or radar bases, etc. This topic continues to attract attention, with further reports of vehicle breakdowns or other problems apparently caused by RFI.

Martin Rolls, M3JZI, sent a cutting from a local paper headlined "Fields of Screams for irate motorists – stranded woman forks out £45 to get immunised car started". This reports an incident where a key fob seems to have been scrambled by RF radiation from a mobile phone antenna mast sited close to a retail car park, backed up by the suggestion that this was a frequent occurrence. An Ofcom spokesman is quoted: "This is certainly not an unknown phenomenon. It is usually that the key fob and the mast are operating on similar frequencies and therefore interfering with each other. If members of the public have concerns about a particular mast we will take it up."

Ian Brodwell, G4EAN, writes: "With reference to 'More on RFI to Automotive Electronics' ('TT' May 2006), I enjoy reading about interactions of electronics systems, such as radio signals and car electronics. I have encountered this myself only once – when, with my handheld GPS, I was on a train halted to the south-east of Manchester. I do not know what electronic system(s) were nearby but my GPS (used to keep track of my journey) was indicating a speed of 1500mph or more!

"Recently, browsing through the handbook for a Mercedes-Benz E220 car, I found the following: The transmit output of a mobile phone or two-way radio must not exceed the following maximum transmission powers: Short-wave, below 50MHz – 100W. 4m waveband – 20W. 2m waveband – 50W. 70cm waveband – 35W. 25cm waveband – 10W. G4EAN adds: "It was interesting to see that the handbook singled out the amateur radio bands (25cm may have been a misprint for 23cm)."

**50 YEARS – LC OSCILLATORS TO DDS.** Fifty years ago, at the dawn of 'TT', amateurs were still searching for better LC VFO stability for HF receivers and transmitters. The classic Hartley, Colpitts and Franklin circuits had sprung many derivatives: the ECO (electron coupled oscillator/doubler), the Clapp/Gouriet, the Vackar, the Seiler, etc. Other techniques included careful temperature compensation, voltage (and occasionally heater-current) regulation, separation of inductors from heat sources, use of crystal-controlled first mixer oscillator with tuneable or broadband IF in double-conversion receivers, the mixer-VFO (an elementary form of frequency synthesis) with crystal and lower-frequency LC oscillators. All these techniques were well established using valves by 1958.

There were innumerable items in those early 'TT's on how LC oscillators could be made more stable, or adapted for bipolar transistors, MOSFETs and later ICs. But in the professional arena, other ideas were stirring and reported in the professional journals or at national or international conferences, not infrequently first reaching the UK amateur press as items in 'TT'.

The historic importance of oscillator developments has received relatively little attention. A notable exception is the excellent contribution to the 1995 international conference "100 Years of Radio" by Dr Mike Underhill, G3LHZ: "From quartz to direct digital synthesis" (IEE Conference Publication No 411, pp 167-176).

While by the 1950s, professional equipment was often capable of good short-term and long-term stability, the growing use in the 1950s and 1960s of 'up-conversion' designs with single-span VHF oscillators, although eliminating the need for band-switching the VFO, introduced to many of us the importance of oscillator phase noise. Up-conversion for an HF receiver implies an IF that is the sum (rather than the difference) of the signal and intermediate frequencies. It thus requires an oscillator working at VHF, typically covering say 42 to 70MHz to provide a fixed first IF of 40MHz for a receiver covering 2 to 30MHz. The difficulty of providing a stable and drift-free LC VHF oscillator encouraged the development of various forms of 'frequency synthesis'. The original high-cost crystal-bank valve synthesiser gave way to the much less expensive solid-state phase-locked-loop (PLL) synthesiser, although usually at the cost of significantly more phase noise.

Until the late 1950s early 1960s few of the standard texts on oscillators even mentioned the presence of oscillator sidebands or ways of reducing them.

A few dates seem called for. The term 'frequency synthesis' had been coined by Plessey engineers during WW2 in connection with the development of a high stability signal generator using a large bank of crystals. Such an approach was applied to a few professional receivers and transmitters in the late 1950s, notably by the Marconi Company, but the solid-state PLL synthesiser was soon a strong rival. I recall being given the chance of trying out a 10W solid-state Hughes manpack in early 1963 when one was brought to the UK for demonstrations by a Hughes professional design engineer who was also a keen amateur. I recall we worked G3PLX, then on the Isle of Wight, on the 3.5MHz band. The manpack provided 1kHz-spaced frequency settings over the HF band with four drum wheel switches. Methods of stabilising LC oscillators in existing receivers as an alternative to frequency synthesisers included the Racalator, developed by Keith Thrower of Racal (see IEE Conference Publication No 31, May 1967, pp122-128) and for amateurs the "Huff & Puff" stabiliser developed by Klaas Spaargaren, PA0KSB and first described in 'TT' (1973).

Dr Wadley began work on his drift-free triple-mixer loop techniques during WW2; it finally reached the professional receiver market in the form of the Racal RA17 in 1957.

Oscillator noise, primarily at UHF, had been discussed by W Edson in his book "Vacuum Tube Oscillators" published in 1953. But its importance to HF receivers, particularly with up-conversion, was not fully appreciated until the late 1960s. I first drew attention to the practical significance of oscillator noise in a series of 'TT' items in 1968 with reference to HF parametric up-converters. These notes were followed two years later by an excellent article "Oscillator noise and its

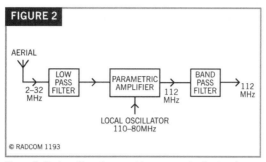

Figure 2: Basic outline of parametric up-converter front-end

effect on receiver performance" by Barry Priestley, G3JGO (*Radio Communication*, July 1970), although he provided no source references. The term "reciprocal mixing" was later coined by B M Sosin, a Polish-born engineer working for the Marconi Company: see 'TT' January 1971.

G3LHZ points out that the concept of DDS was mooted by V Manassewitz in 1980 but its implementation delayed until the 1990s when IC technology had advanced. With a crystal clock the phase noise (jitter) can be very low, but there remained two major deficiencies: a high level of spurii due to DAC inadequacies and, for battery equipment, the power consumption. It has been the practice to combine DDS with PLL in hybrid designs, although the latest generation of DAC chips open the way, as G3SBI has pointed out, for DDS without PLL.

Colin Horrabin, G3SBI, has compiled a useful personalised review (including diagrams) of the progress of oscillators over the past 50 years, including his pioneering work on DDS (direct digital synthesis) in 1986 even before the first DDS chips became available, and his development of the low-noise two-tank LC oscillator. I hope to find space to include at least part of this soon.

**PARAMP UP-CONVERSION HF REVISITED.** Meanwhile it seems worth adding to the recent review of 50 years of mixer progress ('TT', August 2007), first, a note on the largely forgotten low-noise paramp up-conversion mixer that could handle volts of RF input, and, second, advance details of a promising new high-performance mixer developed by RA4LZ.

In my August review I failed to mention the paramp up-conversion mixer as this (as far as I am aware) has never figured in any marketed amateur radio receivers despite its theoretical exceptional performance. I first drew attention to it in January 1968 (further details in the February issue) when I reported that a series of professional/military receivers by the American firm Avco Electronics was using paramp up-converters in conjunction with full frequency synthesis and with first IFs in the region 120 to 180MHz. The local oscillator of a paramp is usually termed the 'pump'.

These reports led to a long letter from Walter Schreuer, K1YZW/G3DCU, who wrote that while at National he may have been the first person to demonstrate the use of a parametric device for the front-end of a tuneable HF receiver (US patent No 3,063,011 of 1962): see 'TT' May, 1968. Collated material on paramp up-converters appeared in *Amateur Radio Techniques* (3rd, 4th and 5th Editions 1970, 1972, 1974).

To quote briefly: "The basic idea is to obtain an HF receiver which is virtually free from the usually disastrous effects of strong off-channel interfering signals, such as blocking, cross-modulation etc. If the desired signal together with the strong signal is converted linearly (in amplitude) to a fixed frequency, the strong interfering signal can be filtered out. A conventional mixer would require excessive local oscillator power which must always be large compared to the signal plus interference power. and is quite noisy. The paramp converter, (**Figure 2**) consisting of a varying capacitor needs only volt-amps, not watts, and being reactive, is almost noiseless. The non-inverting 'up-converter' is the only circuit configuration which is stable with changing source (antenna) impedance and also provides some gain.

"The gain will be larger the higher the 'first IF'. All up-converting receivers suffer from the disadvantage of strong signals at sub-multiples of the 1st IF causing spurious responses. This is particularly serious in a receiver with a very large linear (or dynamic) range, so that a very high 1st IF is again desirable and also helps in minimising local oscillator radiation from the antenna. Against the use of a high IF is the need for a highly selective (crystal) filter with low insertion loss and low noise amplification. As a compromise, in the first receiver design, 112MHz was chosen, being over three times the highest signal frequency of 30MHz....

"I have obtained a useful linear signal input range of about 136dB. In more exact terms, the minimum useful input (10dB above noise) was 0.35µV emf behind 50 ohms, with a detector bandwidth of 3kHz and maximum input of 2 volts which caused a 1dB departure from linearity. That is a noise figure of 6dB. Theoretically, there is no limit to the dynamic range that can be obtained with a single RF signal. However the real value of a large dynamic range is in combating large amplitude off-

# TECHNICAL TOPICS

channel interference, and here we run into a severe limitation which is the noise power associated with the local oscillator. ..."

K1YZW/G3DCU showed that at the high IF (eg 112MHz) of a parametric mixer the oscillator noise power decreased the dynamic range when the interfering signal was spaced even at 100kHz from the weak wanted signal. This was further reduced by the noise sidebands of the strong interfering signal. The parametric upconverter can be considered as a cross between a balanced modulator and a coupled pair of circuits: see **Figure 3**.

Thus although the HF paramp upconverter was, at least theoretically, almost the perfect front-end mixer in offering a low enough noise figure to dispense with the need for pre-amplification, combined with a linearity that could cope with volts of incoming signal, in practice it seems to have soon been superseded by the progress in doubly-balanced hot-carrier diode and JFET mixers as described in the August 'TT'.

However, it played a role in bringing to notice the greater vulnerability of HF up-conversion single-span receivers to oscillator noise than the traditional down conversion receiver configuration (a problem that still affects PLL frequency synthesisers). Another problem is the need for a high dynamic range VHF crystal roofing filter. But one wonders whether in view of the progress in reducing oscillator/synthesiser noise it might be worth revisiting the paramp mixer.

## OSCILLATOR NOISE SIDEBANDS.
In regard to oscillator noise, the 1970 two-page article by Barrie Priestley, G3JGO, referred to above still provides a clear

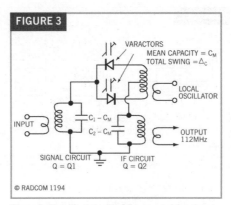

**Figure 3**: Simplified circuit of the parametric mixer as described by Walter Schreuer, K1YZW/G3DCU in 'TT' in 1968

explanation of this mechanism and its effect on receiver performance (later called 'reciprocal mixing'). To quote very briefly from G3JGO's introduction to the mechanism of noise sidebands and 'flicker noise' and his conclusion:

"Any sort of modulation of a receiver local oscillator will degrade receiver performance. 50Hz hum modulation will appear as hum in the audio output, while a sideband at, say, 10kHz would mix with an unwanted signal 10kHz away from a desired one and result in serious interference with a weak signal.

"Obviously, one does not deliberately modulate the local oscillator or let it squegg at 10kHz, while supply frequency hum can be removed by adequate smoothing of the supplies. However, when these obvious sources have been removed there remain noise sidebands which are inherent in any oscillator and are now becoming the limiting factor in receiving weak signals [notably in the presence of strong unwanted off-frequency signals].

"Any oscillator is basically a selective amplifier with positive feedback, ie a Q multiplier. If the gain is sufficient the noise voltages at the centre frequency build up into a continuous sine wave till limiting or AGC reduces the gain to maintain a constant level. Noise voltages near the centre frequency also build up, but the limiting level depends on how near the frequency is to the circuit centre frequency. The result is that the oscillation is accompanied by a roughly triangular spectrum of noise (see **Figure 4**) rising from the flat noise present in any device. The width of the spectrum depends on the tuned circuit bandwidth and the level on the operating conditions. These are factors which the designer can choose.

"On the HF bands .... efforts [in 1970] to produce a receiver capable of withstanding an interfering signal of 100mV 10kHz away from a $1\mu V$ signal (ie a ratio of 100dB) can be frustrated by the noise from a poorly designed VFO or synthesiser!

"*Conclusion:* Oscillator noise is becoming the limiting factor in receiver design due to [strong signal] interference in the HF bands. Amateurs can apply with advantage the basic principles of low noise oscillator design even though the actual measurement of oscillator noise is rather difficult."

A listing of the basic requirements for low-noise oscillators. as taken from Dr Ulrich Rohde, DL2LR/KA2WEU in his 1993 book "Digital PLL Frequency Synthesisers" and largely reprinted in QEX as "All about phase noise in oscillators" Part 1, December 1993, was summarised in 'TT' January 1995 (see also TTS 1995-1999, pp9-11). This item also includes material on reciprocal mixing and notes by G3SBI on his method of measuring phase noise. I would stress that the full benefits of low-noise oscillators depend on their use on both transmitters and receivers.

## RZ4HK's TRANSMISSION LINE TRANSFORMER (TLT) MIXER.
I7SWX has drawn G3SBI's attention to a new high-performance mixer based on broadband transmission line transformers devised by Gannardy Bragin, RZ4HK, as presented on web Page www.cqham.ru/trx86.49.htm (3 pages of Russian text plus 6 pages of enlarged oscillograms, analysis and construction of the transmission line transformers). The performance claims made by RZ4HK have been the subject of controversy among Russian amateurs but it is already clear that the TLT mixer represents an important development.

G3SBI passed the web details of RZ4HK's TLT mixer to Martein Bakker, PA3AKE, who has built and tested the

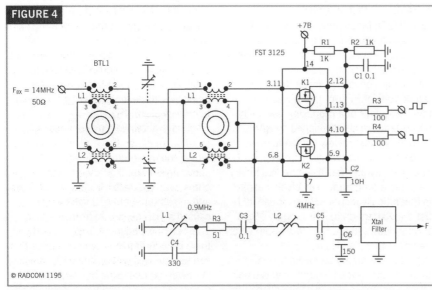

**Figure 4**: RZ4HK's proposed new high-performance Transmission Line Transformer (TLT) HF mixer

mixer as described by RZ4HK and is also trying various modifications etc including some suggested by G3SBI. Colin has also arranged for RA4HK's Russian text to be expertly translated into English by Professor Vasilii Zakharov, G0KGD. G3SBI and PA3AKE are already firmly convinced that this 'TLT' configuration may prove to be a winner, offering a doubly-balanced mixer with very low conversion loss using only two FET switches. RZ4HK used half of an FST3125 four-switch device but it seems likely that the later two-switch FSA3157 would be a better choice. Of particular merit is the low attenuation. While PA3AKE has shown that performance peaks on 7MHz, falling off on lower and higher frequencies, G3SBI believes that this can be overcome, and additionally, with some modification, that good performance could be achieved at VHF and above, making the mixer suitable for up-conversion receivers.

It is intended this month to explain the principles on which the TLT mixer works, quoting from G0GKD's translation, together with a brief look at some preliminary results already achieved by PA3AKE on a TLT mixer based on RZ4HK's circuit.

"Data handbooks show that in broadband transmission line transformers (BTLT1 and 2 see, **Figure 5**) the windings are in the form of homogenous two-conductor lines. In such systems the dominant form of energy transmission is the direct electromagnetic coupling between the windings. In signal transmission along the lines, power is transmitted from the source connected to the transformer input to the load at the output. It is a feature of such a process that the currents at any arbitrary cross-section of each line are equal in magnitude but opposite in direction. In this case the regime that obtains in matching the load to the lines is that of a travelling wave.

"The main difference between a BTLT and a classical impedance transformer is that the BTLT has the circuit property of zero damping within the working frequency range. The characteristics of a BTLT may be improved, in particular symmetry [balance], by connecting an additional matching line to it. An example of a classical form of BTLT is a transformer consisting of two lines connected in parallel on one side but in series on the other. Such a transformer is the basic element used in the proposed mixer shown in **Figure 4.**

"At its input BTLT1 has series connected windings of two 25Ω lines. Hence the input and output impedances of BTLT1 are 50Ω and matched to the input impedance of BTLT2 which uses 50Ω lines. The free output ends of BTLT2 are connected via electronic gates [FET switches] K1 and K2 to ground. The IF signal is derived from the transformed signal at the output ends of the lines connected together.

"To obtain a travelling wave in the mixer circuit, a broadband matching load (diplexer) is connected at the output, and it is to this that the basic (crystal) selectivity filter (X1) is connected.

"Consider the instant when gate K1 is closed and K2 is open. The input ends 1-3 of line L1 of BTLT2 are connected to output connections 4-6 of BTLT1. But line L2 of BTLT2 has ceased being a transmission line because K2 is open. For simplicity, consider that winding 7-8 of BTLT2, at that moment, does not exist. and BTLT2 is then a classical symmetrising [balancing] transformer, with an additional winding or phase-compensating line with 50Ω input and output impedances. At the next moment. winding 1-2 of BTLT2 is disconnected, and now L1 no longer has the property of a transmission line, while winding 3-4 becomes a phase-compensating line. In this case BTLT2 remains all the time a symmetrising transformer with a degree of asymmetry tending to zero. But note that at that moment a phase reversal of 180° has occurred in the output which is indeed essential for normal operation of the mixer.

"Thus, by using BTLTs plus some additional gates we have obtained a well-matched RF mixer having excellent symmetry [balance] with all parameters as good as or better [on 14MHz] than a modern H-mode mixer. ..."

RA4HZ has also made computer simulations of single, double and triple-

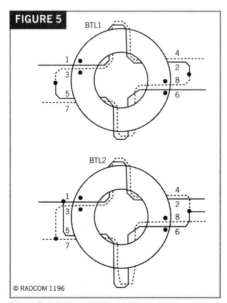

**Figure 5**: Winding details of the broadband transmission line transformers. See text.

balanced variants of this circuit with the results shown on three screen displays on the web. As noted above, RA4HZ's performance figures have been challenged by a number of Russian amateurs.

G3SBI comments: "The circuit of Figure 5 appears to be tweaked only for the 14MHz band with a claimed performance of 4.5dB conversion loss, 150dB blocking. 120dB IP3 dynamic range and 80dB isolation of both signal input and oscillator input. The initial results obtained by PA3AKE for an all HF (3 – 50MHz) version give an insertion loss on 50MHz of 4.3dB – a truly excellent figure since the theoretical loss for a perfect switching mixer is 3.9dB. The IP3 on 7MHz (52dBm) is the same as that achieved by PA3AKE for an H-mode mixer [see 'TT', May 2007] but falls off on the other bands; isolation is quite poor, being below 36dB." Colin and Martein are convinced that isolation can be much improved. Work continues, and PA3AKE has provided G3SBI with some interim results.

G3SBI continues: "In principle the TLT mixer is a much better idea than the H-mode mixer because it should give good results at VHF, UHF and even SHF using strip-line techniques and the right FET switches. So PA3AKE and I are not prepared to write it off. The TLT works like the H-mode mixer in that the FETs switch to ground and it is the transformers that do the mixing. We think that the transformers as made by RZ4HK are the problem [for the fall off of performance above and below 7MHz] and PA3AKE is testing a four-transformer version using 1:1 transformers as baluns."

HERE & THERE. Errors crept into Figure 3 in 'TT', September 2007, page 79 (Controlled Carrier AM): C1 is 10nF not 10μF. R2 is 250Ω not 25Ω. Y1, Y2 cathodes are connected to ground (missing dot).

August 2007 was the lowest month yet of Sunspot Cycle 23. Solar flux was below 70 on 21 days (lowest 67 on 2 days, 68 on 8 days, 69 on 9 days, highest 72 on 4 days). *[September was even lower – Ed]*

Thanks to Jim Lee, G4AEH for providing a CD record of the BBC World Service programme marking the 100th Anniversary of the first radio broadcast as made by Reginald Fessenden on December 24th, 1906 (see July 'TT' pp79-80). As a freelance broadcaster who joined World Service in 1997, he notes that there is now little interest at Bush House in HF broadcasting as the BBC concentrates on FM relays and the internet.

# TECHNICAL TOPICS

PAT HAWKER, MBE
37 DOVERCOURT ROAD,
DULWICH, LONDON SE22 8SS

G3VA

# Technical topics

## 50 YEARS OF AMATEUR ANTENNAS

IS THERE NOTHING NEW IN ANTENNAS? By April 1958, when 'TT' first appeared, the fundamentals of antennas had been well established: the doublet, the dipole and 'folded-dipole', the vertical monopole and 'ground-plane', the 'slot' and 'skeleton-slot', driven (W8JK) and parasitic 'Yagi' close-spaced beams, the helical (axial and normal modes), the rhombic, the Vee beam, the receiving loop antenna, the 'cubical quad' and single-element 'quad loop', the Zepp', the G5RV, the W3EDP, the T2FD the 'Windom' (off-centre-fed dipole), the parabolic reflector dish antenna, 'phased arrays, Marconi inverted L tuned against earth or with counterpoise, the Smith chart etc. It must have seemed then that there was little more to be written on amateur antennas!

Yet, between 1958 and 2000, as shown in *Antenna Topics*, (RSGB, 2002), more than 375 pages of 'TT' were devoted to this topic, providing many new variations such as the HB9CV, Delta loop, ZL-special, ZS6BKW etc. Significant *new* basic developments may have been relatively rare, although I would single out the work of the late Les Moxon on multi-band Quad elements and his endorsement of the VK2ABQ end-coupled wire beam from which has emerged the Moxon rectangle. The professionals developed the 'log-periodic' broadband beam; fractal antennas, the 'active' receiving loop; and the Method of Moments from which emerged the NEC computer programs that now seem to dominate antenna design. The US Military showed the value of small tuneable transmitting loops for jungle warfare leading to their adoption for amateur operation at restricted sites despite their relatively low radiation efficiency.

Less happily, 'TT' became involved in seemingly endless controversy over the efficiency of electrically small loops and other electrically-small transmitting antennas, although my scepticism of some claims appears to have been fully justified (see the contribution from VE2CV below).

Yet there is little doubt that for the newcomer (and even some 'oldies') the antenna remains a mystery that now discourages construction of simple wire antennas supported by trees etc. Consequently, it seems worthwhile to use some 'TT' space to review again some basics, even where these have appeared before in 'TT' or *A Guide to Amateur Radio* (RSGB, out of print):

"Whenever the intensity of an electric current passing through a wire changes, some amount of energy is radiated into space in the form of radio waves. *Any* piece of wire or conductive metal tubing or sheeting will radiate a signal if one can couple RF energy into it. A short piece of wire will radiate energy less readily than a piece one half-wavelength ($\lambda/2$) or more long. But this is primarily because it is more difficult to couple energy into and out of a short length of wire and there will be an increases in energy losses (IR or 'current times resistance' losses) rather than because of any fundamental requirement that an antenna element should always have a (resonant) length of $\lambda/2$ or more.

"Since an antenna radiates power when there is current flowing in it, it can be thought of as possessing *radiation resistance*. This is *not* the same as its 'characteristic impedance' (or 'feed-point impedance') which will vary along its length. The feed-point impedance comprises capacitive or inductive reactance and resistance. At the centre point of a resonant half-wave dipole the impedance is low (about 70$\Omega$ depending on height above ground) and purely resistive, with no inductive or capacitive reactance; considerable use is made by amateurs of this characteristic. The two ends of a resonant half-wave dipole are also purely resistive, but with a feed-point impedance of several thousand ohms; this is used in the 'Zepp'."

In practice an amateur antenna *system* should normally aim at being resonant at the frequency of operation. This does *not* imply that the radiating wire or element itself need always be exactly $1/2\lambda$ long, since the system may be "loaded" with lumped inductance (or its electrical length reduced by means of series capacitors) or by using the earth (or radials or a counterpoise) as part of the resonant antenna systems. Note that the physical length of an element will be less than its electrical length and an end correction of some 5% is usually required for a half-wave dipole: see **Table 1**.

Where an element is itself $1/2\lambda$ long, or a multiple of this, then its operation does not rely upon the presence of the earth. A radiating element should preferably be well removed from any energy absorbing materials, such as buildings, trees, metal fences, metal drainpipes, house electrical wiring, etc. Its height above earth (which acts as both an absorber and reflector of radio waves) has an important bearing on its operation, although for most practical purposes the "higher the better". The electrical conductivity of the earth below and surrounding the antenna also affects its operation, and is especially important at the lower amateur frequencies where the "earth" is likely to form an important part of the antenna system. There may also be significant losses in an ASTU (antenna system tuning unit) or ATU Transmatch unit unless designed with suitable low-loss components of correct value(s).

Some modern transceivers have built-in automatic ATUs but it should be noted that these usually have restricted range and cope only with a limited transmitter/feeder mismatch (SWR). Solid state transmitters normally incorporate automatic power reduction that begins to act when the SWR exceeds about 1.8 – 2.

The post-war use of SWR meters connected between a coaxial feeder line and transmitter output has led to

**TABLE 1: Antenna resonant lengths**

| Frequency (MHz) | $\lambda/2$ (492/f) | End correction (approx. 5%) | $\lambda/2$ resonant length (468/f) |
|---|---|---|---|
| 1.825 | 269ft 6in | 13ft 2in | 256ft 5in (78.21m) |
| 1.900 | 258ft 11in | 12ft 6in | 246ft 4in (75.12m) |
| 3.525 | 139ft 7in | 6ft 11in | 132ft 9in (40.5m) |
| 3.650 | 134ft 9in | 6ft 6in | 128ft 3in (39.11m) |
| 7.020 | 70ft 1in | 3ft 6in | 66ft 8in (20.33m) |
| 10.125 | 48ft 7in | 2ft 5in | 46ft 3in (14.1m) |
| 14.050 | 35ft 0in | 1ft 9in | 33ft 4in (10.16m) |
| 14.200 | 34ft 8in | 1ft 9in | 32ft 11in (10.05m) |
| 18.100 | 27ft 2in | 1ft 4in | 25ft 10in (7.89m) |
| 21.050 | 23ft 4in | 1ft 2in | 22ft 3in (6.78m) |
| 21.200 | 23ft 3in | 1ft 2in | 22ft 1in (6.73m) |
| 21.300 | 23ft 1in | 1ft 2in | 22ft 0in (6.7m) |
| 24.940 | 19ft 9in | 1ft 0in | 18ft 9in (5.72m) |
| 28.050 | 17ft 6in | 0ft 11in | 16ft 8in (5.09m) |
| 28.400 | 17ft 4in | 0ft 10in | 16ft 6in (5.03m) |
| 29.500 | 16ft 8in | 0ft 10in | 15ft 10in (4.84m) |

# TECHNICAL TOPICS

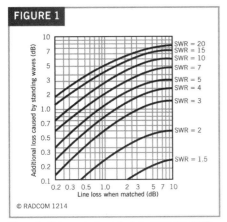

**Figure 1**: This graph shows by how much (or at HF usually how little) additional feeder line loss occurs with a moderate VSWR on a coaxial line compared with the same feeder line having an accurately matched 1:1 (unity) SWR.

misunderstandings. Too many amateurs believe that an SWR meter indicates the "goodness" of an antenna system, equating a low or "unity" VSWR with an ideal and perfectly functioning antenna and significant VSWR with a poor system. It is important to note that an unusually low SWR, particularly when achieved over a wide bandwidth, may often denote a *poor* rather than a good system.

The following notes on SWR facts and fallacies are partly based on a list compiled by Walter Maxwell, W2DU:

- Reflected power does not represent lost power except for the usually modest increase in line attenuation; in a lossless feeder line no power would be lost because of reflection *no matter how high the SWR*. On all HF bands with low-loss cable, reflected power loss is usually insignificant; at VHF it may become significant; at UHF it may be extremely important. Feeder attenuation loss depends primarily on the cable

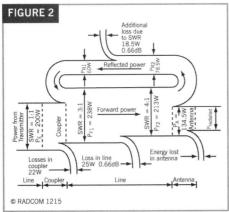

**Figure 2**: An ingenious French diagram that illustrates typical losses between a transmitter and the antenna element. With an SWR of 3:1, the extra loss compared to unity (1:1) SWR amounts to only 0.66dB, in this case 18.6W compared to 22W in the coupler, 25 watts in the line and a small IR loss in the element. Of the 200W transmitter output power, 134W reaches the antenna.

characteristics and its length – at HF it needs a very long or very poor cable for the loss to become at all significant: see **Figures 1** and **2**.

- Reflected power does *not* flow back into the transmitter and cause damage. Damage sometimes ascribed to a high SWR is usually caused by improper output-coupling adjustments and not by the SWR. A transmitter does not 'see' an SWR but only the impedance that results from the SWR; this means that the impedance can be correctly matched (eg by an ATU) without concern for the SWR on the feeder.
- *Efforts to reduce SWR below 2:1 on any coaxial line generally represent wasted efforts from the viewpoint of increasing the radiation from the antenna – unless the transmitter protection circuit begins to cut in below this figure.*
- A low SWR is *not* evidence that an antenna system is a good one. On the contrary *a lower than normal SWR over a significant bandwidth is reason to suspect that a dipole or vertical antenna is being affected by resistance losses*. These may arise from poor connections, poor earthing systems, lossy cable (egress of moisture etc) or other causes.
- The radiator of an antenna system need not be of self-resonant length to achieve maximum resonant current flow, nor need the feeder be of any particular length. A substantial *mismatch* at the junction between feeder line and radiator does not prevent the radiator from absorbing all the real power that is available at the junction. Where suitable matching (ATU etc) cancels out the reactance presented by a non-resonant radiator and a random length of feeder, mismatched at the antenna junction, then the system is *matched* and virtually all the real power may be radiated effectively. Remember few MF broadcasting antennas are of resonant length, yet they radiate efficiently.
- The SWR on the feed line is not affected by any adjustment of an ATU at the transmitter end. A low SWR achieved by this means is usually is usually an indication of a mismatch between the transmitter and the input to the ATU.
- With an effective ATU (balanced output) and good open-wire feeder a 132ft centre-fed dipole does not (contrary to general belief) radiate significantly more power on 3.5MHz than an 80ft dipole fed with the same transmitter power. A dipole self-resonant on say 3750kHz does not radiate more power on 3750kHz than

on 3500kHz or 4000kHz with any normal length of feeder; although it is to be expected that the SWR will rise to about 5:1 at these outer frequencies and that the coaxial cable will then, in effect, be working as a tuned feeder. Proper coupling, in these conditions, requires the use of a matching arrangement (eg ATU) between transmitter and feeder. If the coaxial feeder of any antenna system requires to be a specific length to satisfy a particular matching condition, the same input impedance can be obtained, regardless of the length of the coaxial feeder, by the provision of a simple L network of only two components (either two capacitors, two inductors, or one of each).

- High SWR in a coaxial feeder resulting from a severe mismatch at the antenna junction does not in itself produce common mode currents on the line or cause the line to radiate. At HF a high SWR in any open-wire line caused by a severe mismatch will not produce antenna currents on the line, nor cause the line to radiate provided that the feed currents are balanced, and if the spacing is small relative to the operating wavelength (this is equally true at VHF provided that sharp bends are avoided in the line). Common-mode currents and radiating feeders are most often reduced where the element is balanced against earth with the feeder dropping vertically towards ground; in such cases it is unnecessary to use a balun at the antenna/feeder junction.

SWR meters do not provide a more useful measurement of SWR by being placed at the antenna/feeder junction.

- The SWR in a feeder cannot be adjusted or controlled in any practical manner by varying the line length, If a meter provides significantly different SWR readings when moved along the line, this may indicate 'antenna' (common mode) current flowing on the outside of the coaxial cable, or an unreliable SWR meter, or both, but not that the SWR is varying along the line.
- Any reactance added to an already resonant (resistive) load of any value for the purpose of compensation to reduce the reflection on the line feeding the load will, instead, only increase or worsen the reflection. Lowest feeder SWR occurs at the self-resonant frequency of the radiating element it feeds, completely independent of feeder length.
- Of the various types of dipoles (thin wire, folded, fan, sleeve, trap or coaxial) none will radiate more field than another,

providing that each has insignificant ohmic losses and is fed the same amount of power. However, fat and folded elements have a broader operating bandwidth than a thin wire element.

Note that when an antenna element presents to the transmission line an impedance other than its characteristic impedance, the impedance offered to the transmitter at the input end of the line may be quite different from either the characteristic of the line (unless the line is an exact multiple of an electrical λ/2) or the impedance at the antenna junction. The impedance represented by the line then depends on the length of the feeder (which acts as an impedance transformer). In such cases, *unless a suitable matching network is interposed between transmitter and transmission line*, the impedance may be of a value (in the form R + jX) with which the transmitter output circuit cannot cope; in other words it may not be possible to *load* the transmitter properly without changing the length of the transmission line; it is this factor rather than any *losses* associated with SWR which leads to many misconceptions.

To sum up, it should always be possible to make any centre-fed antenna of any reasonable length, with any type of low-loss feeder, radiate quite effectively provided one has a good, flexible, ATU between a transmitter intended to work into a low impedance (eg 50Ω) and the feeder.

This is why the centre-fed dipole/doublet remains such a dependable multi-band antenna: **Figure 3** is a typical design described in "Lightweight Multi-band Aerial" by W H Hazeldon, G3KBE (*RSGB Bulletin,* June 1959, p573), consisting of a 132ft length of stranded PVC (electric) flex pulled apart to provide two 132ft lengths forming a junction-less radiating element (2 x 58ft) and open-wire feeder (2 x 76ft). The division between element and feeder is not critical. As described. G3KBE found it easy to load on the then existing 1.8, 3.5, 7, 14 and 28MHz bands. With a flexible tuner with balanced output there is no reason why it should not prove effective on all the present 1.8 to 28MHz bands including the 5MHz assignments.

NATO REPORT HIGHLIGHTS IN-HOUSE BPL RFI. Dave Williams, G3CCO, draws attention to an excellent NATO RTO Technical Report, TR-IST-050 "HF Interference, Procedures and Tools" (dated June 2007) that runs to over 150 pages. Although copyrighted, it is available on the web (www.rto.nato.int) and single copies may be made for individual use.

Chapter 2 covers the characteristics and users of HF radio by the military (land, air and maritime forces), broadcasting, and other users including Amateur Radio, noise levels and protection requirements.

Chapter 3 covers the characteristics of PLT and xDSL transmission systems, including the various systems for transmission on power lines (PLT) including medium (MV) and low voltage (LV) systems; access systems; in-house systems; technical characteristics with respect to HF radiation; transmission methods and characteristics (OFDM, DSSS, GMSK); PLT systems in use; example of an Access PLT system. A section covers transmission on telephone lines (xDSL).

Other chapters covers the current and proposed limits for wire-line transmission systems; measurement of PLT systems; propagation path line loss models; modelling of wire-line transmission systems as HF noise; EMC analysis methods.

Chapter 9 provides four pages of "conclusions and recommendations", although it is emphasised that an investigation of this complexity cannot be reduced to a few simple highlights and that the entire report should be consulted for full benefit. This is especially true in picking out a small number of points of importance to amateur radio:

"It is accepted that other studies have indicated that the HF noise level in the vicinity from PLT may affect the reception of low-level signals up to 400m from a single Access PLT installation using overhead power lines. This may require a sensitive military receiving site to have a protection radius of up to 1km. Even then protection against the cumulative long-distance propagation from a large number of PLT installations may be a more serious problem."

It is anticipated that a great number of PLT In-House systems (eg HomePlug) seem likely to be deployed. The report points out that such products are readily available on the market and can be installed by anyone, with no verification of the quality of the installation.

*[From personal enquiries, I gather that HomePlug PLT In-House systems are already proving popular in the UK, handling data at a significantly higher rate than most Wi-Fi systems. Since overhead power lines are relatively rare, PLT In-House systems may prove a major problem at HF in urban areas.]*

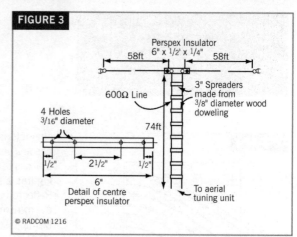

**Figure 3**: G3KBE's 1959 design of a lightweight multi-band antenna requiring no outdoor soldered joints, using two 132ft lengths of wire forming both open-wire line and element. Spreaders are spaced 2ft apart. This type of doublet requires the use of a flexible ATU with a wide range of balanced outputs but with good height should prove reasonably effective on all bands between 1.8 and 28MHz.

The report confirms that ITU-R P.372-8 noise curves (based on measurements carried out in the 1970s) are still valid. Recent measurements in Great Britain and Germany show that there has been no increase of the ambient noise in quiet rural zones within the last 30 years. A suggested *Absolute Protection Requirement* is that the cumulative interference field strengths far away from telecommunications networks should not be higher than –15dBmV/m (9kHz bandwidth) across the entire HF range. This value is in the range of 10 to 1dB below the ITU-R P.372-8 Quiet Rural noise curve (based on median values) across the HF band.

A FINAL, FINAL NOTE ON SMALL LOOP ANTENNAS. Throughout the years 1997 to 2005, controversy about the radiation efficiency of small transmitting loop antennas raged in *RadCom. IEEE Antennas & Propagation Magazine* and at various professional conferences, etc. This followed on from an airing of the views of Professor Mike Underhill, G3LHZ, at an IEE 1997 conference, reported in 'TT' November 1997. Briefly, G3LHZ (supported by his post-graduate students) believed that a very small loop could achieve a high radiation efficiency far in excess of the long-established Chu theory for electrically small antennas.

On the contrary, Dr Jack Belrose, VE2CV, Alan Boswell, G3NOQ, and Dave Gordon-Smith, G3UUR, among others, have consistently argued that the efficiency of G3LHZ's loops was in reality well below his claimed performance. They believed that the measurements made in good faith by G3LHZ and his team were seriously flawed in several respects.

By 2005, I felt that the debate had finally been settled in favour of the Chu theory. However, if there are still doubters, they should study a paper "Performance of Small Tuned Transmitting Loop Antennas Evaluated by Experiment and Simulation" by John S Belrose [VE2CV], *IEEE Antennas and Propagation Magazine,* Vol 49. No 3, June 2007, pp126-132). This paper was

## TECHNICAL TOPICS

written in 2005 but the original manuscript was mislaid by the magazine.

Much of VE2CV's work, based on the use of AMA loops, has already been reported in 'TT', but the paper provides a useful and convincing identification of the loss parameters involved, clearly demonstrates the importance of using very high, low-loss variable capacitors (**Figure 4**); vindicates Chu's theory, and shows that simulation by NEC can be used to predict achievable performance of loops for various operational requirements.

VE2CV concludes *inter alia*: "In spite of the low radiation efficiencies of small transmitting loops (about 1% for $P/\lambda = 0.03$ and 22% for $P/\lambda = 0.18$, where P is the perimeter of the loop), compact loops are very useful for communications from restricted transmitter sites, and for transportable communications since it should be noted that successful communication links are often maintained using relatively small radiated powers. In addition, they are operationally convenient, a particular small loop can be tuned by a low-loss capacitor over a frequency band of up to 6:1 (say, 3.5–21MHz), without the need for an additional (perhaps lossy) antenna system tuning unit (ASTU).

NOVEL VALVE PUSH-PULL CIRCUIT. It is unusual these days to come across a new circuit configuration for valves. An exception would seem to be found in "New Instant Push-Pull" by Chas E Miller, editor of *The Radiophile*, (Issue No 110, Summer 2007, page 9). His application was to add a push-pull output stage to a battery TRF receiver without requiring either a phase-splitting valve or transformer – simply adding a further DL35 and a couple of resistors. However, a similar arrangement might be used to convert a transmitter based on two output pentodes or tetrodes (eg two 807s or even for QRP two DL35s etc) from parallel to push-pull – and possibly offering other possibilities at RF or AF (though none of these ideas has been tested and would require selection of suitable resistor and capacitor values). **Figure 5** shows the circuit as published in *The Radiophile* from which it can be seen that G2 (screen-grid) of the first valve is not decoupled directly to earth but provides a phase-reversed drive to the grid of the second valve. As the author puts it: "Obviously the voltage levels of the two signals will have to be equal for the audio output of the two valves to be balanced and this can be arranged by choosing suitable values for the two resistors."

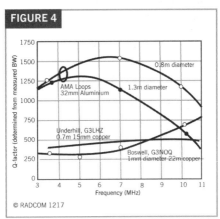

**Figure 4**: The loop Q factors determined from measured swept-frequency bandwidths for two German AMA loops, compared with the experimental loops fabricated by G3NOQ and G3LHZ. (Source: *VE2CV, IEEE Ant & Prop Magazine*)

MORE ON 500kHz. Several members have commented on the item in the October 'TT' on the availability of UK amateur licence NOVs (permitting experimental operation between 501 and 504kHz) with solid evidence of successful contacts, despite the low maximum ERP of 0.1W.

By mid-September, Chris Osborn, G3XIZ, Biggleswade, had made over 200 contacts with some 27 different UK stations and heard several more as beacons. He has compiled a spreadsheet of 500kHz enthusiasts including a number who, at that time, were still awaiting their NOV or were 'interested'. He expresses surprise at my failure (so far) to receive any stations but knows well the limitations caused by local QRN in non-rural areas. He writes: "Many of my contacts would be impossible if it were not for my small active receiving loop antenna, about 2ft square. This is located in a wooden shed as far as possible from the house, I can rotate it remotely and accurately from the shack – another garden shed. Activity is not high but several contacts can usually be heard

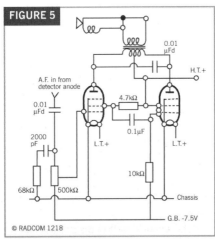

**Figure 5**: Novel form of valve push-pull circuit devised by Chas Miller and described in *The Radiophile*

at weekends and there are several UK and German beacon stations including G0MRF transmitting CW, QRSS and PSK31."

G3XIZ hopes that more UK amateurs will become active on the band, and points to his early efforts: "My NOV arrived quicker than expected and found me without a transmitter. So using a lashed-up direct-conversion receiver, I put my old HP 3310P signal generator in series with a home-brew variometer, Morse key and antenna. I managed to work several stations including G3KEV in Scarborough at a distance of 245km who gave me 559.

"Quite quickly I made a half-decent transceiver. This has a DC receiver with cascode front-end and an IC double-balanced demodulator plus a narrow audio peak/notch filter and recently added an audio-derived (calibrated) S-meter. The power amplifier uses a single IRF500 CMOS FET which, with the 12V supply, gives 4W into a 50Ω dummy load. I manage to generate about 0.6A into my 40m 'inverted-L' antenna of 10m mean height. The earth is a single copper spike driven 2m into ground immediately below the vertical section of the antenna. The local water table is near the surface so I get a good earth. I measure my ERP using a home-built field-strength meter based on PA0SE's design in the "LF Experimenters Guide' (RSGB) calibrated using his recommended design for Helmholtz coils. After spending several happy hours trudging around the neighbourhood, taking readings, I calculated my ERP as 65mW although the scatter of the readings was greater than I had expected."

G3XIZ has also been trying 500kHz transmissions (as he did on 136kHz) using QRSS-3 and DFCW-3, putting out CQ calls several evenings per week. Unfortunately, there seems little activity with these modes and so far (mid-September) he has made only one QSO using QRSS-3 (with G4WGT) although receiving regular reports from Hartmut Wolf in Germany via the 'LF Reflector'. He adds: "Initially there was some opposition to the use of the slower PC modes on this band, with some amateurs expressing the opinion that it should be reserved for 'straight' CW, but the present activity is too low for this to be a real problem."

CORRECTION. "SSB transceiver based on the 7360" ('TT' October, 2007, pp 79 & 82) requires amendment: VA0CJ/AB4O is Adam Farson (*not* Parson). The original articles appeared in *Radio ZS* October, 1964 (*not* 1954) and the frequency offset for cross-modulation measurements was 25kHz (*not* 23kHZ). Apologies!

# Technical topics

**TECHNICAL TOPICS** — PAT HAWKER, MBE, 37 DOVERCOURT ROAD, DULWICH, LONDON SE22 8SS — G3VA

This month a significant anniversary leads to the history of semiconductors

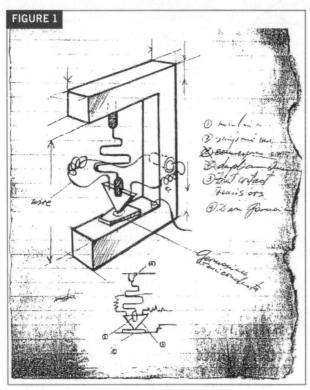

Figure 1: Sketches from the laboratory notebook of Walter H Brattain recording the events of 23 December 1947 when the first point-contact transistor was demonstrated and written up the following day. Brattain, Bardeen and Shockley jointly received the Nobel Prize for Physics in 1956, with Bardeen later to share a second Nobel Prize for his work on superconductivity.

**TRANSISTORS – 60 YEARS OLD.** *[The following item is largely based on a six-page, two-part, article "The Invention that Changed the World" by Pat Hawker that appeared in the bulletin of the British Vintage Wireless Society in 1994].*

In the afternoon of 23 December 1947, three Bell Telephone Laboratories scientists – John Bardeen, Walter Brattain and William Shockley – showed that a small piece of germanium could be made to amplify a speech signal and thus constituted a practical realisation in solid-state form of the thermionic triode valve. The device was duly named the "transistor" although it remains uncertain whether this was derived by Claude Shannon from a contraction of "transfer-resistor", or by John R Pierce who pointed out that for a thermionic valve the important parameter is its transconductance or ratio of output current to input voltage whereas the new solid-state amplifier provided gain by transresistance.

The lab demonstration showed that the device could provide a power gain of 18 or more measured with a 1000Hz tone. As recorded the following day in the laboratory notebook of Walter Brattain (**Figures 1, 2**): "This circuit was actually spoken over and by switching the device in and out a distinct gain in speech level could be heard and seen on the scope presentation with no noticeable change in quality. ... Various people witnessed this test." He listed eight BTL staff who had assisted in setting up the circuit and witnessing the demonstration.

The point-contact transistor was first demonstrated to the press on 30 June 1948 but attracted only a muted response. *The New York Times* ran – on their next to last page – four down-page paragraphs as the final item in the paper's regular 'News of Radio' feature:

"A device called a transistor, which has several applications in radio where a vacuum tube ordinarily is employed, was demonstrated yesterday at Bell Telephone Laboratories, 463 West Street, where it was invented.

"The device was demonstrated in a radio receiver which contained none of the conventional tubes. It also was shown in a telephone system and in a television unit controlled by a receiver on a lower floor. In each case the transistor was employed as an amplifier, although it is claimed that it also can be used as an oscillator in that it will create and send radio waves.

"In the shape of a small metal cylinder about a half-inch long, the transistor contains no vacuum, grid, plate or glass envelope to keep the air away. Its action is instantaneous, there being no warm-up delay since no heat is developed as in a vacuum tube.

"The working parts of the device consist solely of two fine wires that run down to a pinhead of solid semiconductive material soldered to a metal base. The substance on the metal base amplifies the current carried to it by one wire and the other wire carries away the amplified current."

Thus by June 1948, the first crude germanium point-contact transistors of the previous December had been tidied up into manufacturable form and a Patent applied for. They clearly resembled the old crystal detector of the 1920s and the wartime microwave "crystal valve" that played an important role as diode frequency changers in wartime radar but with two instead of one "cat's whisker". But, even to the technical press, the new device seemed more of an interesting curiosity than an invention destined to become one of the greatest inventions of the 20th Century.

There was already a long history of crystals that could function, albeit unreliably, as amplifiers and oscillators. Had not Henry Sutton in Australia, and in the UK Dr W H Eccles (later President of the Wireless Society of London that under his guidance transmogrified into the RSGB), reported oscillating galena experiments (*The Electrician*, 16 December 1910). And there had been a flurry of interest in the 1920s, peaking in 1924, when articles in *Wireless World* reported the work by the Russian engineer O Lossev on oscillating and amplifying zincite crystal detectors. Lossev investigated many circuit possibilities for receivers and even low-power transmitters and apparently achieved transmission over a distance of one mile. On both sides the crystal served simultaneously as a generator and detector, so that even duplex transmission was possible.

In 1925, Captain H J Round and N M Rust of the Marconi Company, in "New Facts about Oscillating Crystals" revealed that, in early 1924, they had decided to repeat Dr

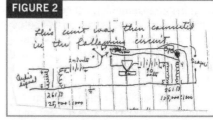

Figure 2: Laboratory notebook sketch of the transistor circuit demonstrated on 23 December.

Eccles's oscillating galena experiments. They found that many other crystals exhibited similar properties, including silicon and 'arzinite' (a pure crystalline form of zincite) but recognised that the publication of Lossev's work had reduced the investigation from one of major importance from the commercial point of view, to one of interest.

In 1928, Julius Lilienfeld of Leipzig described the principles of what in the

## TECHNICAL TOPICS

1950s was developed as the field-effect transistor (FET), and in 1934 O Heil took out a UK patent covering such a device. Prof W Gosling (*The Radio & Electronic Engineer*, Vol 43, No. 1/2) has shown that Lilienfeld secured patents on what we would now call a junction field effect transistor (JFET) and also the insulated-gate FET. Gosling also notes that van Geel of Philips continued to work on a form of bipolar transistor during WW2. But all came to nothing.

A 1938 article by Hilsh & Pohl (abstracted in *The Wireless Engineer*, April 1939) began: "Some experiments are described in which crystals of potassium bromide are used as a model of a barrier-layer rectifier and its control by means of a built-in grid; a 'three-electrode crystal' is devised which behaves in a manner analogous to a three-electrode valve."

The wartime need for a microwave radar detector/mixer led to a revival of interest in crystal detectors. In 1940, among the first problems to be settled was the choice of whisker and semiconductor materials. Among the substances tried were galena, carborundum, silicon, copper pyrites and zincite-bornite. Choice in the UK fell upon silicon with a tungsten whisker.

With the American entry into WW2 in December 1941, an intensive investigation into the properties of silicon and germanium was initiated. Germanium is a relatively rare substance but means were developed for its extraction from flue dusts in gasworks and industrial concerns. Purdue University found that its electrical properties could be predicted from its impurity content, but abandoned solid-state studies in 1945.

However, BTL pressed ahead. In July 1945 an "Authorization for Work" on solid-state physics was issued, which called for "the fundamental investigation of conductors, semiconductors, dielectrics, insulators, piezo-electric and magnetic materials." A solid-state physics group was set up under William Shockley with a sub-group on semiconductors including Walter Brattain, and (soon) John Bardeen.

In the late 1930s Shockley had studied the work in Germany of Schottky and his theories on the space-charge region – the surface layer of a semiconductor near its junction with the metal – and had conceived the idea of an amplifier using semiconductors rather than vacuum tubes. In 1945, the new sub-group began to work on Shockley's field effect concept, but without success. Then Brattain discovered that Schottky's theory was flawed. John Bardeen proposed a new and immensely important hypothesis: the theory of surface states – a semiconductor surface must be in equilibrium even before any electrical

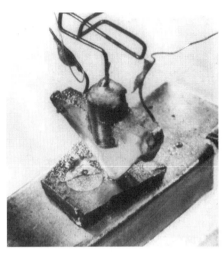

The first point-contact transistor demonstrated at BTL on December 23rd 1947.

contact is made to it – so that any electrons trapped at the surface must be neutralized by a space charge region equal and opposite to the electrons' charge at the surface.

Brattain and Bardeen resumed experiments centred on this idea, with some ideas contributed by Shockley who continued to pursue field-effects concepts until it was appreciated that these might be covered by the German pre-war patents. After disappointments, the group had the idea of placing two electrodes next to each other on a germanium ingot. A current was injected into the base by an emitting point and collected by the second oppositely polarized point; the current could be controlled by a small voltage applied between emitter and the base.

It was this point-contact device that was demonstrated in front of members of BTL staff on December 23, 1947. In January 1948 Shockley abandoned his field-effect quest and began to concentrate on his belief that the transistor effect could be achieved not only on the surface between two whiskers but also within a sandwich of semiconductor provided that an n-type region could be interposed between two p-type regions. Initially there seemed no way of fabricating such a device but continued progress led to the development of the "junction transistor". First described in the *Bell System Technical Journal*, July 1949, it had to wait for the development by Gordon Teal and J B Little of crystal-pulling techniques in 1950 before junction transistors could actually be fabricated. It was to prove far more important than the original point-contact devices.

1948, 1949, 1950, 1951 rolled by with the new transistors causing hardly a ripple on the industrial pond. Bell, in the autumn of 1951, offered manufacturing licences for a $25,000 advance on royalties and even waived all fees for use in hearing-aids.

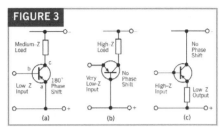

**Figure 3:** The three basic transistor amplifier configurations; (a) Common emitter; (b) Common base (useful at high RF frequencies); and (c) Common collector (useful for impedance transformation).

By Spring 1952 some two dozen firms had taken out licences. But most major manufacturers remained unexcited. The devices were deemed difficult to manufacture and difficult to characterise, with wide variation between individual transistors. But there was also in-built resistance to a new technology that appeared so radically different to the familiar thermionic valves on which all their knowledge and training was based.

It was not until November 1952, when the influential *Proc. IRE* devoted a special issue to transistors, that engineers began to take them seriously. Soon, Raytheon began large scale production of junction transistors and RCA engineers designed AF "complementary" (n-p-n/p-n-p) stages capable of 0.5W output. Work at General Electric, RCA and BTL led to a commercial process for making germanium transistors by alloying techniques. In 1954 diffusion and oxide masking techniques for making p-n junctions were developed.

Texas Instruments, by the end of 1953, were able to produce grown-junction germanium transistors in batches of one thousand or more. By April 1954 they were also producing silicon junction transistors, although initially at a cost of $100 per device. It was not until 1960 that Fairchild Semiconductor introduced a new type of planar geometry for making junction transistors based on the earlier oxide masking and diffusion techniques. That year also saw the epitaxial transistor developed by BTL.

As early as 1952, at a radar conference in the USA, Geoffrey Dummer of RRE, Malvern, suggested that "it seems possible to envisage [semiconductor] equipment in a solid block with no connecting wires. The block may consist of layers of insulating, conducting, rectifying and amplifying materials, the electrical connections being connected directly by cutting out areas of the various layers." This prediction of the integrated circuit later began to be realised independently by Jack Kirby of Texas Instruments and Robert Noyce of Fairchild. By 1962, the first ICs were in production although priced at $100 each. The first low-

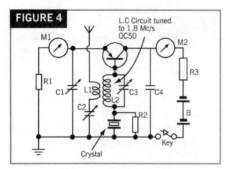

**Figure 4:** Circuit diagram of G3IEE's 1954 QRP transmitter with which he made local contacts early February 1954 using a 30V hearing-aid battery.

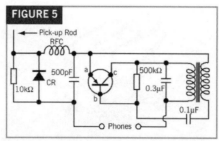

**Figure 5:** The first diagram ever to appear in 'TT' (April 1958), showing a simple CW monitor originally described by a German amateur in DL-QTC.

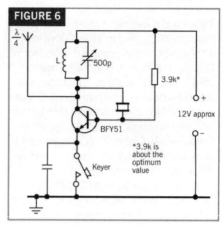

**Figure 6:** By the early 1970s, amateurs were using BFY51 silicon n-p-n devices in oscillator-transmitters providing about 1W output.

cost ICs arrived about 1966 including the dual-gate Fairchild μL912, a small round device (with eight leads) comprising four transistors and six resistors, opening the way for the mass production of digital rather than analogue systems. By 1980, the microprocessor based on large scale integration (LSI) had arrived.

Meanwhile, consumer applications of transistors had appeared; the first, a hybrid transistor/valve hearing-aid, was announced by Sonotone in February 1953 using five transistors plus two valves. Under a 1953 agreement between Texas Instruments and IDEA (Industrial Development Engineering Associates) an all-transistor AM radio was developed in time to reach the 1953 Christmas market. This model, the Regency TR4 shirt-pocket radio, sold some 100,000 units priced $49.95 (although valve table sets were selling at one-third the price). It had eight transistors, weighed two ounces, had an IF of 262kHz, and was powered by a 22.5V hearing-aid battery. The 4mA consumption gave a battery life of some 20-30 hours.

Ratheon began marketing, in February 1954, a six-transistor set with push-pull output, powered by four 1.5V D-cells but weighing some 4lb. In Japan, a small firm Tokyo Telecommunications took out a licence from BTL and by June 1954 had begun to make transistors, although the yield at first was poor. However, in August 1955 a transistor radio was launched by the firm, under the Sony brand. In March 1957, Sony launched its first pocket-sized model, selling more than a half-million sets world wide.

EUROPE AND THE SEMICONDUCTOR. In the UK, germanium point-contact transistors began to appear in research laboratories about 1951. GEC produced GET1, GET2 etc devices using germanium refined from chimney flues. Mullard, owned by Philips, marketed low-cost alloy-junction transistors (probably made in Nymegan) in the mid-1950s as the OC44, OC45 RF types and OC71, OC72 AF types, and OC16 power type. The first MW/LW transistor radio to appear on the UK market, in June 1956, was the Pam Model 710, a brand of Pye, using eight transistors made by Newmarket Transistors Ltd, a subsidiary of Pye, with an IF of 315kHz. The following year, Pye marketed, under their own name, a table receiver (Model 123). The first British firm to concentrate solely on transistor sets, including miniaturised components, was Perdio, formed in 1955. Their first receiver, the PR-1, was shown at the 1957 Earls Court Radio Show but initially suffered component failures, not overcome until about 1958/9.

Although the vast majority of UK amateurs remained committed to all-valve equipment throughout the 1950s and early 1960s, the availability of alloy-junction transistors that could oscillate at MF and low HF soon began to attract experimental interest. Early in February 1954, Tony Cockle, G3IEE, using a 1.8MHz transmitter (**Figure 4**) based on a single Mullard OC50 with an input of 30mW (described in the *RSGB Bulletin*, March 1954) made contact with a number of local stations, the first with G3DAZ at a distance of 2 miles. On 21 February, 1954, the Yeovil Amateur Radio Club station, G3CMH, made an un-arranged 3.5MHz CW sky-wave contact with G3CAZ in Haslemere, at a distance of some 90 miles, using a p-n-p point contact transistor in a crystal-controlled, negative-resistance oscillator designed by C G Banbury, BRS 20100 and described in the *RSGB Bulletin* April 1954, see also *Radio Communication, February 1994, p5*.

Other reports of QRP transistor transmitters soon followed. During the night of April 19-20 1954, G3CCA of Leicester made a number of 1.8MHz contacts, using a transistor oscillator with an input of 50mW, including stations in Harlow, Essex, Littlehampton and Liverpool. During a contact with G3ERN (Harlow) at a distance of 100 miles, G3CCA made possibly the first transistor DX 'phone contact using a modulator consisting of five GET1 transistors. In *A Guide to Amateur Radio* (6th edition, 1958), I included an illustration captioned "Transistors now offer many exciting possibilities for fresh experimental work. Here is a low power, two-stage transmitter with which G3CCA successfully spanned the North Sea in 1954."

By the time that 'TT' began in April 1958, transistors were established as devices that could be used for many purposes. The very first diagram ever in 'TT' was of a simple transistorised CW monitor, using an AF transistor oscillator powered from rectified RF pick-up (**Figure 5**).

The 1950s and 1960s saw the availability of an increasingly wide range of semiconductor devices, including silicon transistors and diodes, JFETs, dual-gate MOSFETS, CMOS, VMOS, Zener diodes, varactors etc. Small-signal p-n-p and n-p-n transistors became available at lower cost than thermionic valves. The silicon BYF51 as an oscillator could deliver 1W RF output (**Figure 6**). Hybrid and all-transistor communications receivers and hybrid transceivers began to appear on the market in the mid-1960s. Bipolar silicon transistors and MOSFETs capable of increasing power output at up to 30MHz became available at some cost; low-cost VMOS devices could provide some watts of RF on the lower part of the HF spectrum. Progress is still continuing. 'TT' has reported on a German solid-state transceiver with an output of 600 watts; a description in QEX by JE1BLI and JA11DJW amateurs of a 1.5kW linear using two ARF1500 devices in push-pull; a similar design is being marketed by Tokyo Hy-Power as the HL-2.5KFX. Freescale Semiconductors has recently introduced their MRF6VP11KH dual LDMOSFET which can deliver 1kW pulsed output (duty cycle 20%) at 130MHz intended for use in MRI medical systems.

Yet, for various reasons, many amateurs, at least until the late 1970s, still preferred to use valves, particularly for home construction. Valves seemed more familiar, less vulnerable to over-voltages; with more consistent and better-defined characteristics.

# TECHNICAL TOPICS

This feeling gradually lessened but, for a few of us, has never entirely disappeared. The difficulty of handling the latest generation of tiny surface-mounted components, with their many pin-outs, has revived interest in discrete components, and even, to some extent, in the 'obsolete' thermionic valve.

In 'TT' (October 2007, p82) I noted that Peter Chadwick, G3RZP, claimed that July 2007 had seen the 50th anniversary of the opening by Plessey Semiconductors of the Cheney Manor plant in Swindon, which was Europe's first purpose-built semiconductor facility. It was a joint effort with Philco for making surface-barrier transistors, the first type of transistor capable of working up to VHF. This device had first appeared in 1954.

This brought the following comment from Gerald Drijver, PG2D: "In my opinion this [claim] is not correct. I claim that the Philips Semiconductor Facility in Nijmegan was the first European purpose-built semiconductor plant."

He explains: "I started my professional career as an electronics engineer in semiconductor development at Philips in Eindhoven, The Netherlands, in April 1954. In October 1953, Philips started a semiconductor production department in the basement of the Twentse Bank (later ABNAMRO) building in the city of Nijmegan and at the same time began building there a new factory for that purpose. The already-running production moved into the new building at the end of 1954. The development laboratory (including me) moved from Eindhoven to the new building in Nijmegan about 19 January 1955, by which time two production buildings, the laboratory and administrative-building were ready. The official opening by the Dutch Minister of Economic Affairs was on 12 July 1955. I retired on 30 November 1988, after almost 35 years in semiconductor development."

Peter Chadwick, G3RZP, in turn, was surprised at the comment ('TT' November, 2007) that DDS chips were not implemented at the time (1986) that G3SBI built a DDS system using standard ICs and discrete components. He recalls a 1987 tour of American firms demonstrating the Plessey SP2001 DDS chip with a 500MHz output and 2GHz clock. The SP2001 and later SP2002 were designed at Plessey Research Caswell Ltd by a design team led by Dr Peter Saul, G8EUX, who produced a number of DDS patent applications and later wrote an article "Direct Digital Synthesis - What is it, and How can I use it?" in *RadCom*, December 1990. These chips used the Plessey high-speed bipolar process and were easily the fastest in the world at the time, and may possibly be still so. They were power hungry. G3SBI may have been thinking of the CMOS DDS chips developed in the 1990s. The Plessey process also resulted in dividers working at 10GHz in the 8-lead mini DIP package."

DATONG PARAMP UP-CONVERTER. In the November, 2007, p78 'TT' item on parametric up-conversion I wrote: "As far as I am aware, this has never figured in any marketed amateur radio receivers despite its theoretical exceptional performance." Several readers have reminded me that in the 1980s, Datong, always innovative, produced and marketed the PC1 general coverage converter: a receiving converter unit that enabled the entire spectrum from 30kHz to 30MHz, in thirty 2MHz bands, to be received on a 144-146MHz receiver or transceiver.

The PC1 had seven RF bandpass filters directly to a balanced parametric mixer using varicap diodes with its drive (pump) signal derived from a frequency synthesiser (113 to 144MHz in 1MHz steps). The PC1 was given a good review by Peter Hart, G3SJX in *Radio Communication*, April, 1982, pp310-1.

Michael O' Beirne, G8MOB writes: "I [still] have a PC1. It's quite a small box but it works well enough. The limiting factor is the VHF receiver or transceiver. My FT221R has the muTek replacement front-end board designed by Chris Bartram, G4TKZ, to a proper professional spec. The new board is streets ahead of the original Japanese board and is very sensitive, yet difficult to overload."

HOME-PLUG IN-HOUSE PLT SYSTEMS. Last month ('TT', December 2007) it was noted that a recent NATO Technical report (TR-IST-050 "HF Interference, Procedures and Tools") suggested that the major problem of interference from Power Line Transmission (PLT) systems may arise from the anticipated very large number of PLT In-House systems (eg HomePlug) likely to be deployed. The report points out that such products are readily available on the market and can be installed by anyone, with no verification of the quality of the installation." I added a note that HomePlug systems are already proving popular in the UK, handling data at a significantly higher rate than most Wi-Fi systems and could become a major HF interference problem in urban areas.

Since writing those notes, I have obtained two HomePlug Powerline Alliance, Inc documents downloaded from the web: 'HomePlug 1.0 Technology White Paper' (2005) and 'HomePlug AV White Paper' (2006). These documents provide detailed technical specifications of the two current HomePlug systems. HomePlug PowerlineAlliance is a non-profit industry association formed in March 2000 by a group of companies to create specifications for home powerline networking products and services and to promote their use worldwide. The Alliance issued a first industry standard (HomePlug 1.0) in June 2001 using Orthogonal Frequency Division Multiplexing (OFDM) system in "the unlicensed frequency band between 4.5 to 21MHz for transmitting the signal over the power line. According to FCC rules, HomePlug is a secondary user in this band. FCC has two primary requirements with respect to emissions performance of such devices. Products must meet FCC part 15 radiation requirements. Product must not cause harmful interference to licensed users of the band." It is recognised that the systems are vulnerable to strong RF signals.

In theory, this should reassure us – but note the NATO report. It is also claimed that HomePlug limits the power spectral density by inserting 30dB notches in amateur bands between 4.5 and 21MHz, and that due to this notching only 76 OFDM carriers are usable in the United States. HomePlug 1.0 systems manufactured in Germany, and possibly elsewhere, are being used in the UK. I do not know if these systems are at present approved by OFCOM. There is also a Spanish specification for in-house PLT.

The later, faster HomePlug AV specification, operates in the frequency range of 2 – 28MHz and provides a 200Mbps physical channel rate and 150Mbps information rate. The specification includes 30dB notches.

Perhaps the best advice that can be given is to ensure that your antenna is as far as possible from any internal or external electricity power lines

HERE & THERE. Jim Readings, G4MVO, writes: "The item on RFI to vehicle electronics ('TT' November, 2007) struck a long forgotten note with its reference to Mercedes Benz cars. My late brother (G3KFT) and I were directors of the local MB dealership. When the 280E model was announced my brother had his mobile rig transferred to a new MB280 by our late friend and colleague Doug Price, G3YZ. Engine problems followed, causing concern at Mercedes Benz, and ending in a new engine management 'box' being provided. I believe the 280 was one of the earliest production cars to be fitted with electronic fuel injection management. Subsequent MB models reverted to mechanical injection."

# Technical Topics

Pat Hawker presents correspondence on DDS, Power Line interference and antennas

**EARLY WORK ON DDS.** The October 2007 'TT' item "50 years – LC oscillators to DDS" included a brief mention of the early work of Colin Horrabin, G3SBI on DDS (direct digital synthesis) during 1986, before the first (CMOS) DDS chips became available in the mid-1990s. Because I wanted to include details of RZ4HK's novel TLT (transmission line transformer) mixer, I omitted to expand on G3SBI's work which was initiated as a result of reading early professional papers on the concept of DDS. One result was that G3RZP reminded us ('TT' January 2008) of the work of Dr Peter Saul, G8EUX in the 1980s at Plessey Research Caswell and his excellent article on DDS basics in Radio Communication, December 1990, pp44-46.

My first 'TT' item on DDS was in December, 1988, (see also Technical Topics Scrapbook, 1985-1989, pp 275-6) based on an article "A direct digital synthesis VFO" by Robert Zarvel, W7SX (Digital RF Solutions of Santa Clara, California) in Ham Radio, October, 1988. W7SX introduced DDS as follows: "The direct digital synthesiser has arrived in amateur radio (the new high cost Icom 781 uses DDS). In the past several months DDS state-of-the-art has progressed to the point where good radio performance is obtainable using DDS. .... Unlike the PLL, DDS doesn't use a VCO, loop filter, phase detector or digital divider and prescaler. Waveform information is generated using digital information only".

A further comment has come from David Guest, GM3TFY (see later) but, first, more detail about G3SBI's early work on DDS. The following notes are taken from G3SBI's description of his personal findings on the progress of amateur radio applications of DDS since the 1980s: "The basic DDS oscillator is shown in **Figure 1**. I first heard about this technique in the late 1970s, but I don't think any suitable chips were around then. If they were, they would have been very expensive.

"How does the system work when a 10-bit DAC only gives you 0.1% control of voltage. The best way of looking at it is that it is a rate effect on the DAC by the phase accumulator; the more bits in the phase accumulator binary adder, the greater the tuning resolution. This is potentially an easy way to generate frequencies that could have fine frequency control and would avoid the

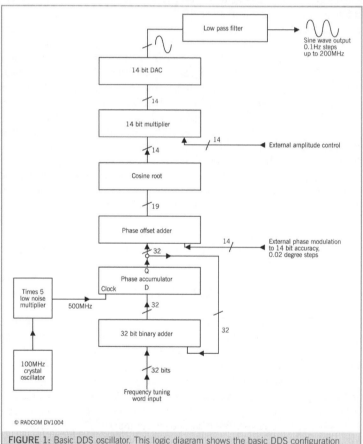

**FIGURE 1:** Basic DDS oscillator. This logic diagram shows the basic DDS configuration such as that used in the AD9951 500MHz part. With a 500MHz clock, the output is usable to 200MHz in 0.1Hz steps. (Source G3SBI)

use of multiple loop (PLL) synthesisers for radio applications. If your 50MHz clock oscillator has good phase noise, the lower the output frequency the better the phase noise performance.

"The snag is that the DDS technique produces 'spurs' - say 50dB down relative to the wanted frequency. Owing to the wide dynamic range of a modern HF receiver, this is bad news if you want to produce DDS directly to generate the frequency in a local oscillator. However, you can use DDS as a variable reference frequency of a single loop PLL to form a combined VCO/DDS where the PLL with a narrow loop bandwidth eliminates the spurs of the DDS. This technique was used in the AOR7030 receiver and the CDG2000 transceiver. With this approach, the narrow 1kHz PLL bandwidth means that the VCO phase noise has to be very good. The double tank grounded-gate J310 FET VCO I developed in 1994 is capable of doing a good job.

"Early on, I soon decided that I needed to learn more about DDS technology. At the time I had access to a development system for the early EPLDs (only 24- and 16-cell devices). Three of these devices were used in a binary adder accumulator about 28 bits long. The 12 most significant bits went to a home made R/2R DAC, with the result that my 1986 DDS (**Figure 2**) could produce a ramp waveform with steps of under 1kHz. The goal was to provide a frequency source tuneable between 80-150kHz as a variable reference for a DDS/PLL with a frequency resolution of a few Hertz.

"A rising ramp from the 12 most significant bits was used to trigger a 74C74 flip-flop used to divide the ramp frequency by two. This gave me a square-wave output and fine control of frequency. This was then used as the reference in a single loop PLL/DDS synthesiser similar to that later used in the AOR7360 receiver and the CDG2000.

"However in 1986, I took the system no further, but after the development of the H-mode mixer (1993), I realised it needed a good low-noise drive order not to degrade its close-in performance. By 1994 a 12-bit Harris DDS chip was available from RS Components and was pressed into service for the AOR7030 and later the CDG2000. I would stress that with this approach a special low-phase-noise VCO is essential owing to the narrow loop-bandwidth used in

# TECHNICAL TOPICS

the PLL system to eliminate the effects of the DDS spurs.

"*Super DDS oscillators*: Since 1996 Analog Devices has been a company with an impeccable record in the development of state-of-the-art semiconductor technology. In 1993 they introduced the AD7008 DDS chip that was one of the first to include a DAC so that it could produce directly a sine-wave at the output frequency. The AD7008 found its way into a lot of QRP equipment, partly due to Steve Weber, KD1JV, who produced a nice little module with shaft encoder and LCD display, all driven by a small Atmel microprocessor. Since 1968, I have used (professionally) quite a lot of Analog Devices and can vouch for the firm's good relationship with their customers, including many amateurs. Recently, they have given away many samples of their new 1GHz chips.

"*DDS – the Italian Way*: A number of amateurs world-wide have used the AD9951 directly as a local oscillator for down-conversion receivers with the device clocked at 500MHz. Most have used a PCB designed by 10CG. These chips have very low spur levels and when a 100MHz crystal oscillator is multiplied to 500MHz it results in good phase noise, and that seems to include the $7 oscillators from Digikey. The resultant outputs at frequencies required for down-conversion receivers can now have excellent phase noise close to the carrier. It is always worth remembering that the lower the output frequency relative to the clock frequency, the lower the phase noise.

"*Further DDS developments*: To quote Jeff Kelp of Analog Devices' High-Speed DDS Development Group: 'DDS is likely to get faster and faster with more bits in the DAC'. Recent parts such as the AD9910 and AD9912 with 1GHz clock promise even better results; of the order of 150dBc/Hz down at 1kHz spacing when generating a 20MHz output frequency – and not much worse at 100MHz. It is likely that their next development will be a 4GHz part with 14-bit DACs using new technology (even though current technology is pretty impressive). Perhaps we are beginning to see the demise of the DDS/PLL approach even for up-conversion HF receivers.

"*Postscript*: During the 1980s there were HF transceivers for which the software appears to have been written by people with little or no experience of using short-wave receivers. For example the Yaesu 757 had a shaft encoder giving 10kHz per revolution in 10Hz steps. The only way to tune to a required frequency was to use the 500kHz button, getting you within 500kHz, and requiring up to 50 turns of the tuning knob

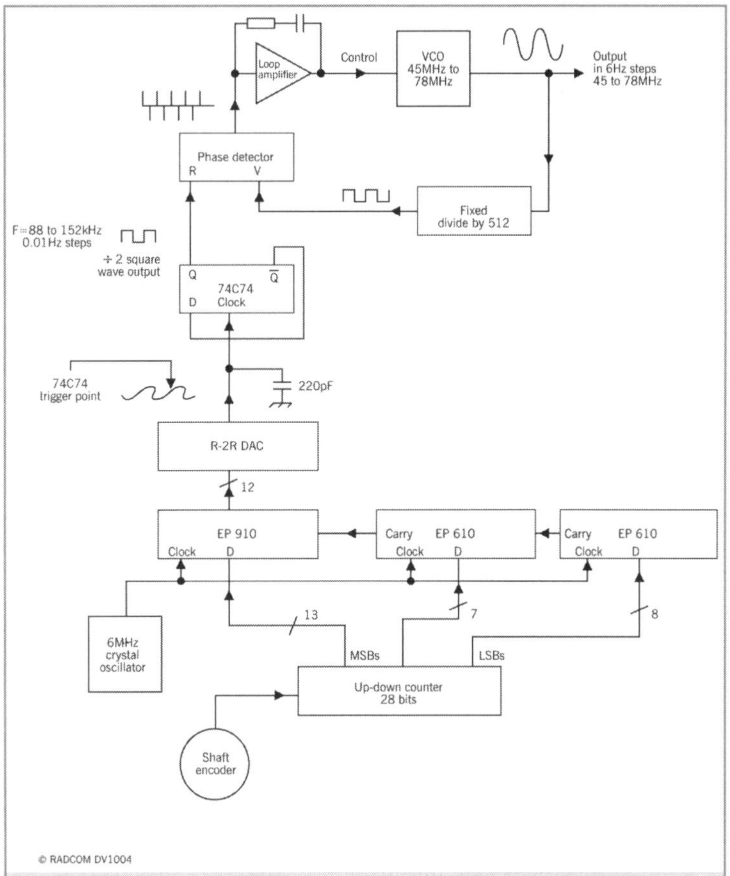

**FIGURE 2:** G3SBI's experimental 12-bit DDS/PLL synthesiser as built in 1986. It was never used in a complete receiver/transceiver but it did work and there are some interesting features. As a result Analog Devices may manufacture a triangular wave DDS because this should have very low spurs particularly close to the carrier. (Source G3SBI)

to reach the actually required frequency. I modified my 757 to be able to tune in 100kHz rather than 10kHz steps, selected by the 500kHz button. So it was then 100kHz per revolution to reach close to a required frequency. Then I went back to 10Hz steps.

"An important design consideration is to know how low should the noise floor be. The phase noise plots of devices such as the AD9910 show that the phase noise rises as the frequency increases. In principle, ECL parts can give a 148dBc/Hz noise floor; with CMOS capable of 165/dBc/Hz. A device such as the AD9910 can give a noise floor of about 165dBc/Hz at 20MHz output, rising to 150dBc/Hz at 100MHz. It seems that it is something of a black art, with combined analog and digital chips, to achieve a really low noise floor at the speeds at which chips are now clocked."

### EARLY DDS – GM3TFY's COMMENTS.

David Guest, GM3TFY, writes: "The dates given in 'TT' (November 2007) require correction. You suggest that the concept of DDS was mooted in 1980 and delayed in implementation until the 1990s. From personal experience, I know that these dates are wildly inaccurate. As a new R&D engineer at Hewlett-Packard (South Queensferry, Scotland), fresh with a shiny degree and endless self-confidence gained from amateur radio, I spotted a paper describing DDS in *IEEE Transactions on Audio & Electroacoustics*, March, 1971. This was probably the first mention of the idea, although it took a couple of decades before the label DDS appeared widely.

"Somehow my managers allowed me to incorporate DDS in a new telecomms test product (measuring audio group delay and level distortion) for which I was largely responsible. I am not sure that many of my colleagues really understood the concept! But somebody trusted me. It worked brilliantly, albeit limited to 20kHz maximum frequency and 10Hz steps. That was all that was needed for the product, but at that time the limited speed of early DACs would have caused difficulty in going much higher in frequency, although it is possible that, with readily available parts, 100kHz might have been possible.

"However, the spectrum was good, very much better than we needed. In those days, before easy integration, the design required rather a lot of discrete logic. With simple real-time trigonometry, the sample look-up was squeezed into 1024 bits of ROM, then a rather unusual new IC device which required explanation for my boss.

"Following prototypes, production runs were shipped in mid-1974. To the best of my knowledge, this was (probably) the first commercial implementation of DDS. I wrote an article about the test product and the DDS technique for the *Hewlett-Packard Journal* (November, 1974). Only many

# TECHNICAL TOPICS

years later did I realise that people were using the name DDS to describe the method.

"Those were great times for young engineers. My spotting the IEEE paper shows the slightly academic environment we worked in. It was good fun. Bill Hewlett himself used to appear from time to time and take a direct interest in what his lab engineers were up to. He talked to the most junior employees in a most approachable and unassuming manner.

"The HP Journals are now on-line. You can read about my DDS and its distortion spectrum in the 1974 article (page 15 onwards): www.hparchive.com/Journals?Low-Resolution/HPJ-1974-11-Low-Resolution.pdf

"True, my DDS was not really RF!"

## MULTIBANDING THE IIDM ANTENNA.

Duncan Telfer, G8ATH/G0SIB, writes: "Most antennas originally designed for single band operation can be 'doctored' to operate successfully on higher frequencies, provided that we are happy with resulting changes in far-field (FF) pattern and VSWR. Ideally, the antenna impedance at the new VSWR minima should be entirely resistive, which is not always the case. However, notable successes in 'teasing out' multiband responses from what is basically a horizontal dipole include the G5RV antenna and its enhancements by Brian Austin, G0GSF/ZS6BKW [see 'TT' July 2007, etc].

"Likewise, my relatively compact 'inwardly inclined dual monopole (IIDM) antenna' for 7MHz NVIS ('TT' Sep 2001 and *RadCom*, Dec 2005), lends itself to investigating responses on the 14MHz and higher bands. I have used EZNEC modelling to optimise firstly a vertically-aligned IIDM wire antenna (2mm copper) for dual-band performance on 7 and 14MHz. A quarter-wave matching section (see **Figure 3**) of characteristic impedance Zm (here 150 Ω) was then added between the antenna and main feeder cable (Zf = 50 Ω) of a length selected for the higher band. The required impedance Zm was found from $Zm = (Zr*Zf)^{1/2}$ where Zr is the antenna radiation resistance at the desired frequency, in this case as determined from the EZNEC plot of minimum VSWR. For simplicity the feeder and matching section were assumed to be of balanced flat twin-conductor of zero loss and unity velocity factor.

"As expected, adding the matching section in turn shifted the lower frequencies response. But because of its small size compared with wavelengths at 7 and 14MHz, the effects were slight and re-adjustments were accomplished with slight (symmetrical) reduction in antenna wire length. It was found that, with care, mutual interactions between configurations, adjustments could be reduced to manageable proportions. A solution was arrived at, the dimensions in **Figure 3** being for assumed EZNEC conditions of poor lossy ground (conductivity 0.001S/in, DC=2) similar to my QTH. Added small lumped inductances were chosen (5μH) to ease the centring of the 14MHz band response.

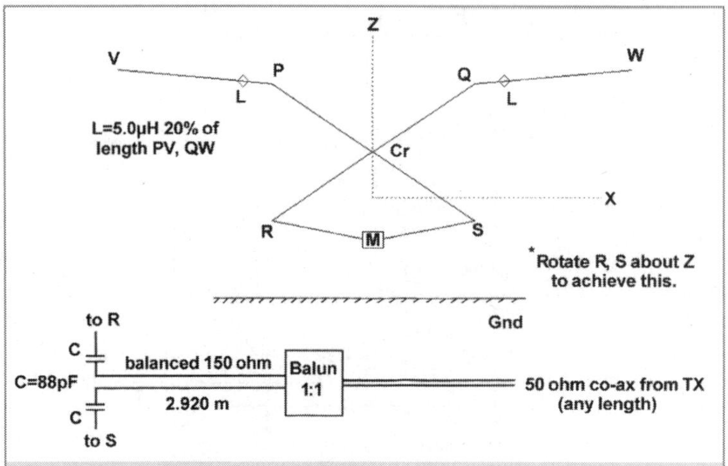

**FIGURE 3:** (a) 3-band IIDM antenna. Dimensions: Lengths PV, QW 4.065m; SP, RQ 6.351m; RM, SM 2.646m. Heights V, W 6.35m; P, Q 6.00m; R, S 2.50m; M 2.00m. Separations: P, Q & R, S 5.30m; Cr (crossover) 1.0m* (b) Detail showing Matching unit M add total of 20cm for connecting capacitors C)

"**Figure 4** shows the resulting VSWR for 50-ohm feed ahead of the matching section over a range 3 to 30MHz. The marked breadth and magnitude of the response at higher frequencies is evident and this peak effectively straddles the 24.9MHz WARC band and the lower part of the 28MHz band. Its position can be shifted by changing the length and impedance of the matching section. However the f3 curve is not a simple one.

"The Smith chart in **Figure 5** shows clearly the loops corresponding to the three principal peaks in the SWR spectrum, along with the frequencies of interest. It also reveals other crossing points on the central (R = ±j0) axis. For example, one of these corresponds to a weak 'shoulder' on the f3 peak at around 22.7MHz, quite near the 21MHz band. But before we start reaching for snips and more cable to make another matching section, the position on the chart indicates the value of R is quite low (it is actually about 10.4 ohms).

"As the FF pattern at a particular frequency is dependent only on the configuration of conductors in the antenna structure and the prevailing conductor and ground losses, the near spherical far-field is preserved, with zenith gains of 3.67dBi predicted for the above loss conditions and 5.23dBi for the less lossy EZNEC 'standard' ground conditions (Conductivity 0.005S/m, DC = 13). At the higher frequency responses, elongation of the FF pattern in the Y-direction (normal to the plane of the antenna) favours longer skip communication. For example, the EZNEC computed maximum gains, even at the losses given above, for 30° elevation are 7.1MHz, 1.06dBi; 14.2MHz, 3.72dBi; 24.9MHz, 2.61dBi; 28.3MHz, 3.35dBi. For EZNEC 'standard' real ground the corresponding gains are 7.1MHz, 1.05dbi; 14.2 MHz, 3.22dBi; 28.2MHz, 4.19MHz. This IIDM scheme is not a unique solution and there is continuing scope for further experimentation."

## POWER LINE INTERFERENCE.

Richard Hankins, G7RVI, adds to the recent 'TT' items on In-house PLT systems (Dec 2007 and Jan 2008). He notes my comment that, in the UK, overhead power lines are relatively rare, but points out that while this may be the case in urban areas, they are common in some rural areas. He writes "Our house in the small market town of Ross-on-Wye is fed by them and they appear common around these parts. Overhead power lines are definitely the norm outside towns.

"The effect of overhead power (and telephone) lines is to make towns like mine a pretty noisy [radio] environment. In such locations operating from home is a constant challenge. I have tried various solutions. The well-known noise-cancelling system has been built and tested; it works exceedingly well when you have a single (or predominant) point source. It is no good when you are surrounded by many sources at different angles or when the source is distributed, as with overhead power lines. Similarly, going to multiple neighbours and getting them to fit ferrite beads, etc is costly and time consuming. Finding a noise source on a long overhead line at HF is a tricky business.

"The effect of all this is obviously frustration for amateurs, but I think there is also a more subtle problem. My observation of 3.5MHz NVIS working is that trying to take part in UK-wide SSB nets is now difficult unless you have an output of 100W PEP or more. My normal experience on such nets is that I simply cannot hear a number of the lower power stations because their signals are below my local noise level. That they are getting out is evident from the reports from stations in more favourable [lower local noise] situations. The effect is to encourage the use of higher output power. But when this is done in locations like mine, the old problem of EMC in the form of interference from my transmitter into my neighbour's TV etc then arises.

"I have an elderly amateur friend who is in just this situation. With overhead power lines across his back garden, his transmissions shut down his neighbour's satellite TV system, but his neighbour is not interested in fixes or compromise. His solution is to threaten my friend with 'a brick through the window' if he ever causes interference. The result is that my friend now rarely ventures on the air unless his neighbours are out.

"Apropos In-house PLT, it is distinctly bad news for amateurs to have anything operating on HF – particularly in houses and in abundance. I know the systems have 'amateur band notches' – however these are not generally real filters, but rather involve the switching off of some of the multiple [OFDM] carriers that would otherwise fall into the amateur bands. That may be effective in reducing the amount of interference suffered by amateurs, but it leaves a wideband receiver at the PLT user's end just waiting to be swamped by our RF. I can't help feeling that the days of amateur radio in densely populated areas may now be numbered!"

**G3LHZ & SMALL LOOPS.** The Editor has passed to me by post the following email from Mike Underhill, G3LHZ, dated before the publication of the December 'TT' with its notes on the IEEE article by VE2CV:

"The remarks of Brian Austin, G0GSF ('TT' October, 2007) are a significant step forward in resolving the so-called 'Loop Controversy' in the minds of the doubters. But perhaps not quite as Brian intended. He says ".... nobody has ever managed to achieve anywhere like the theoretical Q with these antennas ...". Here he quite correctly includes all small tuned loop antennas, not just those with loss in the capacitor rotor mechanism. These can include loops with split-stator capacitors, 'butterfly' capacitors, capacitors with 'pigtail flexible braids' connecting the rotor, loops capacitors with no sliding joints, superconductive loops, etc. So all those loops where there obviously is no rotor loss mechanism still have a Q which

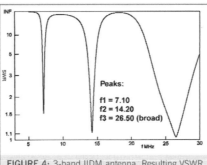

FIGURE 4: 3-band IIDM antenna. Resulting VSWR for 50-ohm feed ahead of the matching section over a range 3 to 30MHz

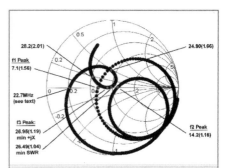

FIGURE 5: 3-band IIDM antenna. Smith Chart trajectory of 3 to 30MHz scan salient frequency (SWR) points seen from 50-ohm source ahead of matching section

is still much lower than the 'theoretical' prediction.

"Surely the logic is that the Q is being lowered by a radiation resistance that is much higher than any of the resistive loss mechanisms? Thus the loop antenna is fundamentally efficient – as proved by Q measurements that now everyone accepts. The loop controversy is therefore finally ended to everyone's satisfaction. Thank goodness for that. This serious misunderstanding has done enormous damage to the take-up and exploitation of loops (and arguably also to other small antennas?). How many amateurs have been put off using small loops by this?

"Theories and their assumptions are practically useless until proved or disproved by practical experiment. (Simulation is no help – it is designed to fit existing theory only!). A 1m-diameter copper loop of 10mm tube rises in temperature by 56 degrees Celsius, or thereabouts, for every 100 watts dissipated in it. Thus 400 watts with an ambient temperature of 20 degrees C would make the loop conductor temperature rise to 244 degrees C. Solder joints and any plastic supports would melt! Such loops would self-destruct. They don't!

"Why not? For 400 watts to such a loop on 160m I measure a 20 degree rise, showing 112 watts dissipated in the loop and a *directly measured* loop efficiency of 72%. On 80m the rise is about 14 degrees C giving a *measured* efficiency of 84%. It is difficult to deny the evidence of the first law of thermodynamics!

"Once the 'canard' of 'low small loop efficiency' has been firmly exposed for what it is, we can move on. We can start to 'invent' and explore new small (tuned) antennas that old theory has declared to be 'impossible'. What do I mean? Perhaps a forthcoming publication will reveal all? Wait and see!"

I have reproduced G3LHZ's email as received by the Editor and posted to me in full and without editorial change. But this does not mean that I am convinced by his claims that it is possible to achieve a radiation efficiency of 72% on 160m or 84% on 3.5MHz with a 1m-diameter loop. In this, I am not denying the first law of thermodynamics but instead would remind readers of the letter from Dave Gordon-Smith, G3UUR, in "Technical correspondence – small loops" (*RadCom*, November, 2004, p100). G3UUR pointed out serious flaws in the thermal balance technique used by G3LHZ and his team to 'measure' RF radiation against power loss. I cannot recall that G3UUR's criticisms have ever been refuted by G3LHZ.

To quote G3UUR's final paragraph: "When the assessment is done properly, I think we will find that there is no discrepancy between his findings and conventional theory. I believe his technique, as it stands, estimates the power loss with a 300 to 400% error. This fully accounts for his claims of small loop radiation resistances that are thousands of times greater than conventional theory predicts. Perhaps we can now lay this topic to rest!"

**NOT SO NOVEL PUSH-PULL.** Richard Hankins, G7RVI, takes me to task for labelling the circuit shown in Figure 5 ('TT', Dec, 2007. p80) "a novel form of valve push-pull circuit, devised by Chas Miller and described by him in *The Radiophile*." He writes: "The circuit is anything but new – it comes directly from the R209 receiver designed during WW2 .... It's a neat solution to the need for a push-pull output when extra valves are to be avoided for cost, complexity, space and current drain reasons. ... The credit should go to the unknown (to me) designer of the R209 not to Chas Miller." All I can say is that it was new to me, and possibly a case of re-inventing the wheel by Chas Miller!

# Technical Topics

Pat Hawker looks at battery technology, receiver front-ends and clandestine equipment

**FAREWELL AND THANKS.** This month marks the completion of 50 years of 'Technical Topics' which first appeared in the RSGB Bulletin in April 1958. For the first 10 years 'TT' appeared bimonthly, then monthly for 40 years: a total of some 600 columns - and an appropriate time to wind it up as a regular monthly feature. Next month, I hope to contribute a special 50th Golden Jubilee 'TT', celebrating some of the items that have had lasting influence on our changing hobby. And to take the opportunity of thanking the hundreds of members – both living and now SK – who have contributed items or suggestions - and the many periodicals and publications from which ideas have been abstracted. Thereafter, it may prove possible for me to contribute an occasional Technical Topics Extra column free from the pressure of deadlines, etc but since April will also see my 86th birthday (with arthritis-limited mobility) there can be no guarantees!

### 140 YEARS OF BATTERY DEVELOPMENT.

The continued development of portable consumer devices such as laptop computers, iPods, mobile phones with imaging facilities, digital cameras with flash and many other 'cordless' products has brought about a major increase in battery research. Traditionally this has been a rather slow moving and unexciting field of activity in an apparently mature branch of electro-chemistry in which new developments add to rather than driving out those that came before. This may soon change. A European directive this year puts stringent limits on the amount of highly toxic mercury and cadmium permissible in batteries, and the disposing of them.

The year marks the 140th anniversary of the development by Georges Leclanché of the type of cell that bears his name. It was soon appreciated that a major advantage over other, even earlier, cells was that little action occurred until the external circuit was applied, thus extending its shelf life when not in use. Essentially it comprised a zinc plate (cathode), a solution of ammonium chloride as electrolyte and a positive pole (anode) made by packing powdered manganese dioxide into a porous pot round a centrally place carbon rod. I still remember the large liquid Leclanché cells used to power the bells in the house where I was born. Such cells being "wet" cells were not readily portable.

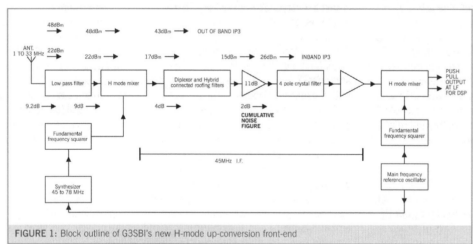

**FIGURE 1:** Block outline of G3SBI's new H-mode up-conversion front-end

It was twenty years after this invention - 1888 - that the first "dry" Leclanché cell appeared, generally attributed to the German scientist, Dr Gassner. He formed the zinc element into a cup that contained all the other elements and added zinc chloride to the electrolyte to improve shelf life. The round-cell became a commercial product. With various improvements it still - 120 years later - forms the basis of the standard carbon-zinc torch battery. Production was greatly increased during WW1 and in WW2 the "layer-type" construction reduced very significantly the space required. Layer-type HT batteries were widely used for the lowest power clandestine transmitters and for such compact receivers as the MCR1 (miniature communications [superhet] receiver) and the "Sweetheart" straight receiver.

Both round and layer-type carbon-zinc batteries have continued to find wide application in the miniature-valve and the later transistor era, but the increasing use of cordless portable devices has encouraged the development or re-evaluation of many other primary (disposable) cells. "Leakproof"; "High Energy"; "combined HT + LT layer type batteries", all based on the carbon-zinc Lelanché cell, were widely used in the post-war period.

In the 1950s, the mercury button cell appeared on the market with a zinc anode, a caustic alkali (potash) electrolyte and a cathode of compressed mercuric oxide-graphite in contact with a steel container forming the negative electrode. It had an on-load voltage of about 1.2V which remains steady throughout most of its useful life. However, it does not recover during rest periods and shows little difference in the total energy supplied during continuous or intermittent use.

The 1960s and 70s saw increasing use of the still popular alkaline manganese cell which performs well under relatively heavy loads that would quickly exhaust a carbon-zinc cell. It requires the use of good quality materials and more sophisticated construction. The anode is formed of zinc particles, combined with a little mercury or other heavy metals to suppress gassing. The alkaline electrolyte is a solution of potassium hydroxide (KOH). The entire cell is enclosed in a steel case, providing a considerably stronger and more secure casing than the zinc used for carbon zinc cells, and does not form part of the working system but is in close contact with the cathode material known as electrolytic manganese dioxide.

The major recent development has been the lithium-ion and lithium-polymer cells, providing a working voltage of some 3V, double that of the carbon zinc or alkaline cell and with a much higher energy/weight content.

The performance of a primary cell for a specific application depends on a number of factors including: (a) the physical size of the cell; (b) method of construction and skill of

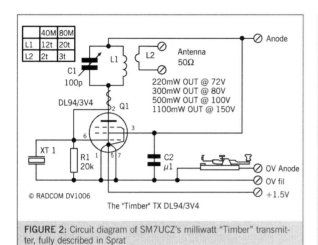

**FIGURE 2:** Circuit diagram of SM7UCZ's milliwatt "Timber" transmitter, fully described in Sprat

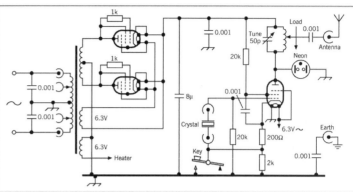

**FIGURE 3:** Circuit diagram of the 3W SOE "pocket transmitter" Type 51/1 as introduced in 1945 and possibly the smallest Allied HF transmitter of WW2. The outer case measured only 5 by 4 by 1 inches. The unit weighed only about 1lb 11oz. Component values shown include some modern replacements in the example held by John Lawrence, GW3JGA

the manufacturer; (c) the rate at which the cell is discharged; (d) period of time per day for which it is used; (e) the voltage output to which it is required to work; (f) the temperature; and (g) the age of the cell.

All primary cells are to some extent reversible, that is to say they can be recharged to some degree when fed with a carefully controlled current. However the number of charge-discharge cycles is limited and there can be safety hazards unless care is taken. Chargers suitable for a mixture of popular sizes of primary cells are available on the market, although I have no personal experience of their use.

The disposal or recycling of batteries has become a major concern. In the UK alone it is estimated that some 30,000 tonnes of used batteries are thrown away annually. There is at present no such thing as an environmentally friendly cell: each contains some toxic materials such as traces of heavy metals including mercury, cadmium etc. Hence the new European Law.

Battery R&D is increasingly aimed at rechargeable (secondary) cells and making them more energy efficient, more suitable for recycling and more safe. It is not so long since Sony had to recall some ten million of its lithium high energy cells when it was found that small metallic particles could cause internal short-circuits, leading to overheating and even fires or explosions.

The key to advance is seen in finding new combinations of materials for the anode/cathode/electrolyte and the engineering of the cells to meet specific requirements. Some applications involving only occasional use of the battery need to put emphasis on long shelf-life; others need to produce low-currents over prolonged periods; others heavy bursts of energy over short periods. It has for long been recognised by amateurs that 12-volt vehicle lead-acid batteries come in two types with the so-called "deep discharge" variety more suitable for powering transceivers than the conventional vehicle battery which is required to supply very high currents over short periods for the starter motor.

For low-current applications, new forms of low-cost, easily disposable cells are currently being researched in Finland and the UK. In these, an anode paste is printed on one side of a sheet of paper which acts as separator and electrolyte, a cathode paste printed on the other side, and then enclosed in a laminated plastic cover.

### G3SBI's UP-CONVERSION FRONT-END.

During the past decade, 'TT' has featured several high-performance front-ends based on the H-mode mixer, originally developed by Colin Horrabin, G3SBI, and later used in the AOR AR7060 receiver and the CDG2000 transceiver as described in a series of *RadCom* constructional articles. The CDG2000, with its down-conversion front-end and amateur-band RF signal-frequency band input filters, remains a state-of-the-art transceiver.

However, G3SBI has recently been concentrating on designing and building an H-mode mixer front-end using up-conversion. He writes: "The mixer termination used in the CDG2000 transceiver at 9MHz can also be used with good results in an up-converter with a 45MHz IF. As with the 9MHz IF version, two hybrid-connected roofing filters are used to present a 50Ω impedance to the mixer for close-in signals. As for the CDG2000, in order to achieve a sensitive receiver with a preamplifier requires low-loss roofing filters, in this case at 45MHz, that also have a good in-band and out-of-band IP3.

"A block diagram of a complete front-end is shown in **Figure 1**. This delivers an amplified signal down converted to a DSP IF system. This configuration will easily outperform the current up-conversion radios and is not unduly complicated.

"The FSA3157 switch has been investigated in detail as an H-mode Mixer by Martein Bakker, PA3AKE, and is well suited to be driven by a simple fundamental-frequency squarer up to 75MHz. It is a fast switching device providing a single-pole, double-throw switch with break-before-make action so that it does not require external logic to develop both the Q and not-Q signals that would have been needed with an FT3125 switch.

"A complete front-end has been built and tested - and has been given to my friend John Thorpe, the designer of the AOR7030 receiver, who is now working on the development of a new high-performance receiver for AOR UK. Richard Hillier, G4NAD, and Mark Summer, G7KNY, of AOR UK have provided components to both PA3AKE and myself (G3SBI) for evaluation and for use in the experimental front-end. It has been possible to capitalise on the excellent work done by Martein on switches and transformers to further simplify the design of the front-end.

"Overall performance of the experimental front-end on 7MHz is as follows:

"IP3 at 15kHz spacing is 45dBm. In-band IP3 (within the 15kHz bandwidth of the roofing filter) is 22dBm at 1kHz spacing. Receiver noise figure is around 10dB without a preamplifier so that the overall performance compares favourably with the CDG2000. The main difference is that the SSB bandwidth of the 9MHz roofing filter used in the CDG2000 will have a better close-in IP3 performance and also lower local-oscillator phase-noise owing to the lower frequency of its local oscillator. However, John Thorpe has come up with an interesting idea that should significantly improve the already excellent synthesiser used on the AOR 7030.

"The four-pole fundamental-mode crystal filters (15kHz bandwidth at 45MHz) have an out-of-band IP3 at 15kHz spacing of 40dBm and within their bandwidth 26dBm. Insertion loss is 1.8dB. A 7kHz bandwidth unit from the same manufacturer had an IP3

# TECHNICAL TOPICS

of 9dBm within the bandwidth, and it was necessary to go out much further to get an out-of-band IP3 of 40dBm. Thus, from a linearity point of view, the wider filter was better.

"An additional four-pole filter is used after the first IF amplifier so that image-rejection is not required at the second H-mode mixer. The in-band IP3 of the second filter sets the in-band IP3 of the front-end. The stop-band of the filter system is over 100dB within 40kHz. It would have been possible to improve the in-band IP3 (to 34dBm at 1kHz) by not having a second filter and using image-rejection circuitry.

"The first amplifier at 45MHz is a modified version of the four times J310 FET amplifier used in the CDG2000; in a 50Ω system, this has a gain of 11dB, a noise figure of 1.5dB and an output IP3 of 40dBm. This is adequate performance to achieve the overall front-end performance detailed above.

"The second H-mode mixer gives an impedance step-up and push-pull output to the low-frequency DSP system."

Although G3SBI first developed the H-mode mixer some 14 years ago, there are still opportunities for further refinement and experimentation. It has become clear that the actual mixing process occurs in the associated transformers and it is these and their cores rather than the switches that finally determine the performance in terms of intermodulation performance over specific frequency ranges.

## QRP – 'TIMBER' & CLANDESTINE VALVE
**TXs.** As a 'Christmas Project', Johnny Apell, SM7UCZ, has described in detail the construction of "The Timber Transmitter" (Sprat, Nr 133, Winter 2007/8, pp3-7). This simple valve milliwatt transmitter is built on a 1in x 4in piece of wood that forms not only the baseboard but also acts as a former for the tank circuit and low impedance coupling coil. The associated Morse key, mounted on the board, is fashioned out of a small piece of oak plus small brass pieces.

**Figure 2** shows the simple crystal oscillator circuit as used by SM7UCZ with a DL94/3V4 'miniature' battery-operated valve providing about 220mW RF output when fed from 3V LT and 72V HT (8xPP3) batteries. Other valves in this once-popular series or their equivalents could be used.

SM7UCZ provides a series of illustrations showing in detail the construction of this little milliwatt transmitter. His rig reminds me strongly of some of the low power wartime "midget" transmitters used for clandestine links to occupied countries during the final months of the war in Europe.

For example, SOE's Midget Transmitter Type 51/1, with an RF output of about 3W,

**FIGURE 4:** Circuit diagram of the Polish NP3 battery-powered midget transmitter as found in the London Polish Archives in the 1990s

Replica Paraset transmitter-receiver as built by G3WXI

used a single X136 or CV136 (EL91) 6.3V 'miniature' valve as a power oscillator but with an integral PSU suitable for AC mains supplies using two X136/CV136 valves as rectifiers. If the oscillator valve failed it could be replaced by one of the rectifier valves, the other then functioning as a half-wave rectifier. For full details see Wireless for the Warrior - Volume 4 - Clandestine Radio by Louis Meulstee and Rudolf F Staritz (Radio Bygones, 2004) or Secret Warfare by Pierre Lorain, F2WL, English translation by David Kahn (Orbis, 1983, originally with French text in 1972, pp60, 61 and 178).

'TT' January 1987 (also Technical Topics Scrapbook, 1985-89, pp148-9) included a circuit diagram (**Figure 3**) and photograph of a Type 51/1 transmitter as held by John Lawrence, GW3JGA. It was used with a simple regenerative "straight" (1-v-1 ie RF- det-AF) receiver Type 53/1 using three 9002 (or 1T4) valves with a separate AC Mains PSU.

Milliwatt transmitters operated from dry batteries were seldom used. Exceptions included the Whaddon (MI6/SIS Section VIII) Mark XXI with a two-stage transmitter (CO-PA) with 1S4 and 3A4 valves, providing about 750mW RF output, used with an associated (separate) regenerative 1-V-1 receiver using three 1T4 valves. I recall seeing this equipment in connection with the "Sussex" 1944 operation although most of the Sussex teams were equipped with the more powerful Mk VII transmitter-receivers and the 35MHz "Ascension" R/T equipment.

Another dry-battery midget transmitter was designed by Ing T Heftman at the London-Polish unit at Stanmore. This was the NP3 (later NP3A) transmitter intended as a companion unit to the OP3 "hip-flask" midget superhet receiver (1R5 –1Y4 –1S5 - 1T4) both operated from a dry-battery power container. Large numbers of OP3 receivers were made and examples can still be seen in various collections, but I cannot recall having ever coming across an NP3 or NP3A transmitter and my details are taken from files seen at the London Polish Archives in 1992. The NP3/NP3A does not appear in any of the publications listed above and I would be interested to know if they went into production, possibly intended for internal use by the Polish Home Army in occupied Poland. The NP3 (**Figure 4**) used two 1J6-G glass valves providing about 0.15W output with 130V HT. The NP3A apparently used a single dual-triode 3A5.

Of all the wartime clandestine sets, the one that seems to have most caught the imagination of present-day amateurs is the "Paraset" (Mark VII and VII/2) as built by MI6/SIS Section VIII at Whaddon Hall and Little Horwood (and possibly elsewhere for the Auxiliary (UK stay-behind) Units. There is now an active "The Paraset" club, with some 50 members, founded a couple of years ago by Rev Adrian Heath, G4GDR, and the late Tom Smith, G3EFY, who sadly became a silent key in the week before Christmas. Alan Strong, G3WXI, is now Membership Secretary (3 Ellorslie Drive, Stocksbridge, Sheffield, S36 2BB) and on behalf of G4GDR has kindly invited me to

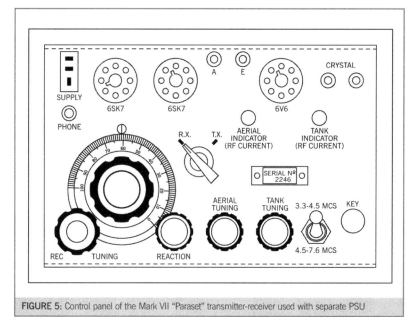

FIGURE 5: Control panel of the Mark VII "Paraset" transmitter-receiver used with separate PSU

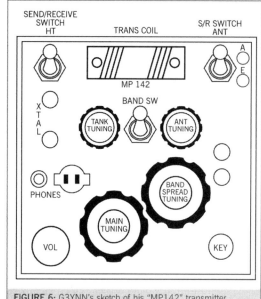

FIGURE 6: G3YNN's sketch of his "MP142" transmitter

become the first Honorary Member.

The prime objective of the Club is to honour the memory of those who designed, built and operated the equipment during WWII but also to facilitate and encourage members to construct accurate replicas, operated with simple wire aerials, using only the set's internal tank and aerial tuning controls. **Figure 5** shows the standard layout of the controls.

The Club's "Aims and Conditions" has reminded me of one of my still unresolved puzzles about the Mark VII which has been featured a number of times in 'TT'. While this set with two 6SK7 valves in a regenerative 0-v-1 receiver and 6V6 power crystal oscillator providing some 4W RF output was a simple but practical equipment, tuning the receiver did call for "safecracker's fingers". The single receiver tuning span of about 3 to 7.6MHz did mean that tuning a signal above about 6MHz required fine adjustment of the Muirhead slow-motion drive. A "Bandspread" (fine) tuning control would have been an advantage but was not provided. The Club's rules specify that "There should be no bandspread devices built into the tuning circuitry".

Yet curiously, many years ago, G3YNN sent me a sketch (reproduced here as Figure 6) of a clandestine-type transmitter-receiver using the Mk VII line-up of 6V6 crystal oscillator, two 6SK7 receiver, in-built Morse key, separate PSU (5.5 x 4.25 x 4 inches) etc but built into a square wooden box with paxolin panel. G3YNN's equipment had come from India, and the only clue to its origin is "MP142". This equipment had clearly been based on the Mk VII, but does not appear to have come from a Whaddon production line. It lacks the usual Muirhead drive but instead has a band-spread tuning knob. I believe that some immediate postwar SOE units had MP designations but have never found any reference to a MP142 unit or workshop. I wonder if any reader can provide information on this equipment?

**POWER MOSFET PRE-AMPLIFIERS.** The use of medium power devices in the front-end of a receiver system has been discussed in 'TT' several times over the years. A notable early example was the use by 'Dud' Charman, G6CJ, of 807 medium power transmitting valves in the distribution amplifiers that he developed in 1941, originally for use at the Hanslope Park SCU3 Radio Security Service's special intercept station, but later widely used elsewhere. These broadband low-noise amplifiers could accept the multitude of strong signals fed from rhombics or Vee beams and distribute them to up to eight receivers without any trace of intermodulation. I understand they remained in use at the Park, without failures, for many years.

In 'TT', September 1996 (see also Technical Topics Scrapbook 1995 to 1999, pp 114-116) in an item 'Towards the Super-linear Receiver', Colin Horabin, G3SBI described some of his experimental work, aided/inspired by American collaborators in using power MOSFETs with source feedback as a low noise and highly linear amplifier. He investigated a number of devices including the (discontinued) Siliconix VMP4, the MRF136, the DU2810s FET from the Phi Division of Macom etc.

G3SBI now writes: "I investigated these techniques because the H-mode mixer of the day used the SD5000 and its insertion loss in a receiver meant that a preamplifier was really necessary above 14MHz. The results were excellent, but when we jointly designed the CDG2000 transceiver this technique was not used, partly because the fast bus switches were a marked improvement on the SD5000. The post-mixer amplifier used the 4 x J310 design originally developed by Bill Carver, W7AAZ.

"One of the reasons that the power MOSFET approach was not used for the CDG2000 was that excessive RF at the receiver input could have produced watts of power that would have burnt out other devices and components. Additionally, the CDG2000 was not was not designed to work on the 50MHz band where a better than 10dB receiver noise figure is required.

"Martein Bakker, PA3AKE, wants his new HF/VHF receiver to cover 50MHz and potentially the 70MHz band so I sent him a copy of the September 1996 'TT' item. He has come up with an interesting device for such an application: the Philips BLF202, and intends to investigate its use as a pre-amplifier once he has built his AD9910 DDS local oscillator. The BLF202 is a surface-mount part and could prove to be ideal as a preamplifier in a 1.8 to 70MHz transceiver. I intend to do some experimental work to see how it compares with my original 1996 work with the VMP4 etc. One reason is that the BLF202 is designed for 12V operation and the device capacitances appear to flatten out at 12V whereas the VMP4 required 22V on the device to achieve this.

"Further, it seems worth trying this device to replace the 4 x J310 post-mixer amplifier used in the CDG2000. This preliminary mention of this device could also encourage other readers to do some development work with it."

# TECHNICAL TOPICS

50 Years in retrospect – Pat Hawker looks back over the half century of developments covered by TT

ANNIVERSARY THOUGHTS. For this 50th Anniversary *"Technical Topics"* it seems appropriate to look back and recall how the column came into existence, its original objectives, how it developed, a few of the many diverse topics and innovations it introduced, and then to contemplate if only briefly what the future may hold for amateur radio.

HOW DID *TT* BEGIN? In 1958 I was a member of the Society's "Technical & Publications Committee" (T&P) that met quite often at the then HQ – the fifth floor of 28 Little Russell Street, WC1, near the British Museum. In 1958, the membership had fallen from its high peak of the immediate post-war period, largely due to the problem of TVI. There was also some criticism by members of the monthly *RSGB Bulletin*, popularly known as *"The Bull"*, with some members comparing it unfavourably with *"Short Wave Magazine"* edited with a professional touch by Austin Forsyth, G6FO.

The *Bull* of the 1950s had some excellent innovational and constructional articles but by 1958 seemed to be running out of steam. With the exception of the HF, VHF, SSB and Mobile columns, articles were subject to a slow-moving peer-review process and were expected to reflect only work based on the practical experience of the author.

There was even a suggestion that the Society should approach G6FO and see if *SWM* could become the Society's journal. At the T&P Committee meeting there was discussion on how the *Bull* could be made to appeal to more members, but nothing firm emerged. So after the meeting, Roy Stevens, G2BVN (an active Council Member), John Rouise, G2AHL (Deputy Editor to John Clarricoats, G6CL) and I adjourned to a nearby hostelry to chew over the problem and try to think up some practical ways of improving the appeal of the *Bull*.

At the back of my mind was a monthly one-page feature in the pre-war "Television and Short Wave World" compiled by its Short Wave Editor, Ken Jowers, G5ZJ comprising abstracts (including diagrams) of articles published in overseas magazines. Again, during the years (1947-51) that I had been a staff member of HQ (finally as Assistant Editor), I had abstracted a few single pieces from overseas journals: for example 'Simple CW/Phone monitor' from *Radio Revista*; "Parallel Cathode Modulation" from *Amateur Radio (WIA)*; 'A Variable Frequency Crystal Oscillator" from *Amateur Radio (WIA)* (the first amateur description for what is now known as a VXO); and the 'cascode' low-noise amplifier from the original paper in *'Proc. IRE'* and since widely used in the pre-amplifier stages of HF and VHF receivers. There had also been an attractive once-only feature "Bright Ideas" (by a contributor who soon fell out with G6CL) that had been intended as the first of a regular series but never reappeared.

I tentatively suggested that I should try my hand at a new monthly column based on digests and short hints and tips. G2BVN and G2AHL welcomed my suggestion. I duly wrote to Clarry sending him a draft of what became the first *"Technical Topics"* of April 1958, sub-headed "A survey of recent Amateur Radio development – the first of a new regular series". It was by-lined "By Pat Hawker (G3VA)", the first time I had used the name 'Pat' in print. For previous published material in the *Bull* and elsewhere I had used my initials 'J. P.' or a nom de plume – a deliberate attempt to lighten the rather formal approach then standard practice in the *Bull*.

Not being a professional radio engineer I realised that I was taking the risk of making technical bloomers, particularly when abstracting material from foreign-language texts. I have to confess that once, in the early days, I took a circuit diagram of an "audio-filter" from *'Radio-REF'* that turned out to be an April fool joke. The filter placed a virtual short-circuit across the audio! A few of the antennas abstracted from professional as well as amateur journals have made exaggerated claims. I have also learned to be careful in using the word "first" which seems inevitably to remind one or more readers of some long-forgotten but basically similar idea. This is but one reason why it is worth reading old as well as current magazines and books, and taking an interest in how practices evolved, only to be overlooked; yet remain well worth reviving. I remain in awe at the ingenuity and inventiveness of the pioneers who worked with spark or the early valves. There is a tendency for each new generation of amateurs to disregard the continued value of past work.

In that first *'TT'* I set out my stall as follows: "To keep abreast of current technical progress and practice in the amateur radio field has never been an easy task. New ideas and circuits are constantly being introduced and old ones revived. Some have a short life, others are absorbed into the main stream of amateur practice. Yet often, unless one has read the original article in a British or overseas journal, it may be many months before one meets someone able to pass along sufficient details to find out what the latest technical trend may be, and to make it possible to try a new aerial or circuit device which may be just what the station needs.

"We cannot promise that this new *Bulletin* feature will solve all these difficulties. All we can hope to do is to survey from time to time a few ideas from the Amateur Radio press of the world; a few hints and tips that have come to our notice; with perhaps an occasional comment thrown in for good measure."

While these aims have been adhered to over the years, they have been extended in two significant ways. After a few issues, comments and suggestions began to come in from UK and overseas members/readers; in the second place I began to find information in the professional journals that had not previously appeared in RSGB publications yet seemed highly relevant to amateurs. So increasingly *'TT'* became recognised by members and even some non-Amateur professional engineers as a forum providing early publication for technical ideas and new developments.

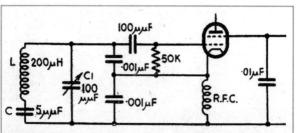

FIGURE 1: A real 3.5MHz Clapp variable frequency oscillator described by G3PL in 1949 in the RSGB Bulletin. Note the recommended values for C an L (see text). Bandspread is about 100kHz when C is about 5pF.

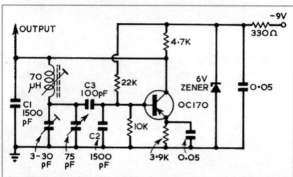

FIGURE 2: Transistorised high-stability Vackar oscillator as developed by BRS25769.

SOURCES. Many of the *'TT'* items have come from such sources as *QST, CQ, 73 Magazine, Electronics, Electronics World (USA), G-E Ham Notes, RCA Ham Tips, Electronics Australia, Australian EEB, CQ-DL, Radio-REF, Radio-ZS, DL-QTC, Sprat, UBA, Electron, Break-in, Interadio – 4U1ITU, Radio (Moscow), Wireless World (later Electronics & Wireless World), Electronic Engineering, Television, EBU Technical Review, Signal, Electronics Weekly* (on which I worked from 1963 to 1968), *Ham Radio, Communications Quarterly, QEX,* etc, and house journals such as *Point-to-Point Communication, Marconi Review,* etc. Institutional journals have included *Proc IRE* until it merged with *Proc IEEE,* IEEE's *'Transactions Antennas & Propagation',* IEEE's *'Antennas and Propagation Magazine',* IERE and IEE journals, IEE's *'Electronics Letters', Proc IREE Australia,* etc. Book sources have included ARRL publications; *"The Radio Handbook";* some of the books which I compiled or edited while with George Newnes Technical Books (1952-63), or from technical publications of the ITA/IBA's Engineering Information Service where I worked from 1968 to 1987, or the Conference Books of the International Broadcasting Conferences that I attended in Montreux, London and Brighton and the National Broadcasting Conventions I attended in Chicago, Washington DC and Dallas. Not all these sources now survive or remain unchanged.

I have always tried when abstracting material to provide full reference and acknowledgment to the original author and publication and have found that in 99.9% of cases both author and journal have been pleased to have their work recognised in *RadCom*. There was one threat of a trade libel, but it was withdrawn.

Although most early digests were from overseas amateur radio publications, I found that technical developments reported in the professional journals, trade and house magazines, available to me at libraries or at work, included interesting material of potential interest to amateurs. Again, after joining *"Electronics Weekly"* in early 1963, I attended many IEE and IERE meetings, seminars and conferences as well as press visits to UK and overseas firms, research establishments and international conferences. This continued after I joined the Independent Television (later Broadcasting) Authority (ITA/IBA) in October 1968, becoming also for some years a spare-time Editor of *"The Royal Television Society Journal"*. Many books and journals were scanned at The Science Museum Library (now largely merged with an Imperial College Library); the Patent Office Library (now part of the British Library); the IEE Library (now IET Library); the IBA Library (now dispersed). Unfortunately most of these libraries have seen changes that either make them less useful as *'TT'* sources or in the past three years less reachable due to my arthritis. And increasingly RF analogue HF and VHF engineering material is now swamped by digital information technology based on software.

All these sources provided *'TT'* useful information on new developments and new components during the years when much professional as well as amateur effort was still being put into HF and VHF radio communications equipment, antennas and propagation, using analogue and subsequently digital systems. Much of the professional HF research and development has of late been cut back, affected by the ending of the Cold War, the development of global satellite communications, short-range mobile phones, the internet and allied information technology. The London libraries have ceased holding on their shelves copies of many of the amateur radio-based journals that they formerly held. Years ago the IERE merged with the IEE; the IEE is now part of the IET and for some years have held fewer evening meetings and largely abandoned their one-day seminars at Savoy Place. The British radio communications industry has changed profoundly with the disappearance of such major firms as Plessey, AEI, Marconi, EMI Electronics, Racal, Redifon, BCC, GEC and so on.

The BBC and IBA no longer own the UK radio and television transmitters, with the IBA becoming first ITC and now part of OFCOM. The Government hopes that broadcasting will eventually be based solely on the complex digital systems and allow it to sell off much of the spectrum, although I suspect that AM and FM radio will not disappear for many years. The public is accepting DTV and anticipating that HDTV will flourish. But DAB and DRM are struggling – and HF SSB broadcasting which should have been mandatory by now has virtually disappeared off the radar. The growth of cable distribution has been slowed by terrestrial and satellite Freeview. Broadcasting seems to be seen increasingly as just part of IT, distributed in digital form over the web or by satellite to laptops, i-Pods, mobile phones etc.

All these changes have made the compilation of *'TT'* more difficult and more time-consuming; indeed the need for it is more questionable in these days of factory-built equipment and with so much information available on the web.

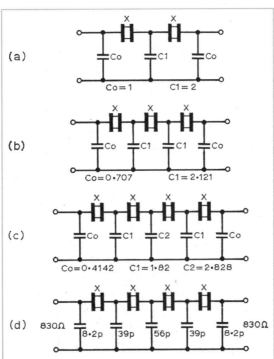

FIGURE 3: Early crystal ladder filters investigated by F6BQP. All crystals (X) are of the same resonant frequency and preferably between 8 and 10MHz. To calculate values for the capacitors multiply the coefficients given above by $1/(2\pi f R)$, where F is the crystal frequency in Hertz, R is the input and output termination impedance and $2\pi$ is roughly 6.28. (a) Two crystal unit with relatively poor shape factor. (B) Three crystal filter can give good results. (d) Four crystal unit capable of excellent results. (d) Practical realisation of four-crystal unit using 8314kHz crystals, 10% preferred value capacitors and termination impedance of 820 ohms. Note that for crystals between 8 and 10MHz termination impedance should be between about 800 and 1000 ohms for SSB. At lower crystal frequencies use higher design impedances to obtain sufficient SSB bandwidth.

**THANKS.** I had intended at this point to list and thank some of the many who have contributed novel circuits, antennas, oscillators, filters (crystal, mechanical, LC etc). But there have been so many contributors to *'TT'*, both UK and overseas – literally hundreds – that it would be invidious to single out and select what could only represent a limited selection. Nevertheless such names as 'Dud' Charman, G6CJ (SK), Les Moxon, G6XN (SK), Dick Rollema, PA0SE, Peter Martinez, G3PLX, Klaas Spaargaren, PA0KSB (SK), Dave Gordon-Smith, G3UUR, Jan-Martin Noeding, LA8SK (SK), Brian Austin, G0GSF, Jack Belrose, VE2CV, Colin

Horrabin, G3SBI and Gian Moda, I7SWX spring to mind, but there are many, many others to whom *'TT'* is indebted. It has always pleased me that *'TT'* reaches and is read by so many overseas amateurs and professionals. Much of my time has been in dealing with and responding to the many letters and queries that have reached me. I take this opportunity to apologise to those readers who have submitted ideas that for one reason or another (including the occasional mislaying of them) have not appeared, and to some correspondents to whose letters, even after extensive searches, I have been unable to answer satisfactorily, or not at all.

So now, in this 50th Anniversary *'TT'*, just a tiny selection of some early items that stick in my memory as being developments that have (or should have) subsequently become accepted as part of standard practice; concentrating on topics that did not appear in the recent "50 year reviews" or are not otherwise available in the *"Technical Topics Scrapbooks"*.

## CLAPP & VACKAR OSCILLATORS

The immediate post-war decades saw intensive search for VFOs of improved stability. Voltage regulation, temperature compensation, high-Q coils, careful choice of components and their placement away from heat sources led to improvement of the classic Hartley, Colpitts and Franklin circuits. Then, in the late forties, J K Clapp described an oscillator based on a crystal-equivalent circuit (it was later shown that the BBC engineer, Geoffrey Gouriet, had developed and used professionally a similar arrangement during WW2). Amateurs soon adopted this oscillator although often failing to observe the correct choice of component values, even after A G Dunn, G3PL ("Clapp or Colpitts?", RSGB Bulletin, June, 1949) pointed out: "Many versions of the Clapp oscillator have been published in recent months, but it is evident that the main difference between the Clapp and Colpitts have escaped the attention of the designers of some of these so-called Clapp circuits. The majority are found on close inspection to be little, if-at-all, better than the original Colpitts circuit ....". After discussion of what constitutes a crystal equivalent circuit, he provided an example of a real Clapp VFO for 3.5MHz: **Figure 1**.

G3PL stressed the low value (5pF) of C and the high value (200µH) of L. He noted that C is often shown as a variable capacitor for tuning purposes but this is not essential. As a result the value has often been given as 400 or 500pF in some published circuits. L is of a value more usually associated with medium-wave circuits, and is some twenty times as large as the values commonly used in VFO circuits for 3.5MHz.

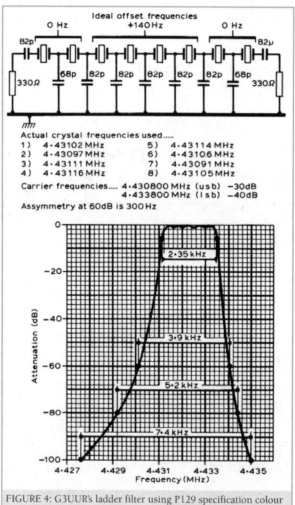

FIGURE 4: G3UUR's ladder filter using P129 specification colour TV 4.43MHz crystals showing the excellent shape factor etc that can be achieved.

Perhaps it was the misuse of the Gouriet/Clapp circuit that led to further investigation of stable variable-frequency oscillators. In *'TT'* ("Transistorised Vackar Oscillator" July, 1966) I wrote: "Thanks to L Williams, BRS25769, of Birmingham, we are able to include a circuit which appears to have real promise for both transmitters and receivers but which, as far as we are aware, is described here for the first time."

BRS25769 wrote: "I have done some work on a high stability oscillator circuit (**Figure 2**) which is a transistor version of the Vackar-Colpitts (Tesla) circuit. I am sure someone must have done it before but have never seen it in print. The circuit arose from a search for a VFO for a high stability receiver... Published circuits seem to be either of the Hartley or Clapp form, usually with buffer amplifiers. The Clapp has the feature of the output varying with frequency and transistor buffers do not give as high a degree of isolation as a well-designed valve stage.

"The Vackar circuit has the transistor terminals shunted with very large lumps of capacitance, as are also the output terminals. The values given cover 2.0 – 2.5MHz, and the prototype will stay zero beat with a crystal standard for hours. With C1 and C2 polystyrene and C3 silvered-mica, temperature drift is plus 10Hz per degree Centigrade; this could be improved by making C3 mixed mica and ceramic. The amplitude of oscillation is controlled by C2. Increasing C2 reduces amplitude without very much effect on frequency. For good stability

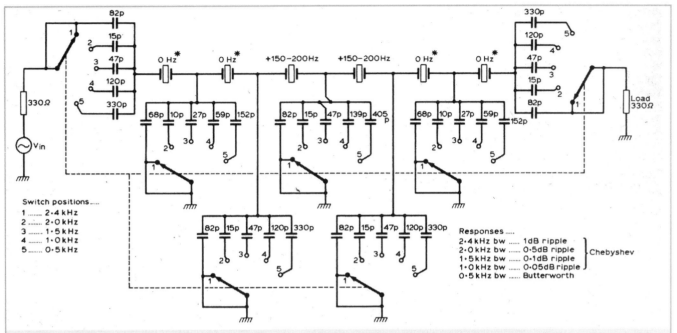

FIGURE 5: G3UUR's design for a switched variable-bandwidth filter using colour TV crystals. Note crystals shown as 0Hz offset can in practice be +/- 50Hz without too detrimental effect on the passband ripple.

the amplitude [with an OC170 pnp device] should be limited to a few hundred millivolts. ..."

CRYSTAL LADDER FILTERS. An item "Making Crystal ladder filters!" ('*TT*', September 1976) noted that for years most bandpass crystal filters had been based on the half-lattice or lattice configuration: "Such filters require the use of a number of crystals of carefully defined (and different) frequencies, and often the use of centre-tapped inductive components - inevitably expensive items to buy and at HF very difficult to construct.

"Relatively little has been published in amateur journals about an alternative filter configuration – the ladder network – that appears to offer very useful features to those who wish to save money by building their own filters. The little that has been published – for instance in current advertisements for the Atlas solid-state transceiver – suggests that very high ultimate rejection figures can be achieved....

"The absence of information on building ladder-type crystal filters has to some extent been rectified by a most useful article by J Pochet, F6BQP in '*Radio-REF*' (May, 1976) covering filters using two, three and four crystals *all of identical frequency*: **Figure 3**.

"F6BQP shows clearly that very useful SSB filters can be made, seemingly with few problems, by anyone having on hand, say, four identical crystals. His prototype filters were based on 8314kHz crystals ...but can be made at any frequency from roughly 5 to 20MHz, although the capacitances and impedances favour the use of 8 to 10MHz."

The November, 1976 '*TT*' showed a CW ladder filter using four 8794kHz and a 50-ohm input and output impedance, adding that J A Hardcastle, G3JIR was preparing a detailed article following extensive work on the design and construction of ladder filters that had begun before my reporting that of F6BQP. His four-part article "Some experiments with high-frequency ladder crystal filters" ('*Radio Communication*' December, 1976, January, February and September, 1977) remains a classic. A shorter version was published in '*QST*'.

For '*TT*' (June 1977) Hans Kreuzer, DL1AN reported on ladder crystal filters based on the low-cost 4.33618MHz crystals used in colour television receivers to provide the colour sub-carrier reference frequency for the synchronous demodulation of colour signals. Since then PAL and NTSC (3.5796875MHz) crystals have been widely used for low-cost CW and SSB ladder filters. Years later, in April 1999, Jan-Martin Noeding, LA8AK showed that useful LF and MF ladder filters could be constructed using ceramic resonators.

G3UUR showed ('*TT*', December 1980) how ladder filters with excellent characteristics and/or switched bandwidths could be implemented, although requiring rather more careful selection of PAL crystals: **Figures 4 & 5**.

More complex than a ladder filter, attention was drawn in '*TT*' (December 1969) to the continuously variable-bandwidth symmetrical IF filter used in the Rohde & Schwartz receiver Type EK07-80 using two high-slope low-pass filters and four mixers: **Figure 6**. Despite its exceptional performance, most readers seem to have decided it was too complex to copy. However, quite recently (see '*TT*', July 2002 and October 2004) Dick Rollema, PA0SE reported his use of this approach (which he called a "sliding doors" system) with two 10kHz low-pass filters in a high-performance, home-built HF transceiver. Unlike the R/S filter, his could be adjusted for

LSB, USB or symmetrical response, each of continuously variable bandwidth: **Figure 7**.

COMPONENTS, REPAIRS & SAFETY. Over the years, *'TT'* has provided early information on many new components and active devices, plus a great number of hints and tips on checking, servicing and constructing equipment. This has included simple test instruments, making of and fault finding on printed circuit boards, soldering techniques and 'third-hand' devices, soldering jigs, handling surface-mount devices and components etc. Some of the ideas have been derived from trade and industry magazines or overseas amateur journals, many others been contributed by readers. Unfortunately, it is now impracticable to recommend home servicing the current generation of factory-built amateur transceivers with their surface mounted components, except possibly to those with a solid background of professional or extensive amateur experience.

However, older hard-wired valve or solid-state equipment based on discrete devices and/or boards using standard size ICs and discrete components can still, with care, be tackled at home with a limited number of test instruments and tools, with some experience. I have to confess that I have spent some months trying, so far unsuccessfully, to trace a fault on an ancient AR88 - the main problem is proving its sheer size and weight, leading to my virtual inability now to move it around!.

Although today one sometimes feels that there is too much attention paid to safety and too little to experimentation, *'TT'* has highlighted the potential hazards arising from electric shock, toxic chemicals in components, servicing aids, asthma inducing solder-flux fumes, and eye protection while soldering or metal working, cadmium in transistor casings etc, erecting and dismantling antennas and masts or using ladders; wartime luminous paint in surplus equipment; and some critical looks at the banning of lead in solder.

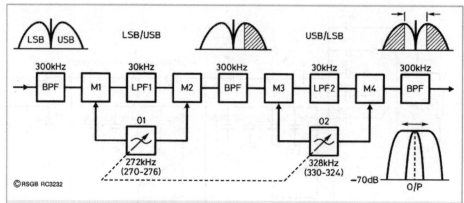

FIGURE 6: Basic principles of the 1969 R/S EK07-80 filter which uses two 30kHz low-pass filters in conjunction with ganged oscillators tuning in opposite directions. Variable from about +/-150Hz to about +/-6kHz with similar slope at all settings down to -70dB.

DIRECT-CONVERSION RECEIVERS. *'TT'* has always shown interest in receivers that can be easily constructed yet capable of providing practical CW and SSB reception. This may reflect my own use pre-war of a simple two-valve (0-v-1) receiver with regenerative detector and a transmitter limited by my licence to ten watts DC input which enabled me to work some DX including North & South America, Australia, and Southern Rhodesia (Zimbabwe). The regenerative detector can be extremely sensitive and fairly selective when just oscillating (good for CW and not too bad for SSB) but the gain and selectivity suffer badly when out-of-oscillation and used for AM or in the presence of strong AM signals. Simple regenerative and VHF super-regenerative receivers have been featured a number of times.

In March 1967 I expressed what has always been a tenet in compiling *'TT'*: "One way of developing something new is to look back at ideas that have been around for a long time but which have never been widely used. Often basic principles are developed many years in advance of the materials and devices that make them practical or economical. Poulsen's magnetic recording of 1889 had to wait until the 1940s brought forth the practical domestic tape recorder; Blumlein's stereo disc techniques of 1929-31 were not used until the past decade [1950s]; Robinson's 'stenode' tone correction with a single-crystal filter was largely forgotten until G6XN revived the idea in 1962".

As an example, in November 1967, I wrote: "In a recent *Electronics Weekly* we drew upon these histories to

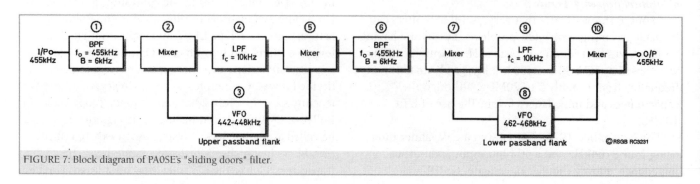

FIGURE 7: Block diagram of PA0SE's "sliding doors" filter.

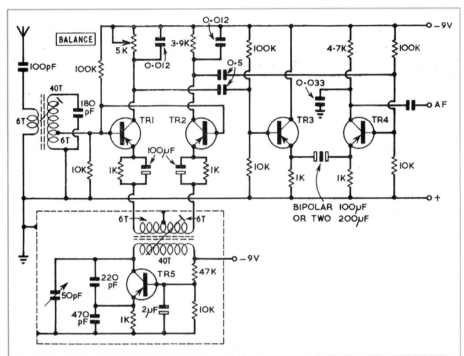

FIGURE 8: Circuit diagram of PA0KSB's homodyne-type direct-conversion 3.5MHz receiver, first published in the Dutch journal 'Electron' and then in 'TT' (March 1967) was one of if not the earliest transistorised receivers of this type. Transistors were pnp AF124s in the original but almost any RF types would be suitable (with npn types voltages should be reversed). The coils were wound on Philips-T formers.

– this time with semiconductors – is to be found in the Dutch journal 'Electron' (January, 1967) where [Klaas Spaargaren] PA0KSB describes the front-end of a simple 3.5MHz receiver for CW or SSB: **Figure 8**. Two transistors form a balanced detector followed by a form of differential amplifier to produce an unbalanced AF output suitable for feeding into a conventional AF amplifier, with the fifth transistor used as a heterodyne oscillator. Using a 7m long aerial wire, PA0KSB says that he has been able to hear VE1 stations on 3.5MHz in the winter evenings…"

Quite soon, this form of receiver received a major boost with the publication in 'QST' November 1968 (summarised, in 'TT', February 1969) of an article "Direct Conversion – A neglected technique" by Wes Heyward, W7ZOI and Dick Bingham, W7WKR, with full constructional details of a 3.5MHz receiver capable of copying CW signals down to below the microvolt level and of sufficient stability to give reasonably good SSB performance: **Figure 9**."

Throughout the following years, this form of direct-conversion has been widely adopted, with many variations, as the basis for low-cost, home-built receivers, despite the presence of the audio image. Later on, they included the addition of image-rejection techniques using phasing-type filters (including those based on digital ICs, 'third-method' and polyphase networks). This approach, although adding complexity, opened the way for direct-conversion receivers with a performance as good as, or even better, than conventional superhet receivers, as found for example in the Elecraft K2 transceiver kit. Direct-conversion also forms the basis for some of the "software defined radios" currently attracting attention.

question whether it is not time to look again at synchrodyne or homodyne receivers. The synchrodyne can be regarded either as a superhet with an IF of 0kHz or as a straight [direct-conversion] receiver with a balanced linear heterodyne detector. The true synchrodyne is the simplest form of phase-locked receiver, and in essence consists simply of a balanced mixer (product detector), a local oscillator locked to the incoming carrier, and an audio amplifier: no IF strip or second detector, no images or spurious responses, and – at least in theory – the ability to govern selectivity entirely by the bandwidth of the [high-gain] AF circuits, a far cheaper way of achieving high selectivity than with a crystal or mechanical filter.

"Of course, if it were all as simple as the last sentence suggests we would all have thrown away our superhets years ago. We recall, back in the 'forties, in conjunction with [Charles Bryant] GW3SB trying to get a Tucker synchrodyne circuit to work on AM broadcast stations without much success. But then, a lot more is known today about linear balanced mixers, and about phase-locking. And again, for SSB and CW reception there is no need for the local oscillator to be phase-locked to the incoming signal. Some time ago, 'TT' drew attention to W2WBI's 3.5/7MHz receiver of this type ('QST', May 1961) using two well-balanced 6SB7Y valves as heterodyne detector, with a separate well-screened heterodyne oscillator – but the item did not attract much attention. ...

"A positive sign that some amateurs are still considering the possibility of using such an arrangement

POLYPHASE FILTERS. While the vast majority of circuit developments reported in 'TT' have represented the work and ideas of amateurs and/or professional radio engineers, I can claim to have played a small part in the application of 'polyphase' filters to HF SSB transmitters and receivers. In October 1973 I wrote: "All amateur transmitters employ one of three systems for generating SSB: (a) the filter method; (b) phasing ('outphasing' or quadrature) method in analogue or digital form (**Figure 10**); and (c) the 'third method' (Weaver or Barber). .... It would be exciting to report that some entirely new method of SSB generation has been developed which combines all the

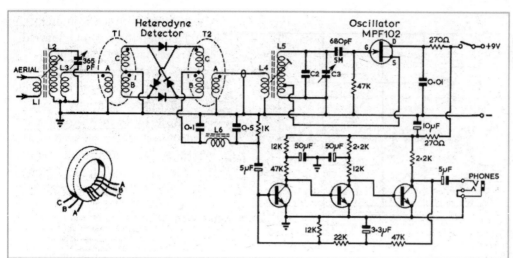

FIGURE 9: Circuit diagram of the 3.5 or 7MHz homodyne-type direct-conversion receiver by W7ZOI in 'QST' (November 1968) and in 'TT' (January 1969) that set the pattern for the home construction of simple but effective receivers. FET oscillator C2 is 470pF S.M. for 3.5MHz, 120pF S.M. for 7MHz. C4 is 140pf to cover 3.5 to 4.0MHz, 40pF for 7MHz. The three bipolar transistors were all RCA 40233 but other AF types could be used. L1, L3 3-turns, L2 40 turns No 28 enam. on 0.680-in toroid; L4 5 turn link; L5 22 turns No 28 enam. on 0.680 toroid with tap 5 turns from earthy end. L8 is 88mH toroid (a small smoothing choke might prove reasonably suitable). Since 1967 a large number of variations on this general type of receiver and hints on improvements etc have been published in 'TT' and RadCom.

virtues and none of the vices of all three systems, providing, say 60dB of sideband suppression at low cost and without unduly critical component values. It would be even more sensational to publish a circuit diagram showing exactly how to assemble such a system.

"I cannot do the second – but it is possible to draw attention to an article which explains the principles of a system which seems to fill most of the requirements: "Single-sideband modulation using sequence asymmetric polyphase networks" by M J Gingell, of STL, in *Electrical Communication*, Vol 48, No 1 – 2 (combined issue) 1973, pp21-25. Gingell's basic polyphase network is shown as in **Figure 11**.

"My difficulty is that even after reading it through several times and phoning the author, I find it difficult to attempt to translate a pretty involved and complex paper into terms which I, and I suspect many readers, would understand. ... It is claimed that a sideband suppression of 60dB can be achieved consistently in a polyphase network of capacitors and resistors with tolerances as much as +/- 2.5% at the input section to about +/- 0.2% at the output. In fact by adding or taking away extra sections one can design for varying degrees of performance and component tolerance. Up to 70dB of suppression has been achieved."

Michael Gingell's work at STL was in connection with line telecommunications at around 100kHz but in our telephone conversation he could see no reason why a polyphase network should not be used in HF transmitters. Years later, I learned that he had been disappointed that STC (ITT) had decided to use conventional line filters and his work had been shelved.

I concluded the item by writing: "I hope that by drawing attention to this novel technique I may stimulate somebody who really understands or can grasp how a sequence asymmetrical filter really works and how it could be designed into an SSB generator suitable for amateurs to build. It would be possible to make a bit of communications history by doing so, since I gather that no commercial applications have yet appeared. So how about it somebody?".

Peter Martinez, G3PLX was quick to respond. The December, 1973 *'TT'* included an item "More on polyphase SSB" which began: "Peter Martinez, G3PLX, has done a valiant and valuable follow-up to the October item on the polyphase system for SSB generation. Although he has not had time to build a complete SSB generator he has gone back to the original paper by M J Gingell and translated the mathematics into practical terms, and has built several polyphase networks with results well in line with expectations: for example **Figure 12**.

He sees the polyphase network as, in effect, a new way of making a wide-band audio phase-shift network – but with the advantage of being much more tolerant of component values than the well-known approaches. With a six-stage network, which can be realised using readily available components, he believes that 10% tolerance can be used in the first five stages and two per cent in the final stage, yet this should be capable of providing some 40dB sideband suppression. .... To complete the SSB generator, Michael Gingell feeds the a, b, c, d outputs to four modulators, and combines the outputs; each modulator being fed with RF carriers in the sequence 0, 90, 180, 270 degrees. This has the advantage of cancelling out harmonic sidebands which occur on either side of the harmonics of the carrier frequency. But a simpler arrangement would be just to feed a and b to two balanced modulators fed with carriers at 0 and 90 degrees. ..."

G3PLX provided a most useful simplified explanation of this type of network including the advantages of using a six-section network, permitting the use of readily available preferred-value components.

It was a pity that by the time that Gingell's polyphase system was published in *'TT'* the era of large scale home-building of SSB generators had subsided, with the result that it has not made a serious impact on amateur transmitter practice. However there is little doubt that it can be an effective and useful configuration, now recognised by both amateur and professional designers.

As I reported in TT, July 2001, Michael Gingell worked on polyphase and related techniques for some ten years, from about 1966, by the end of which time STL's interest had switched to digital filter techniques. In 1974 he received a PhD from London University for his work on polyphase filters and SSB. He later moved to the States and became KN4BS.

'TT' March and April 2001 included an account of some subsequent developments in the use of polyphase networks in receivers, including the experimental design of a 14MHz high-performance, single-signal, direct-conversion receiver by G3OGW using a four-phase polyphase filter as an SSB demodulator together with the use of FST3253 switching mixers. An outline of G3OGW's experimental design is shown in **Figure 13**. Unfortunately G3OGW became a Silent Key before converting his prototype into a fully operational multiband receiver – but not before showing that this could become a valuable approach still to be fully exploited in receivers or transceivers. This is an area that could well form the basis for further experimentation, possibly linked with DSP and other software defined techniques.

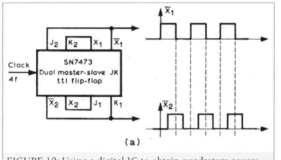
FIGURE 10: Using a digital IC to obtain quadrature square-wave signals.

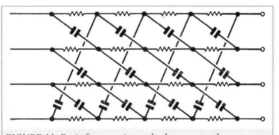
FIGURE 11: Basic four-section polyphase network.

## ANOMALOUS PROPAGATION.

'TT' has always shown interest in reporting and commenting on anomalous propagation conditions that can extend the range of LF, MF, HF, VHF and microwave signals. Such modes may permit communications above the MUF on HF or well beyond the horizon on VHF and above. Examples include G6XN's early work on extremely low angles of elevation by which he achieved SSB contacts with Australia using only 1 watt from a sloping site adjacent to a salt-water loch; Albrecht's chordal hop; whispering galleries; gray-lines; forward- and back-scatter, antipodal enhancement, round the world echoes, etc at HF and 50MHz.

Anomalous modes include Sporadic-E, meteor scatter, troposcatter, ionospheric scatter, moonbounce, trans-equatorial, rain scatter, "aircraft-enhanced" modes at VHF and above. Because many of these VHF modes are fleeting or unpredictable they tend to be of more interest to amateurs than to professional communicators, other than broadcasters who often regard them as sources of unwanted interference.

One of the more significant post-war discoveries in the field of HF propagation was that made by Dr H J Albrecht, VK3AHH/DL3EC, when he realised that the signal strength and reliability of European amateur signals received in Australia on 3.5, 7 and 14MHz could not be accounted for by the conventional theories of multihop propagation, leading him to propose the theory of long-distance 'chordal hop' HF propagation without intermediate ground reflection losses. This concept, together with similar work by Fenwick and Stein on "around-the-world echoes" and the satellite-orientated work on "whispering galleries" have received many mentions in 'TT' and has gradually led some of us to believe that the *majority* of long-distance HF contacts made by amateurs using low power are made without intermediate ground reflections rather than conventional multi-hop mode. Sea reflections attenuate signals less than ground reflections so that multi-hop paths are not uncommon. Albrecht's work was brought to my notice in the early 'fifties by Les Moxon, G6XN.

A 1979 article in *'Telecommunication Journal'* by German External Service broadcast engineers showed in detail how a superior service to Australia and New Zealand could be provided (Australian evenings, European mornings) by utilising the 24,000km long path across South America on about 8MHz rather than via the 16,000km short path, often up to 25dB stronger than that calculated from CCIR formulae for higher frequencies nearer the MUF.

Less successful have been the attempts to investigate and/or explain the "long delay echoes" (LDEs) that have been credibly if rarely reported since the 1920s.

## COMING UP TO DATE AND THE FUTURE.

Two relatively recent developments that have appeared first in 'TT' have been the wide dynamic range H-mode mixer and the low-phase noise twin-tank oscillator, both representing significant state-of-the-art circuit configurations. These were originally developed by Colin Horrabin, G3SBI, in the mid-1990s but since further improved with the aid of the latest IC devices. Since these topics have been covered in recent issues and in the series of *'Technical Topics Scrapbooks'* they are noted here only as examples of the continued role of amateur radio developments contributed by those with both professional and amateur experience based on analogue and digital RF engineering.

It is clear that future developments will increasingly be governed by improvements in IC devices in which more and more bits work at increasing speed at ever lower

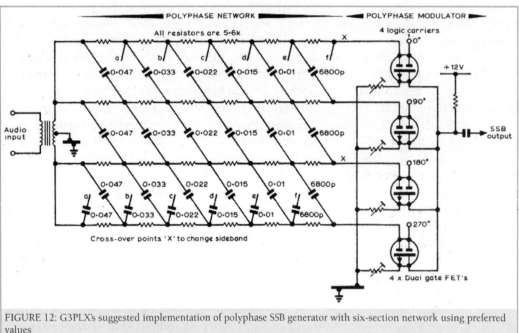

FIGURE 12: G3PLX's suggested implementation of polyphase SSB generator with six-section network using preferred values

voltages. But even here there are many who believe that we are already approaching the limits imposed by silicon and even gallium arsenide, not to mention the problem that devices can demand all the skills of a watch-repairer equipped with powerful optical binocular magnifiers.

This perhaps is why many of those amateurs who believe that home construction should still form at least part of the hobby are turning back, as Peter Chadwick, G3RZP recently suggested, to the use of discrete components and to replicas or refurbishment of valve equipment. I make no apology for the space devoted since the 1970s to the clandestine radios of the Underground groups and Prisoner of War receivers. There is also still room for experimental work with crystal sets, as noted in 'TT' as recently as January 2007.

Many years ago I wrote: "Digital techniques utilising microprocessor intelligence and either with or without substantial electronic memory can do many useful things, Such systems can sense, can control, can encode or decode, can convert one code into another, can change the rate of flow, can retrieve information in many different arrangements, can display information – all with accuracy, and unlimited endurance. Unlike the human operator, the microprocessor does not grow tired and careless, and once the program has been fully debugged does not itself introduce mistakes".

By now, we have learned (though not necessarily used) that effective filtering, mixing, demodulation can all be achieved by digital processing. With modes such as PSK51, signals can be received successfully from below the noise level. The human brain, however, can still hold its own in being more flexible and adaptable and amazingly good at solving those problems that depend on some form of pattern recognition. I believe, or at least hope, that there is still a role for the human operator –
and for the home constructor or informed critic of factory-built equipment stemming from an appreciation of the role of the individual components and their limitations. There are still controversies to settle: the emergencies of global warming seem likely to call for thin-line systems, including the use of NVIS HF propagation, supplied from improvised power sources. There will still be enthusiasts for restoring and repairing vintage equipment. Nostalgia can be a powerful learning tool.

Nor should we overlook or forget the pioneers of radio and electronics. As someone born before the first transmissions of the British Broadcasting *Company*, I feel privileged that I have either met or attended lectures by such 'greats' as Dr Zworykin, Alec Reeves, Dr George Brown (RCA), Sir Robert Watson Watt, Sir Bernard Lovell, Sir Martin Ryle, G3CY and may possibly have brushed shoulders at Hanslope Park with Alan Turing.

But I have to admit that, looking back for this Jubilee issue to the early days of compiling 'TT' in my limited spare time, I cannot help feeling that the first 25 years were the most rewarding and personally satisfying, writing then to readers who had known the joys and frustrations of home construction, who saw that HF radio communication still played a useful, unchallenged role in global communications. Today, well, *sic transit gloria* – but let us believe with some certainty that there will in future still be a role for the technically-minded, experimentally-minded Radio Amateur!

ADDENDA. I had intended to end this special retrospective Anniversary 'TT' at this point but there are still some matters outstanding, arising from recent issues that require clarification and/or amendment rather than being held over for a possible *"Technical Topics Extra"* in a few months time.

ANTENNA TRANSMISSION LINES. In the long leading item: "Is there nothing new in antennas" (December 2007, pp77-79), in the section on SWR, I inadvertently added in parenthesis some misleading advice. As printed (1st column, p79) the sentence read: "Note that when an antenna element presents to the transmission line an impedance other than its characteristic impedance, the

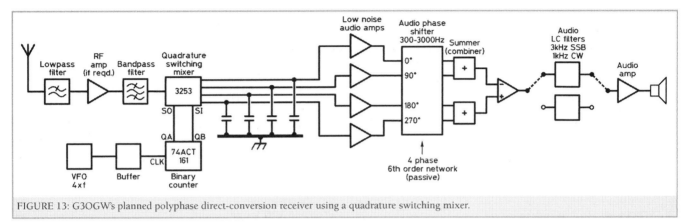

FIGURE 13: G3OGW's planned polyphase direct-conversion receiver using a quadrature switching mixer.

impedance offered to the transmitter at the input end of the line may be quite different from either the characteristic of the line (unless the line is an exact multiple of an electrical $\lambda/2$) or the impedance at the antenna junction."

The note (added as an afterthought) "(unless the line is an exact multiple of an electrical $\lambda/2$)" is incorrect. Stewart Rolfe, GW0ETF writes: "Forgive me if I am misinterpreting your sentence as published but it is my understanding that in such situations the impedance offered to the transmitter will 'never' equal the characteristic impedance of the line. At an electrical half wavelength from the antenna feedpoint, the (lossless) line impedance will be the same as that at the feedpoint itself, being in effect once round the line circle on a Smith Chart where the (normalised) impedance is the circle's centre point. The Smith Chart thus illustrates how the line characteristic impedance will only ever be presented on a flat line (ie no reflections or standing waves).

"I realise how easily meanings are 'lost in translation' when discussing antennas and transmission lines; I have been in friendly discussion with a fellow club member who insists SWR is a function of line length. This is indeed the case in practice due to inevitable imbalance in most antenna systems which results in the feeder itself acting as a third radiating element to a greater or lesser extent. Thus by altering feeder length the dimensions of one (the minor) radiating element is being changed which therefore will vary the antenna feedpoint impedance and in turn the SWR. Trying to find the words to describe this in straightforward terms can be a challenge. My response to those who argue their antenna is 'good' because the SWR is nice and flat over a wide bandwidth is to point out that the system with the best 1:1 SWR bandwidth is a dummy load."

EARLY TRANSISTORS. The December 2007 review of the early development of transistors continues to stir readers' memories. John Teague, G3TGJ recalls that about 1949 while an engineering trainee at EMI, Hayes a friend, Gil Trafford, in charge of the Valve Applications Laboratory, rang him: "Come over here, I've got something to show you". G3TGJ writes: "When I arrived he revealed a padded package he had received that morning which contained some small white-painted devices with three wires [one from each end, the third from a central rectangular block, 5 x 4 x 4mm, overall length about 15-20mm, cylindrical end tubes 2.5mm diameter]. These were the first transistors the company had obtained from RCA with which EMI was then closely connected. Their configuration was entirely different from any transistor I have seen since."

Derek Slater, G3FOZ recalls an article in 'Short Wave Magazine' in the mid-1950s by John Osborne, G3HMO, describing how to make a point-contact transistor from a germanium diode. G3FOZ writes: "I remember destroying several diodes, with inconclusive results. For me, as a chemist, the best bit was putting a point on a tungsten wire by dipping it into molten sodium nitrite. I also remember that a part of the process was to discharge a capacitor through one of the junctions. Quite why, I can't remember.'

SMALL LOOPS. I must also report receiving two letters from Canada in connection with G3LHZ's small loop claims. Alan Goodacre, VE3HX writes in support of Professor Underhill. He puts forward a hypothesis explaining why he believes it is possible to achieve high radiation efficiency. He backs this up with thermal measurements made on 14MHz with a 1m diameter copper tubing loop, comparing them with those on a 10m mobile whip. Ingenious, but one suspects that his methodology may be even more error-prone than that of G3LHZ. Dr John Belrose, VE2CV, on the other hand, strongly reiterates his considered opinion, based on theory and experiment, that the radiation efficiency claims made by G3LHZ cannot be substantiated and fly in the face of long-established theory supported by virtually all professional antenna engineers and physicists.

FINALLY. May I express my gratitude to those UK and overseas readers who in letters or during QSOs have conveyed their appreciation of TT and the pleasure it has given them. *Thanks!*

# Index

250kHz FM generator ... 140
3V MOSFET shunt regulator ... 15
500kHz ... 136, 147
7360 beam-switching mixer ... 22
Alternative power sources ... 72, 81
Aluminium ... 40, 42
AM 68, 77, 85, 98, 107, 133
Antennas ... 144, 170
    50MHz halo ... 22
    Antenna current probe ... 117
    Beverage ... 79, 85
    Electrically small self-resonant antennas ... 57
    Ground plane ... 69
    Handset folded antennas ... 22
    IIDM antenna ... 154
    Large horizontal loops ... 103
    Low dipoles ... 6
    Monopoles ... 69
    Quad loops ... 101
    Radials ... 58, 69, 80
    Ships' aerials ... 90
    Short-span multiwire folded dipole ... 32
    Small loops, G3LHZ loops ... 9, 44, 55, 146, 155, 171
    T antenna ... 102
    Verticals with compact radials ... 80
    Windom antenna ... 64, 81
    Zepp antennas ... 62, 75, 92
    ZS6BKW antenna ... 119
ATU with balanced output ... 79
Audio triode centennial ... 100
Automotive electronics ... 55, 71, 140
AVO meter ... 115
Balanced feeders ... 84
Baluns ... 94, 125, 135
Batteries, chargers ... 107, 156
Binaural cocktail parties ... 35
BPL / PLT ... 146, 151
Bright LEDs & conductive polymers ... 123
Buying overseas – caveat emptor ... 32
Bypassing: one or multiple capacitors? ... 16
Cascode / cascade amplifiers ... 53, 64
Ceramic resonators ... 70
Circuit breaker ... 80
Clandestine valve TXs ... 158
Coax ... 33
Conductive polymers ... 123
Conjugate match ... 86
DC-DC converters ... 15, 30
DDS, DDS chips ... 66, 68, 75, 76, 88, 140, 152, 153
Demodulation – product & envelope ... 1
Dielectric capacitors ... 137
Diode mixers ... 19
Double-balanced beam-switching mixers ... 128
Earth-rod performance ... 62
Ergonomic controls ... 5, 108, 120, 133
Europe and the semiconductor ... 150
Fessenden ... 126, 127
Filters (crystal, bandpass, polyphase) ... 48, 64, 139, 165, 167
Ground losses ... 55
Henn-Collins, Lt Col, ex-GU5ZC / G5ZC, SK ... 99
High sensitivity crystal set ... 116
High voltage & overload shut-down ... 105
High-level mixers using LEDs ... 2, 19
H-mode mixer ... 28, 46, 117, 132
Home construction ... 104
How did 'TT' begin? ... 161
'Huff & puff' ... 37, 105
IMD in digital receivers ... 131
Improving solid-state front-end ... 129
Inventions & inventors ... 126
Jigs 85, 88
Keeping the PA working ... 50
Law of unexpected consequences ... 88
LC balun matching networks ... 94
LC oscillators to DDS ... 140
Lead-free solder ... 42, 53

Low-cost huff & puff stabilised VFO ... 37
Marconi ... 114, 126, 135
Measurements, professional and amateur ... 28
Mixers ... 97, 139, 142
Mobile phone risks ... 45
Morse code ... 134
MOSFETs ... 15, 159
Neat earth connections ... 14
New low-spurious DDS chip ... 88
Nøding, Jan-Martin, LA8AK, SK ... 27
Nylon washers as VHF toroids ... 14
Oscillators ... 39, 142, 164
Passive squarer for H-mode mixer ... 46
Petrol-electric generators ... 73, 132
Phase noise ... 18
PIC stabiliser ... 122
Polyphase networks ... 83
Portable energy source ... 94
Potentially useful chips ... 96
Power line interference ... 154
Power MOSFET pre-amplifiers ... 159
Power reducer using two transformers ... 68, 78
Power sources ... 72, 81, 100
Precision time / frequency ... 16
Propagation, Anomalous ... 169
    Beacons ... 98
    Chordal hop ... 47, 74
    Grey-line ... 47
    NVIS ... 89
    Sporadic E ... 89
    'Supermodes' ... 47, 60, 74
    Sunspot cycle 24 soon? ... 113
    Trans-equatorial (TEP) ... 47, 52
Protecting radios' DC output sockets ... 93
Push-pull circuit ... 147, 155
Quartz crystal contamination ... 120
Radials ... 58, 69, 80
Receivers
    Buccaneer ... 64
    Direct-conversion ... 166
    Digital receivers' IMD ... 131
    Direct Digital Receivers ... 116
    GEC navy receiver model CJA / CJC ... 25
    German & Russian HF receivers ... 4
    JA9MAT's hybrid 'autodyne' ... 11
    KK7B's Micro-R2, T2 ... 104, 111
    Premium receivers ... 11
    R&S EM510 ... 116
    Specifications ... 17
    Wadley loop ... 13
Safety ... 84, 166
Secret listeners of WWII ... 91
Software defined radio ... 21, 32, 43, 51
Soldering to aluminium ... 42
Solid-state front-end ... 129
SSB transceiver based on the 7360 ... 138
Temperature controller ... 61
Tracing unwanted signals on SMD boards ... 52
Transistors – 60 years old ... 148, 171
Transmission Line Transformer (TLT) mixer ... 142
Transmitters
    HF transmitter limitations ... 27
    KK7B's Micro-R2, T2 ... 104, 111
    Micro-power AM 'transmitter' ... 6
    PDM transmitters ... 95
    QRP transmitters ... 95, 111, 158
    Unlicensed TXs & Ofcom ... 124
Up-conversion techniques ... 141, 151, 157
Valve diode harmonic mixers & product detectors ... 37
Variable capacitors ... 20, 29, 137
Vector network analyser (VNA) ... 6
VHF oscillators with inverted-mesa quartz resonators ... 40
VHF squarer ... 97
VXO ... 38
Wide-band-gap diode & LED mixers ... 10
Wide-range tuneable oscillator with stable output ... 60